# Trigonometry Update

# Trigonometry
# Update

**Marvin L. Bittinger**
Indiana University — Purdue University at Indianapolis


**Judith A. Beecher**
Indiana University — Purdue University at Indianapolis

 **Addison-Wesley Publishing Company**

Reading, Massachusetts • Menlo Park, California
New York • Don Mills, Ontario • Wokingham, England
Amsterdam • Bonn • Sydney • Singapore • Tokyo
Madrid • San Juan • Milan • Paris

# Contents

# 6

## THE TRIGONOMETRIC FUNCTIONS

**353**

6.1    Trigonometric Functions of Acute Angles    353

6.2    Application of Right Triangles    367

6.3    Trigonometric Functions of Any Angle    373

6.4    Radians, Arc Length, and Angular Speed    391

6.5    Graphs of Basic Circular Functions    405

6.6    Graphs of Transferred Sine and Cosine Functions    428

6.7    Transformed Graphs of Other Trigonometric Functions    441

Summary and Review    447

Chapter Test    448

# 7

## TRIGONOMETRIC IDENTITIES, INVERSE FUNCTIONS, AND EQUATIONS

**451**

7.1    Trigonometric Manipulations and Identities    451

7.2    Sum and Difference Identities    458

7.3    Cofunction and Related Identities    465

7.4    Some Important Identities    471

7.5    Proving Trigonometric Identities    478

7.6    Inverses of the Trigonometric Functions    486

7.7    Composition of Trigonometric Functions and Their Inverses    493

7.8    Trigonometric Equations    498

7.9    Identities in Solving Trigonometric Equations    503

Summary and Review    508

Chapter Test    509

# 8

## TRIANGLES, VECTORS, AND APPLICATIONS    511

8.1    The Law of Sines    511

8.2    The Law of Cosines    519

8.3    Vectors and Applications    526

8.4    Polar Coordinates    534

8.5    Complex Numbers    541

8.6    Complex Numbers:  DeMoivre's Theorem and $n$th
       Roots of Complex Numbers    547

       Summary and Review    551

       Chapter Test    553


## ANSWERS                                      554

# Trigonometry Update

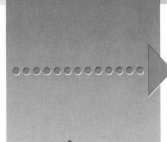

In this chapter, we consider an important class of functions called *trigonometric*, or *circular*, *functions*. Historically, these functions arose from a study of triangles, hence the name trigonometric. ● We will begin our study with right triangles and degree measure, and solve applications involving right triangles. We then consider trigonometric functions of angles or rotations of any size with both degree and radian measure. ● We also define the six basic functions by using a circle of radius 1 (a unit circle, hence the name circular functions). The domains and ranges of these functions are real numbers. ●

# The Trigonometric Functions

## 6.1

## TRIGONOMETRIC FUNCTIONS OF ACUTE ANGLES

We begin our study of trigonometry by considering right triangles and acute angles measured in degrees.

### A  The Trigonometric Ratios

In a right triangle, it is customary to label the right angle *C*. The side opposite the 90° right angle is called the **hypotenuse.** Its length is *c*. The other two sides are called the **legs.** The *acute* angles are labeled *A* and *B*, and the lengths of the sides opposite them are *a* and *b*, respectively. We recall that an **acute angle** is an angle whose measure is between 0° and 90°.

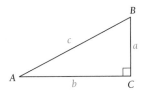

**OBJECTIVES**

You should be able to:

A  Determine the six trigonometric ratios for a given acute angle of a right triangle.

B  State the function values for 30°, 45°, and 60°.

C  Find function values for any acute angle and given a function value for an acute angle, find the angle.

D  Given the function values for an acute angle, find the function values for its complement.

E  Given a function value for an acute angle, find the other five function values.

F  Solve right triangles.

We are now ready to define trigonometric ratios. Let us begin by considering angle $A$. The length of the side **opposite** angle $A$ is $a$, and the length of the side **adjacent** to angle $A$ is $b$.

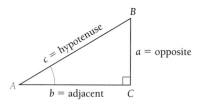

The *sine* of angle $A$ is the length of the side opposite $A$ divided by the length of the hypotenuse:

$$\sin A = \frac{a}{c}.$$

The ratio $a/c$ depends on angle $A$, and thus is a function of angle $A$.

Similarly, the *cosine* of angle $A$ is the length of the side adjacent to $A$ divided by the length of the hypotenuse:

$$\cos A = \frac{b}{c}.$$

There are other such ratios, or trigonometric functions, defined as follows. The symbol "$\theta$" is the Greek letter "theta." It will often be used to denote an angle.

---

**DEFINITION**      **Trigonometric Function Values of an Acute Angle $\theta$**

$$\sin \theta = \frac{\text{side opposite } \theta}{\text{hypotenuse}} \qquad \csc \theta = \frac{\text{hypotenuse}}{\text{side opposite } \theta}$$

$$\cos \theta = \frac{\text{side adjacent } \theta}{\text{hypotenuse}} \qquad \sec \theta = \frac{\text{hypotenuse}}{\text{side adjacent } \theta}$$

$$\tan \theta = \frac{\text{side opposite } \theta}{\text{side adjacent } \theta} \qquad \cot \theta = \frac{\text{side adjacent } \theta}{\text{side opposite } \theta}$$

---

**Example 1**   In this triangle, find each of the following:

a) The trigonometric function values for $\theta$.
b) The trigonometric function values for $\phi$ (the Greek letter "phi").

**Solution**

$$\sin \theta = \frac{\text{side opposite } \theta}{\text{hypotenuse}} = \frac{3}{5} \qquad \sin \phi = \frac{\text{side opposite } \phi}{\text{hypotenuse}} = \frac{4}{5}$$

$$\cos \theta = \frac{\text{side adjacent } \theta}{\text{hypotenuse}} = \frac{4}{5} \qquad \cos \phi = \frac{\text{side adjacent } \phi}{\text{hypotenuse}} = \frac{3}{5}$$

$$\tan \theta = \frac{\text{side opposite } \theta}{\text{side adjacent } \theta} = \frac{3}{4} \qquad \tan \phi = \frac{\text{side opposite } \phi}{\text{side adjacent } \phi} = \frac{4}{3}$$

$$\cot \theta = \frac{\text{side adjacent } \theta}{\text{side opposite } \theta} = \frac{4}{3} \qquad \cot \phi = \frac{\text{side adjacent } \phi}{\text{side opposite } \phi} = \frac{3}{4}$$

$$\sec \theta = \frac{\text{hypotenuse}}{\text{side adjacent } \theta} = \frac{5}{4} \qquad \sec \phi = \frac{\text{hypotenuse}}{\text{side adjacent } \phi} = \frac{5}{3}$$

$$\csc \theta = \frac{\text{hypotenuse}}{\text{side opposite } \theta} = \frac{5}{3} \qquad \csc \phi = \frac{\text{hypotenuse}}{\text{side opposite } \phi} = \frac{5}{4} \quad \blacktriangleleft$$

**DO EXERCISE 1.** _____

In Example 1, we note that the value of $\cot \theta$, $\frac{4}{3}$, is the reciprocal of $\frac{3}{4}$, the value of $\tan \theta$. Likewise, we see the same reciprocal relationship between the values of $\sec \theta$ and $\cos \theta$ and between the values of $\csc \theta$ and $\sin \theta$. For any angle, the cotangent, secant, and cosecant function values are the respective reciprocals of the tangent, cosine, and sine function values.

$$\cot \theta = \frac{1}{\tan \theta}, \qquad \sec \theta = \frac{1}{\cos \theta}, \qquad \csc \theta = \frac{1}{\sin \theta}$$

## Similar Triangles

Triangles are similar if their corresponding angles have the same measure. In the following right triangles, if we know that $\angle A$ and $\angle A'$ are the same size, then we know that all corresponding angles have the same measure; thus the triangles are similar.

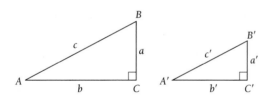

In similar triangles, corresponding sides are in the same ratio (are proportional). For the triangles above, this means the following:

$$\frac{a}{a'} = \frac{b}{b'} = \frac{c}{c'}.$$

**Example 2** In each of the following triangles, find $\sin \theta$, $\cos \theta$, and $\tan \theta$. The second right triangle is similar to the first, but the length of each side is 6 times the length of the corresponding side in the first triangle.

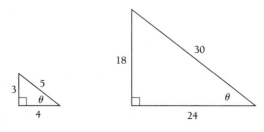

1. Find the following.

a) $\sin \theta$, $\cos \theta$, $\tan \theta$, $\cot \theta$, $\sec \theta$, $\csc \theta$

b) $\sin \phi$, $\cos \phi$, $\tan \phi$, $\cot \phi$, $\sec \phi$, $\csc \phi$

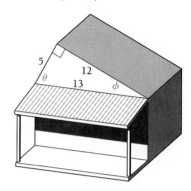

2. Find $\sin \theta$, $\cos \theta$, and $\tan \theta$ for the following triangle, and compare the values with those obtained in Margin Exercise 1(a).

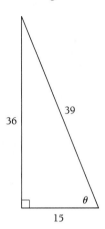

**Solution**

$$\sin \theta = \frac{3}{5} \qquad \sin \theta = \frac{18}{30} = \frac{3}{5}$$

$$\cos \theta = \frac{4}{5} \qquad \cos \theta = \frac{24}{30} = \frac{4}{5}$$

$$\tan \theta = \frac{3}{4} \qquad \tan \theta = \frac{18}{24} = \frac{3}{4}$$

For both triangles, the corresponding values of $\sin \theta$, $\cos \theta$, and $\tan \theta$ are the same.

> The trigonometric function values of $\theta$ depend only on the size of the angle, not on the size of the triangle.

**DO EXERCISE 2.**

Similar triangles can be used to determine distances without measuring directly.

**Example 3** Find the height of the flagpole using a rod and a measuring device, as shown.

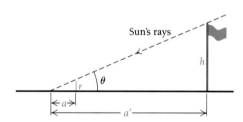

**Solution** Place the rod of length $r$ so that the top of its shadow coincides with the shadow of the top of the flagpole. We know that we have similar triangles, because the acute angle $\theta$ at the ground is the same in both triangles. We then measure and find that $a = 3$ ft, that $a' = 30$ ft, and that $r = 2$ ft.

Now $h/r = a'/a$ by similar triangles, so

$$h = r\frac{a'}{a} = 2\left(\frac{30}{3}\right)$$

$$= 20 \text{ ft.}$$

**DO EXERCISE 3.**

3. Find the height of this building.

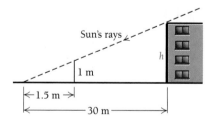

## ▶ B Function Values for Some Special Angles

We can determine the function values for certain angles using our knowledge of geometry. First, recall the Pythagorean theorem. It says that in any right triangle, $a^2 + b^2 = c^2$, where $c$ is the length of the hypotenuse.

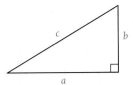

A right triangle with a 45° angle actually has two 45° angles. Thus the triangle is isosceles, and the legs are the same length. Let us consider such a triangle whose legs have length 1. Then its hypotenuse has length $c$:

$$1^2 + 1^2 = c^2, \quad \text{or} \quad c^2 = 2, \quad \text{or} \quad c = \sqrt{2}.$$

Such a triangle is shown below. From this diagram, we can easily determine the trigonometric function values for 45°.

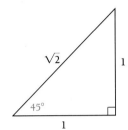

$$\sin 45° = \frac{\text{opposite}}{\text{hypotenuse}} = \frac{1}{\sqrt{2}} = \frac{\sqrt{2}}{2} \approx 0.707$$

$$\cos 45° = \frac{\text{adjacent}}{\text{hypotenuse}} = \frac{1}{\sqrt{2}} = \frac{\sqrt{2}}{2} \approx 0.707$$

$$\tan 45° = \frac{\text{opposite}}{\text{adjacent}} = \frac{1}{1} = 1$$

$$\cot 45° = \frac{\text{adjacent}}{\text{opposite}} = \frac{1}{1} = 1$$

$$\sec 45° = \frac{\text{hypotenuse}}{\text{adjacent}} = \frac{\sqrt{2}}{1} = \sqrt{2} \approx 1.414$$

$$\csc 45° = \frac{\text{hypotenuse}}{\text{opposite}} = \frac{\sqrt{2}}{1} = \sqrt{2} \approx 1.414$$

In a similar way, we can determine function values for 30° and 60°. A right triangle with 30° and 60° acute angles is half of an equilateral triangle, as shown in the following diagram. Thus if we choose an equilateral triangle whose sides have length 2 and take half of it, we obtain a right triangle that has a hypotenuse of length 2 and a leg of length 1. The other leg has length $a$, which can be found using the Pythagorean theorem as follows:

$$a^2 + 1^2 = 2^2$$
$$a^2 = 3$$
$$a = \sqrt{3}.$$

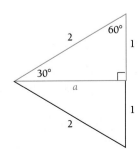

We can now determine function values for 30° and 60°.

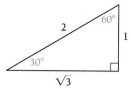

$$\sin 30° = \frac{1}{2} \qquad\qquad \sin 60° = \frac{\sqrt{3}}{2}$$

$$\cos 30° = \frac{\sqrt{3}}{2} \qquad\qquad \cos 60° = \frac{1}{2}$$

$$\tan 30° = \frac{1}{\sqrt{3}} = \frac{\sqrt{3}}{3} \qquad\qquad \tan 60° = \frac{\sqrt{3}}{1} = \sqrt{3}$$

$$\cot 30° = \frac{\sqrt{3}}{1} = \sqrt{3} \qquad\qquad \cot 60° = \frac{1}{\sqrt{3}} = \frac{\sqrt{3}}{3}$$

$$\sec 30° = \frac{2}{\sqrt{3}} = \frac{2\sqrt{3}}{3} \qquad\qquad \sec 60° = \frac{2}{1} = 2$$

$$\csc 30° = \frac{2}{1} = 2 \qquad\qquad \csc 60° = \frac{2}{\sqrt{3}} = \frac{2\sqrt{3}}{3}$$

Since we will often use the function values for 45°, 30°, and 60°, they should be memorized. It is sufficient to learn only those for the sine, cosine, and tangent, since the others are their reciprocals.

$$\sin 45° = \frac{\sqrt{2}}{2} \qquad \cos 45° = \frac{\sqrt{2}}{2} \qquad \tan 45° = 1$$

$$\sin 30° = \frac{1}{2} \qquad \cos 30° = \frac{\sqrt{3}}{2} \qquad \tan 30° = \frac{\sqrt{3}}{3}$$

$$\sin 60° = \frac{\sqrt{3}}{2} \qquad \cos 60° = \frac{1}{2} \qquad \tan 60° = \sqrt{3}$$

We can now use what we have learned about trigonometric functions of special angles to solve problems. We will consider such applications in much greater detail in Section 6.2.

**Example 4**    The rays of the sun over the top of a building cast a 40-ft shadow and form an angle of 60° with the ground. Find the height $h$ of the building.

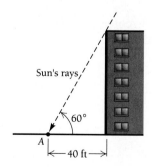

**Solution**    We will typically refer to an angle $A$ and its measure by the same notation. Thus, $A = 60°$. We know the side adjacent to $\angle A$ and also the measure of $\angle A$. Since we want to determine the length of the opposite side, we can use the tangent ratio or the cotangent ratio. Here we use the tangent:

$$\tan 60° = \frac{\text{opposite}}{\text{adjacent}} = \frac{h}{40}$$

$$\sqrt{3} = \frac{h}{40} \qquad \text{Substituting}$$

$$40\sqrt{3} = h$$

$$69.3 \text{ ft} \approx h.$$

The height of the building is about 69.3 ft.

**DO EXERCISE 4.**

  **Function Values of Any Acute Angle**

We were able to determine trigonometric function values of special angles (30°, 45°, and 60°) geometrically. To find function values for other angles, we can use either a table of values or a scientific calculator. With the aid of certain theoretically determined formulas, tables of values of the trigonometric functions can be constructed. Table 3 at the back of the book is such a table with four-digit accuracy. Your scientific calculator contains such a table, with even greater accuracy.

### Minutes Versus Tenths and Hundredths

It is the general practice to express the measure of an angle in *degrees, minutes,* and *seconds*. One **minute** is denoted 1′ and is such that

$$60 \text{ minutes} = 60' = 1°.$$

One **second** is denoted 1″ and is such that

$$60 \text{ seconds} = 60'' = 1'.$$

Then

$$34°42'28'' = 34 \text{ degrees, } 42 \text{ minutes, } 28 \text{ seconds} = 34° + 42' + 28''.$$

When working with a calculator, we will need to have the measure in degrees, tenths, hundredths, and so on. Thus we will need to know how to convert between these.

Converting a measure like 34°41′52″ to decimal parts of degrees, and vice versa, can be done quite easily with a calculator. Some calculators convert directly. With others, we can convert as shown in the following examples.

Example 5  Convert 34°41′52″ to degrees and decimal parts of degrees. Round to two decimal places.

Solution   Using $1' = \left(\frac{1}{60}\right)°$ and $1'' = \left(\frac{1}{3600}\right)°$, we have

$$34°41'52'' = 34° + 41' + 52''$$

$$= 34° + \left(\tfrac{41}{60}\right)° + \left(\tfrac{52}{3600}\right)°$$

$$\approx 34° + 0.6833 + 0.0144$$

$$\approx 34.70°.$$

4. A baseball diamond is really a square 90 ft on a side. If a line is drawn from third base to first base, then a right triangle *QPH* is formed, where ∠*QPH* is 45°. Use the sine function to find the length *PQ* from third base to first base.

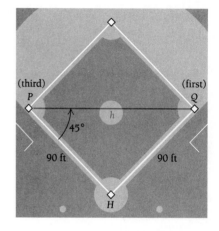

5. Convert 67°13′16″ to degrees and decimal parts of degrees. Round to four decimal places.

6. Convert 37.45° to degrees and minutes.

Find each of the following.

7. sin 15°

8. tan 22.6°

9. cos 75°

**Example 6**   Convert 16.35° to degrees and minutes.

**Solution**   We have

$$16.35° = 16° + 0.35 × 1°.$$

Substituting 60′ for 1° gives us

$$0.35 × 1° = 0.35 × 60′ = 21′.$$

Thus,

$$16.35° = 16°21′.$$   ◀

**DO EXERCISES 5 AND 6.** _____

## Finding Function Values on a Calculator

So far we have measured angles using degrees. Another useful unit for angle measure is the radian, which we will study in Section 6.4. Some calculators require angles to be entered in degrees. Others work with either degrees or radians. Be sure to use whichever mode is appropriate. In this section, we use the degree mode.

When finding values on a calculator, remember that different calculators round to different decimal places, so there may be variance in answers found on your calculator. Also keep in mind the difference between an exact answer and an approximation. For example,

$$\sin 60° = \frac{\sqrt{3}}{2}.$$   **This is exact!**

But on a calculator, you might get an answer like

$$\sin 60° ≈ 0.866025404.$$   **This is approximate!**

Scientific calculators usually provide values of the sine, the cosine, and the tangent functions only. You can find values of the cotangent, the secant, and the cosecant by taking reciprocals of the tangent, the cosine, and the sine function, respectively. Since calculators are different, we suggest that you carefully read the directions for the calculator you are using.

**Example 7**   Find tan 81°.

**Solution**   We check to be sure that the calculator is in degree mode. Then we enter 81 and press the $\boxed{\text{TAN}}$ key. We find that

$$\tan 81° ≈ 6.3138.$$

We rounded the answer to four decimal places.   ◀

**DO EXERCISES 7–9.** _____

**Example 8**   Find sec 43°. Round to nine decimal places.

**Solution**   The secant function value can be found by taking the reciprocal of the cosine function value. We enter 43 and press the $\boxed{\text{COS}}$ key:

$$\cos 43° ≈ 0.731353702.$$

We then press the reciprocal key $\boxed{1/x}$, or divide 1 by 0.731353702. You need not round before taking the reciprocal. We find that

$$\sec 43° = \frac{1}{\cos 43°} \approx 1.367327461.$$ ◄

**DO EXERCISE 10.**

**Example 9** Find $\sin 38°25'57''$. Round to four decimal places.

Solution   We first convert to degrees and decimal parts of degrees. We get

$$38°25'57'' = 38.4325°.$$

We then press the $\boxed{\text{SIN}}$ key and get

$$\sin 38°25'57'' \approx 0.6216.$$ ◄

**DO EXERCISES 11–13.**

To use a calculator in reverse—for example, to find an angle whose sine is given—we first enter the sine value. Then we press the key marked either $\text{SIN}^{-1}$ (on some calculators) or ARCSIN (on other calculators). On still others, two keys must be pressed in sequence: ARC and then SIN, INV and then SIN, or $f^{-1}$ and then SIN. Be sure to read the instructions for your calculator.

**Example 10** Find the acute angle $\theta$, to the nearest hundredth of a degree, for which $\cos \theta = 0.5437$.

Solution   We check to be certain that the calculator is in degree mode. We enter 0.5437 and press the $\boxed{\text{COS}^{-1}}$ key. We find that

$$\theta = \cos^{-1} 0.5437 \approx 57.06°.$$

Thus, $\theta \approx 57.06°$. ◄

**DO EXERCISES 14–16.**

▸ **Cofunctions and Complements**

We recall that two angles are **complementary** whenever the sum of their measures is 90°. Each is the complement of the other. In a right triangle, the acute angles are complementary, since the sum of all three angle measures is 180° and the right angle accounts for 90° of this total. Thus if one acute angle of a right triangle is $\theta$, the other is $90° - \theta$. Note that the sine of $\angle A$ is also the cosine of $\angle B$, its complement:

$$\sin \theta = \frac{a}{c},$$

$$\cos (90° - \theta) = \frac{a}{c}.$$

10. Find cot 29°.

Find each of the following to four decimal places.

11. cos 61°54′

12. tan 12°57′22″

13. sec 55°13′43″

Find the acute angle $\theta$, to the nearest hundredth of a degree, for each of the following.

14. $\tan \theta = 0.4621$

15. $\sin \theta = 0.7660$

16. $\cos \theta = 0.7009$

**17.** Given that

sin 75° = 0.9659,   cos 75° = 0.2588,

tan 75° = 3.732,    cot 75° = 0.2679,

sec 75° = 3.864,    csc 75° = 1.035,

find the function values for the complement of 75°.

Similarly, the tangent of $\angle A$ is the cotangent of its complement and the secant of $\angle A$ is the cosecant of its complement.

These pairs of functions are called **cofunctions**. The name **cosine** originally meant the sine of the complement. The name **cotangent** meant the tangent of the complement and **cosecant** meant the secant of the complement. A complete list of the cofunction properties is as follows.

$$\sin \theta = \cos (90° - \theta) \qquad \cos \theta = \sin (90° - \theta)$$
$$\tan \theta = \cot (90° - \theta) \qquad \cot \theta = \tan (90° - \theta)$$
$$\sec \theta = \csc (90° - \theta) \qquad \csc \theta = \sec (90° - \theta)$$

**Example 11**   Given that

$$\sin 18° = 0.3090, \qquad \cos 18° = 0.9511,$$
$$\tan 18° = 0.3249, \qquad \cot 18° = 3.078,$$
$$\sec 18° = 1.051, \qquad \csc 18° = 3.236,$$

find the six function values for 72°.

**Solution**   Since 72° and 18° are complements, we have sin 72° = cos 18°, and so on. Thus the function values are

$$\sin 72° = 0.9511, \qquad \cos 72° = 0.3090,$$
$$\tan 72° = 3.078, \qquad \cot 72° = 0.3249,$$
$$\sec 72° = 3.236, \qquad \csc 72° = 1.051.$$

**DO EXERCISE 17.**

### **E**  The Six Functions Related

Using the Pythagorean theorem, we can find all six trigonometric function values of an acute angle when one of the function-value ratios is known.

**Example 12**   If $\sin \theta = \frac{6}{7}$ and $\theta$ is an acute angle, find the other five trigonometric function values for $\theta$.

**Solution**   We know from the definition of the sine function that the ratio

$$\frac{6}{7} \quad \text{is} \quad \frac{\text{side opposite } \theta}{\text{hypotenuse}}.$$

Using this information, let us consider a right triangle in which the hypotenuse has length 7 and the side opposite $\theta$ has length 6.

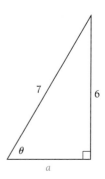

To find the length of the side adjacent to $\theta$, we can use the Pythagorean theorem:

$$a^2 + 6^2 = 7^2$$
$$a^2 = 49 - 36 = 13$$
$$a = \sqrt{13}. \quad \text{\small Choosing the positive square root, since we are finding length}$$

We can use $a = \sqrt{13}$, $b = 6$, and $c = 7$ to find the other five ratios in the original triangle:

$$\sin \theta = \frac{6}{7}, \qquad\qquad \csc \theta = \frac{7}{6},$$

$$\cos \theta = \frac{\sqrt{13}}{7}, \qquad\qquad \sec \theta = \frac{7}{\sqrt{13}}, \text{ or } \frac{7\sqrt{13}}{13},$$

$$\tan \theta = \frac{6}{\sqrt{13}}, \text{ or } \frac{6\sqrt{13}}{13}, \qquad \cot \theta = \frac{\sqrt{13}}{6}.$$

◀

**DO EXERCISE 18.**

  **Solving Right Triangles**

Now that we know how to find function values for any acute angle, we can begin to solve right triangles. Triangles that are not right triangles will be studied later. To **solve** a triangle means to find the lengths of all its sides and the measures of all its angles, provided they are not already known.

**Example 13**  Find $m \angle B$.

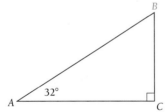

**Solution**  Since the angle measures of any triangle add up to 180° and the right angle measures 90°, the acute angles must add up to 90°:

$$m \angle A + m \angle B = 90°.$$

Thus,

$$m \angle B = 90° - 32° = 58°.$$

◀

**Example 14**  In this right triangle, find $a$ and $b$.

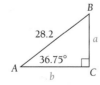

**Solution**  By the definition of the sine ratio, we know that

$$\sin 36.75° = \frac{a}{28.2}.$$

**18.** If $\cos \theta = \frac{7}{8}$, find the other five trigonometric function values.

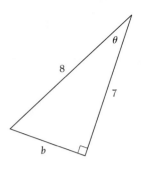

**19.** Find $m\angle A$.

**20.** In this right triangle, find $a$ and $b$.

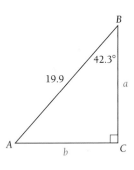

**21.** Solve this triangle.

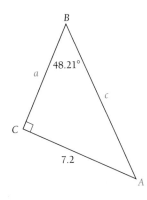

**22.** Solve this right triangle. Standard lettering has been used.

$$A = 37.8°, \qquad b = 8.62$$

Solving for $a$, we have

$$a = 28.2 \sin 36.75°$$
$$\approx 16.9.$$

To find $b$, we use the cosine ratio:

$$\cos 36.75° = \frac{b}{28.2}.$$

We solve for $b$:

$$b = 28.2 \cos 36.75°$$
$$\approx 22.6.$$

◀

**DO EXERCISES 19 AND 20.**

Henceforth, for simplicity, we will use an equals sign, $=$, even when an approximation symbol, $\approx$, might be appropriate.

**Example 15** In this right triangle, find $A$ and $B$.* Then find $a$. That is, *solve* the triangle.

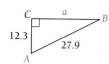

**Solution** We know the side adjacent to $A$ and also the hypotenuse. That suggests the use of the cosine ratio:

$$\cos A = \frac{12.3}{27.9}$$
$$= 0.4409. \qquad \text{Dividing}$$

We now use a calculator to find the inverse cosine of 0.4409, which is $A$. We find that

$$A = 63.84°.$$

Now, since the measures of $A$ and $B$ must total $90°$, we can find $B$ by subtracting:

$$B = 90° - 63.84°$$
$$= 26.16°.$$

To find $a$, we have a choice of how to begin. We could use $\cos B$, or we could use $\sin A$. We could also use the tangent or cotangent ratios for either $A$ or $B$. Let's use $\tan A$:

$$\tan A = \frac{a}{12.3}, \quad \text{or}$$

$$\tan 63.84° = \frac{a}{12.3}.$$

---

* We will often shorten our writing or speaking, saying that we find $B$, rather than $m\angle B$.

Then

$$a = 12.3 \tan 63.84°$$  Solving for *a*

$$= 25.0.$$  Using a calculator  ◀

**DO EXERCISES 21 AND 22 ON THE PRECEDING PAGE.**

● **EXERCISE SET** **6.1**

**A** Find the six trigonometric function values for the specified angle.

**1.**

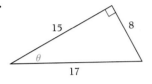

**2.**

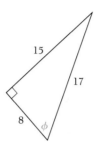

**3.**

**4.**

**5.**

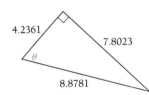

**6.**

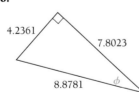

**B**

**7.** Find the distance *a* across the river.

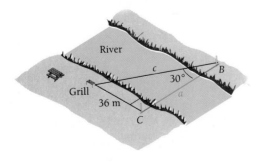

**8.** Find the length *L* from point *A* to the top of the pole.

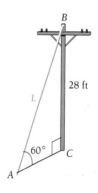

In Exercises 9 and 10, use the fact that $\sqrt{2} \approx 1.414$ and $\sqrt{3} \approx 1.732$. Do not use the trigonometric function keys on a calculator.

**9.** Find the decimal values for the six trigonometric functions of 30°.

**10.** Find the decimal values for the six trigonometric functions of 60°.

In Exercises 11 and 12, use the trigonometric function keys on a calculator.

**11.** Find the six trigonometric function values of 30°.

**12.** Find the six trigonometric function values of 60°.

**C** Convert to degrees and minutes.

**13.** 8.6°  **14.** 47.8°

**15.** 72.25°  **16.** 11.75°

**17.** 46.38°  **18.** 85.21°

**19.** 67.84°  **20.** 38.48°

Convert to degrees, minutes, and seconds.

**21.** 87.3456°  **22.** 11.0256°

**23.** 48.02498°  **24.** 27.899461°

Convert to degrees and decimal parts of degrees. Round to two decimal places.

**25.** 9°45′  **26.** 52°15′

**27.** 35°50′  **28.** 64°40′

**29.** 80°33′  **30.** 27°19′

**31.** 3°02′

**32.** 10°08′

Convert to degrees and decimal parts of degrees. Round to four decimal places.

**33.** 19°47′23″

**34.** 49°38′46″

**35.** 31°57′55″

**36.** 76°11′34″

Use a calculator to find each of the following function values. Round to four decimal places.

**37.** cos 18°

**38.** sin 37°

**39.** tan 2.6°

**40.** cos 34.8°

**41.** sin 62°20′13″

**42.** tan 55°30′40″

**43.** cos 15°35′

**44.** sin 40°55′

**45.** csc 29°

**46.** cot 54°

**47.** sec 10°30′24″

**48.** csc 45°15′51″

Use a calculator to find acute angle $\theta$, to the nearest hundredth of a degree, when the function value is given.

**49.** sin $\theta$ = 0.5125

**50.** cos $\theta$ = 0.8241

**51.** cos $\theta$ = 0.6512

**52.** sin $\theta$ = 0.8220

**53.** tan $\theta$ = 3.163

**54.** tan $\theta$ = 7.425

**55.** csc $\theta$ = 6.277

**56.** sec $\theta$ = 1.175

**57.** Given that

$$\sin 65° = 0.9063, \qquad \cos 65° = 0.4226,$$
$$\tan 65° = 2.145, \qquad \cot 65° = 0.4663,$$
$$\sec 65° = 2.366, \qquad \csc 65° = 1.103,$$

find the six function values for 25°.

**58.** Given that

$$\sin 32° = 0.5299, \qquad \cos 32° = 0.8480,$$
$$\tan 32° = 0.6249, \qquad \cot 32° = 1.600,$$
$$\sec 32° = 1.179, \qquad \csc 32° = 1.887,$$

find the six function values for 58°.

**E** In Exercises 59–62, assume that $\theta$ is an acute angle.

**59.** Given that sin $\theta$ = $\frac{24}{25}$, find the other five trigonometric function values for $\theta$.

**60.** Given that cos $\theta$ = 0.7, find the other five trigonometric function values for $\theta$.

**61.** Given that tan $\phi$ = 2, find the other five trigonometric function values for $\phi$.

**62.** Given that sec $\phi$ = $\sqrt{17}$, find the other five trigonometric function values for $\phi$.

**F** Solve each of the following right triangles. (Standard lettering has been used.)

**63.**

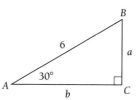

**64.**

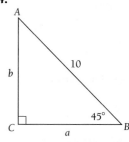

**65.**

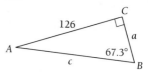

**66.**

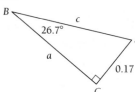

**67.**

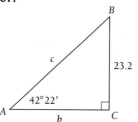

**68.**

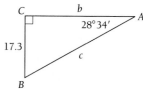

**69.** $A = 36°12′$, $a = 27.2$

**70.** $A = 87°43′$, $a = 9.73$

**71.** $B = 12.65°$, $b = 98.1$

**72.** $B = 69.88°$, $b = 127$

**73.** $A = 17.4°$, $b = 13.6$

**74.** $A = 78.7°$, $b = 1340$

**75.** $B = 23°12′$, $a = 350$

**76.** $B = 69°22′$, $a = 240$

**77.** $A = 47.58°$, $c = 48.3$

**78.** $A = 88.92°$, $c = 3950$

**79.** $B = 82.3°$, $c = 0.982$

**80.** $B = 56.5°$, $c = 0.0447$

**81.** $a = 12.5$, $b = 18.5$

**82.** $a = 10.2$, $b = 20.4$

**83.** $a = 16.0$, $c = 20.0$

**84.** $a = 15.0$, $c = 45.0$

**85.** $b = 1.86$, $c = 4.02$

**86.** $b = 100$, $c = 450$

● **SYNTHESIS**

**87.** Find $h$, to the nearest tenth.

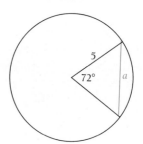

**88.** Find $a$, to the nearest tenth.

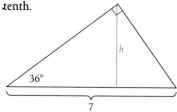

**89.** Show that the area of a right triangle is
$\frac{1}{2}bc \sin A$.

● CHALLENGE _____

**90.** Show that the area of this triangle is $\frac{1}{2}ab \sin \theta$.

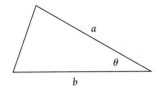

# 6.2

# APPLICATIONS OF RIGHT TRIANGLES

**OBJECTIVE**

You should be able to:

Solve problems involving the use of right triangles and the trigonometric functions.

▲  Right triangles have many applications. To solve a problem, we locate a right triangle and then solve, or at least partially solve, it.

**Example 1**   An observer stands on level ground, 200 m from the base of a television tower, and looks up at an angle of 26.5° to see the top of the tower.

**a)** How high is the tower above the observer's eye level?
**b)** How far is it from the observer's eye to the top of the tower?

**Solution**   We draw a diagram and see that a right triangle is formed.

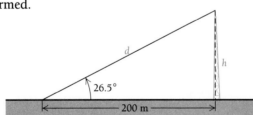

**a)** We use one of the trigonometric functions. The tangent function is most convenient. From the definition of the tangent function, we have

$$\tan 26.5° = \frac{\text{opposite}}{\text{adjacent}} = \frac{h}{200}.$$

Solving for $h$ gives us

$$h = 200 \tan 26.5° = 100.$$

The height of the tower is about 100 m.

**b)** To find the distance $d$ from the observer's eye to the top of the tower, we use a different trigonometric function, one that involves the hypotenuse. The cosine function gives us

$$\cos 26.5° = \frac{\text{adjacent}}{\text{hypotenuse}} = \frac{200}{d}.$$

1. An observer stands 120 m from a tree and finds that the line of sight to the top of the tree is 32.3° above the horizontal.

   a) Find the height of the tree above eye level.

   b) Find the distance from the observer to the top of the tree.

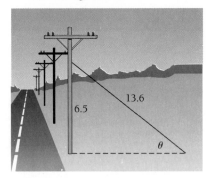

Then solving for $d$, we get

$$d = \frac{200}{\cos 26.5°} = 223.$$

The distance from the observer's eye to the top of the tower is about 223 m.    ◀

**DO EXERCISE 1.** _____

**Example 2    *Safe angle for ladders.*** Suppose that a ladder has length $L$. It has been determined that the ladder is at its safest position on a wall when it is pulled out a distance $D$ from the wall, where $L = 4D$. What angle does this safest position determine with the ground?

**Solution** We draw a diagram and then use the most convenient trigonometric function. From the definition of the cosine function, we have

$$\cos \theta = \frac{\text{adjacent}}{\text{hypotenuse}} = \frac{D}{L} = \frac{D}{4D} = \frac{1}{4}.$$

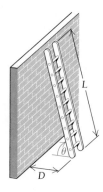

We find that $\theta$ is about 75.52°. Thus the ladder is at its safest position when it makes an angle of about 75.52° with the ground.    ◀

**DO EXERCISE 2.** _____

2. A guy wire is 13.6 m long and is fastened from the ground to a pole 6.5 m above the ground. What angle does the wire make with the ground?

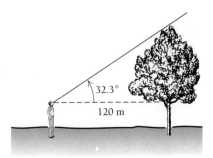

Many applications with right triangles involve an *angle of elevation* or an *angle of depression*.

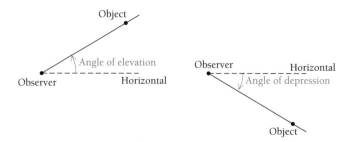

The angle between the horizontal and a line of sight *above* the horizontal is called an **angle of elevation.** The angle between the horizontal and a line of sight *below* the horizontal is called an **angle of depression.** For example, suppose that you were looking straight ahead and then you moved your eyes upward toward an oncoming airplane. The angle your eyes pass through is an *angle of elevation.* If the pilot of the plane is looking forward and then moves his eyes down toward you, his eyes pass through an *angle of depression.*

**Example 3   *Finding a cloud height.***   A device for measuring cloud height at night consists of a vertical beam of light, which makes a spot on the clouds. The spot is viewed from a point 135 ft away. Using a surveying device, the angle of elevation is found to be 67°40′. Find the height of the clouds.

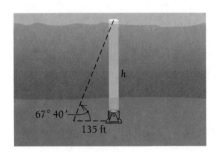

**Solution**   From the figure, we have

$$\frac{h}{135} = \tan 67°40′$$

$$h = 135 \times \tan 67°40′ = 329 \text{ ft.} \qquad \blacktriangleleft$$

The following is a general procedure for solving problems involving triangles.

---

To solve a triangle problem:

1. Draw a sketch of the problem situation.
2. Look for triangles and sketch them in.
3. Mark the known and unknown sides and angles.
4. Express the desired side or angle in terms of known trigonometric ratios. Then solve.

---

**DO EXERCISE 3.**

**Example 4   *Surveying.***   Horizontal distances must often be measured even though terrain is not level. One way of doing so is as follows. Distance down a slope is measured with a surveyor's tape, and the distance $d$ is measured by making a level sighting from $A$ to a pole held vertically at $B$, or the angle $\alpha$ is measured by an instrument placed at $A$. Suppose that a slope distance $L$ is measured to be 121.3 ft and the angle $\alpha$ is measured to be 3°25′. Find the horizontal distance $H$.

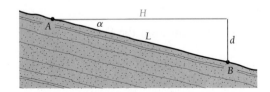

3. The length of a guy wire to a pole is 37.7 ft. It makes an angle of 71°20′ with the ground, which is horizontal. How high above the ground is the guy wire attached to the pole?

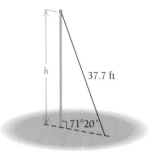

4. A downslope distance is measured to be 241.3 ft, and the angle of depression $\alpha$ is measured to be 5°15'. Find the horizontal distance.

**Solution**   From the figure, we see that $H/L = \cos \alpha$. Thus, $H = L \cos \alpha$, and in this case.

$$H = 121.3 \times \cos 3°25' = 121.1 \text{ ft.} \quad \blacktriangleleft$$

**DO EXERCISE 4.**

Some applications of trigonometry involve the idea of **direction**, or **bearing.** In this text we present two ways of giving direction, the first in Example 5 below and the second in Section 6.3.

**Bearing: First-type.**   One method of giving direction, or bearing, involves reference to a north–south line using an acute angle. For example, N43°W means 43° west of north and S30°E means 30° east of south. Several bearings of this type are shown below.

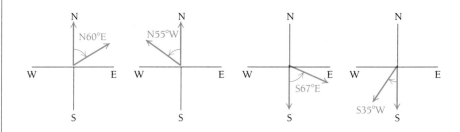

5. Directly east of a lookout station, there is a small forest fire. The bearing of this fire from a station 12.5 km south of the first is N 57°19' E. How far is the fire from the southerly lookout station?

**Example 5**   A forest ranger at point $A$ sights a fire directly south. A second ranger at point $B$, 7.5 mi east, sights the same fire at a bearing of S27°23'W. How far from $A$ is the fire?

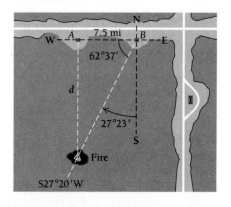

$$\angle B = 90° - 27°23'$$
$$= 89°60' - 27°23'$$
$$= 62°37'$$

**Solution**   From the figure, we see that the desired distance $d$ is part of a right triangle, as shown. We have

$$\frac{d}{7.5} = \tan 62°37'$$

$$d = 7.5 \tan 62°37' = 14.5 \text{ mi.}$$

The forest ranger at $A$ is 14.5 mi from the fire.   $\blacktriangleleft$

**DO EXERCISE 5.**

**Example 6**   From an observation tower, two markers are viewed on the ground. The markers and the base of the tower are on a line, and the observer's eye is 65.3 ft above the ground. The angles of depression to the markers are 53.2° and 27.8°. How far is it from one marker to the other?

Solution   We first make a sketch, looking for right triangles to sketch in. Then we mark the known information on the sketch. The distance we seek is $d$, which is $d_1 - d_2$.

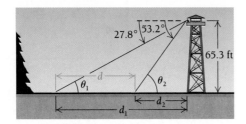

From the right angles in the figure, we have

$$\frac{d_1}{65.3} = \cot \theta_1 \quad \text{and} \quad \frac{d_2}{65.3} = \cot \theta_2,$$

or

$$d_1 = 65.3 \cot \theta_1 \quad \text{and} \quad d_2 = 65.3 \cot \theta_2.$$

When parallel lines are cut by a transversal, alternate interior angles are equal. Thus,

$$\theta_1 = 27.8° \quad \text{and} \quad \theta_2 = 53.2°.$$

Then

$$d_1 = 65.3 \cot 27.8° \quad \text{and} \quad d_2 = 65.3 \cot 53.2°.$$

We could calculate these and subtract, but the use of a calculator is more efficient if we leave all calculations until the end:

$$\begin{aligned} d = d_1 - d_2 &= 65.3 \cot 27.8° - 65.3 \cot 53.2° \\ &= 65.3(\cot 27.8° - \cot 53.2°) \\ &= 75.0 \text{ ft.} \end{aligned}$$

◀

**DO EXERCISE 6.**

6. From an airplane flying 7500 ft above level ground, one can see two towns directly to the east. The angles of depression to the towns are 5.3° and 77.5°. How far apart are the towns, to the nearest mile?

 **EXERCISE SET   6.2**

A   Solve.

**1.** A guy wire to a pole makes an angle of 73.2° with the level ground and is 14.5 ft from the pole at the ground. How far above the ground is the wire attached to the pole?

**2.** A guy wire to a pole makes an angle of 74.4° with the level ground and is attached to the pole 34.2 ft above the ground. How far from the base of the pole is the wire attached to the ground?

**3.** A kite string makes an angle of 31.7° with the (level) ground when 455 ft of string is out. How high is the kite?

**4.** A kite string makes an angle of 41.3° with the (level) ground when the kite is 114 ft high. How long is the string?

**5.** A road rises 3 m per 100 horizontal m. What angle does it make with the horizontal?

6. A kite is 120 ft high when 670 ft of string is out. What angle does the kite make with the ground?

7. What is the angle of elevation of the sun when a 6-ft man casts a 10.3-ft shadow?

8. What is the angle of elevation of the sun when a 35-ft mast casts a 20-ft shadow?

9. From a balloon 2500 ft high, a command post is seen with an angle of depression of 7.7°. How far is it from a point on the ground below the balloon to the command post, to the nearest tenth of a mile?

10. From a lighthouse 55 ft above sea level, the angle of depression to a small boat is 11.3°. How far from the foot of the lighthouse is the boat?

11. An observer at a command post sights a balloon that is 2500 ft high at an angle of elevation of 8°20′. How far is it from the command post to a point directly under the balloon?

12. An observer sights the top of a building 173 ft higher than the eye, at an angle of elevation of 27°50′. How far is it from the observer to the building?

13. Ship $A$ is due west of a lighthouse. Ship $B$ is 12 km south of ship $A$. From ship $B$, the bearing to the lighthouse is N 63°20′ E. How far is ship $A$ from the lighthouse?

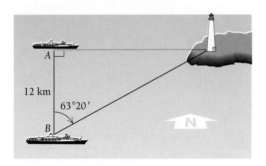

14. Lookout station $A$ is 15 km west of station $B$. The bearing from $A$ to a fire directly south of $B$ is S 37°50′ E. How far is the fire from $B$?

15. A regular pentagon has sides 30.5 cm long. Find the radius of the circumscribed circle.

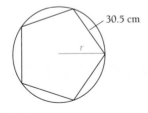

16. A regular pentagon has sides 42.8 cm long. Find the radius of the inscribed circle.

17. A regular hexagon has a perimeter of 50 cm and is inscribed in a circle. Find the radius of the circle.

18. A regular octagon is inscribed in a circle of radius 15.8 cm. Find the perimeter of the octagon.

19. A vertical antenna is mounted on top of a 50-ft pole. From a point on the level ground 75 ft from the base of the pole, the antenna subtends an angle of 10.5°. Find the length of the antenna.

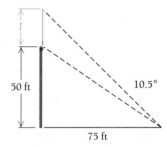

20. An observer on a ladder looks at a building 100 ft away, noting that the angle of elevation of the top of the building is 18°40′ and the angle of depression of the bottom of the building is 6°20′. How tall is the building?

21. From a balloon 2 km high, the angles of depression to two towns, in line with the balloon, are 81.2° and 13.5°. How far apart are the towns?

22. From a balloon 1000 m high, the angles of depression to two artillery posts, in line with the balloon, are 11.8° and 84.1°. How far apart are the artillery posts?

23. A weather balloon is directly west of two observing stations that are 10 km apart. The angles of elevation of the balloon from the two stations are 17°50′ and 78°10′. How high is the balloon?

24. From two points south of a hill on level ground and 1000 ft apart, the angles of elevation of the hill are 12°20′ and 82°40′. How high is the hill?

● SYNTHESIS _____

25. Find a formula for the distance to the horizon as a function of the height of the observer above the earth. Calculate the distance, to the nearest tenth of a mile, to the horizon from an airplane flying at an altitude of 1000 ft.

26. In finding horizontal distance from slope distance (see Example 4), we have $H = L - C$, where $C$ is a correction. Show that a good approximation to $C$ is $d^2/2L$.

27. *Carpentry.* A carpenter is constructing picnic pavilions in parks, as shown in the figure. The rafter ends are to be

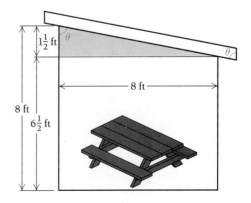

sawed in such a way that they will be vertical when in place. The front wall is 8 ft high, the back wall is $6\frac{1}{2}$ ft high, and the distance between walls is 8 ft. At what angle should the rafters be cut?

**28.** *Carpentry.* A V-gauge is used to find diameters of pipes. The advantage of such a device is that it is rugged, it is accurate, and it has no moving parts to break down. In the figure, the measure of angle *AVB* is 54°. A pipe is placed in the V-shaped slot and the distance *VP* is used to predict the diameter.

**a)** Suppose that the diameter of a pipe is 2 cm. What is the distance *VP*?
**b)** Suppose that the distance *VP* is 3.93 cm. What is the diameter of the pipe?
**c)** Find the formula for *d* in terms of *VP*.
**d)** Find a formula for *VP* in terms of *d*.

The line *VP* is calibrated by listing as its units the corresponding diameters. This, in effect, establishes a function between *VP* and *d*.

● CHALLENGE _____

**29.** *Sound of an airplane.* It is a common experience to hear the sound of a low flying airplane, and look at the wrong place in the sky to see the plane. Suppose that a plane is traveling directly at you at a speed of 200 mph and an altitude of 3000 ft, and you hear the sound at what seems to be an angle of inclination of 20°. At what angle $\theta$ should you actually look in order to see the plane? Consider the speed of sound to be 1100 ft/sec.

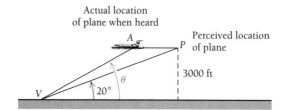

**30.** *Measuring the radius of the earth.* One way to measure the radius of the earth is to climb to the top of a mountain whose height above sea level is known and measure the angle between a vertical line to the center of the earth from the top of the mountain and a line drawn from the top of the mountain to the horizon, as shown in the figure. The height of Mt. Shasta in California is 14,162 ft. From the top of Mt. Shasta, one can see the horizon on the Pacific Ocean. The angle formed between a line to the horizon and the vertical is found to be 87°53′. Use this information to estimate the radius of the earth, in miles.

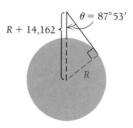

# 6.3

# TRIGONOMETRIC FUNCTIONS OF ANY ANGLE

## ▲ Angles, Rotations, and Degree Measure

An *angle* is a familiar figure.

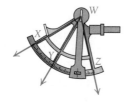
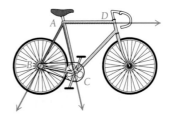

**OBJECTIVES**

You should be able to:

**A** Given the measure of an angle or rotation in degrees, tell in which quadrant the terminal side lies; find positive and negative angles that are coterminal with a given angle; and find the complement and the supplement of a given angle.

**B** Determine the six trigonometric function values for any angle in standard position when the coordinates of a point on the terminal side are given and when the equation of the line that is the

*(continued)*

terminal side is given; and given a rotation, determine the signs of the six trigonometric function values.

C  State the trigonometric function values (when defined) for any angle whose terminal side lies on an axis.

D  Find the trigonometric function values for any angle whose terminal side makes an angle of 30°, 45°, or 60° with the *x*-axis. Using a calculator, find function values for any angle, and given a function value for any angle, find the angle.

E  Given a trigonometric function value for an angle and the quadrant in which the terminal side lies, find the other five function values.

1. **a)** Draw ∠*ABC*.

• *B*

*A* •

• *C*

**b)** Draw ∠*BAC*.

• *B*

*A* •

• *C*

An **angle** is the union of two rays with a common endpoint. Each ray is a **side** of the angle. If *B* is the common endpoint, it is called the **vertex.** If *A* is on one ray and *C* is on the other, and *B* is the vertex, we can name the angle as ∠*ABC*, where it is understood that the vertex is the middle letter.

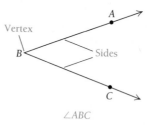

∠*ABC*

**DO EXERCISE 1.**

In trigonometry, we often think of an angle as a **rotation.** To do so, think of locating a ray along the positive *x*-axis with its endpoint at the origin. This ray is called the **initial side** of an angle. Though we leave that ray fixed, think of making a copy of it and rotating it. A rotation *counter-clockwise* is a **positive rotation,** and a rotation *clockwise* is a **negative rotation.** The ray at the end of the rotation is called the **terminal side** of the angle. The common endpoint is called the *vertex.* An angle so formed is said to be in **standard position.**

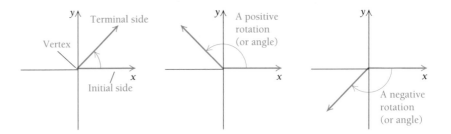

It is common to refer to an angle or rotation by Greek letters such as α (alpha), β (beta), γ (gamma), θ (theta), and φ (phi). A rotation need not stop after it goes around to the initial side. It can continue for more and more revolutions. If two rotations have the same terminal side, they are said to be **coterminal.** The following are two examples of coterminal angles.

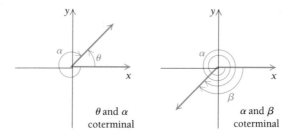

The measure of an angle or rotation may be given in **degrees.** The Babylonians developed the idea of dividing the circumference of a circle into 360 equal parts. Thus one complete positive revolution or rotation has a measure of 360 *degrees,* which we denote with a degree symbol as 360°. One half of a revolution has a measure of 180°, one fourth of a revolution has a measure of 90°, and so on. We also speak of an angle of measure 60°, 135°, 415°, and so on. Measures of negative rotations can be −38°, −270°, −750°, and so on.

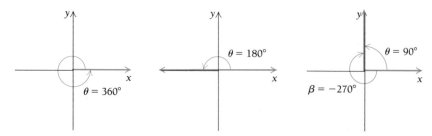

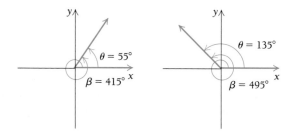

When the measure of an angle is greater than 360° or less than −360°, the rotating ray has gone through more than one complete revolution. For example, 415°, shown above, will have the same terminal side as 55°, since 415° = 360° + 55°. Thus the terminal side will be in the first quadrant. Note that angles of measure 415° and 55° are coterminal. We will often speak of an angle by its measure, rather than the more cumbersome terminology "an angle whose measure is . . . ." Thus we might say "the 45° angle," rather than "the angle whose measure is 45°."

**Examples**   In which quadrant does the terminal side of each angle lie?

1. 53°

   The terminal side lies in the *first* quadrant.

2. −126°

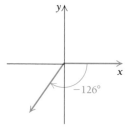

   The terminal side lies in the *third* quadrant.

3. 460°

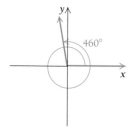

   The terminal side lies in the *second* quadrant.

2. How many degrees are there in:

   a) one revolution?

   b) one half of a revolution?

   c) one fourth of a revolution?

   d) one eighth of a revolution?

   e) one sixth of a revolution?

   f) one twelfth of a revolution?

3. In which quadrant does the terminal side of each angle lie?

   a) 47°

   b) 212°

   c) −43°

   d) −145°

   e) 365°

   f) −365°

   g) 740°

4. Find two positive angles and two negative angles that are coterminal with a 45° angle.

5. Classify each of the following angles as right, acute, obtuse, or straight.

   a)

   b)

   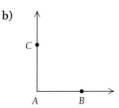

   c)

   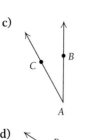

   d)

   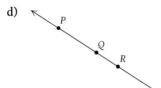

4. 253°

   The terminal side lies in the *third* quadrant.

5. −373°

   The terminal side lies in the *fourth* quadrant.    ◀

**DO EXERCISES 2 AND 3.** _____

Example 6    Find two positive angles and two negative angles that are coterminal with a 30° angle.

Solution    Consider a 30° angle or rotation. If we have a rotation that goes around another 360°, we get another angle, coterminal with 30°. If we go around twice, or 720°, we get a second angle coterminal with a 30° angle. Since

$$30° + 360° = 390° \quad \text{and} \quad 30° + 720° = 750°,$$

we know that 390° and 750° are two positive angles that are coterminal with a 30° angle. Many other answers are possible by adding multiples of 360°. Similarly, we get the negative coterminal angles by adding −360° and −720°:

$$30° + (−360°) = −330° \quad \text{and} \quad 30° + (−720°) = −690°.$$    ◀

**DO EXERCISE 4.** _____

| **DEFINITION** | **Classifications of Angles** |
|---|---|

*Right angle:*    An angle whose measure is 90°.

*Acute angle:*    An angle whose measure is between 0° and 90°.

*Obtuse angle:*    An angle whose measure is between 90° and 180°.

*Straight angle:*    An angle whose measure is 180°.

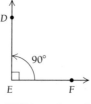

∠*DEF* is a right angle.

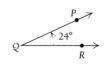

∠*PQR* is an acute angle.

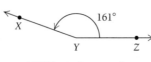

∠*XYZ* is an obtuse angle.

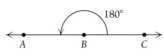

∠*ABC* is a straight angle.

**DO EXERCISE 5.** _____

| **DEFINITION** | **Complementary and Supplementary Angles** |
|---|---|

Two acute angles are *complementary* if their sum is 90°. Two positive angles are *supplementary* if their sum is 180°.

**Example 7**   Find the complement and the supplement of 56°23′16″.

**Solution**

$$90° = \begin{array}{r} 89°59'60'' \\ -56°23'16'' \\ \hline 33°36'44'' \end{array} \qquad 180° = \begin{array}{r} 179°59'60'' \\ -56°23'16'' \\ \hline 123°36'44'' \end{array}$$

Thus the complement of 56°23′16″ is 33°36′44″ and the supplement is 123°36′44″.   ◀

**DO EXERCISE 6.**

**Example 8**   Find the complement and the supplement of 87.46°.

**Solution**   We have

$$90° - 87.46° = 2.54°,$$
$$180° - 87.46° = 92.54°.$$

Thus the complement of 87.46° is 2.54° and the supplement is 92.54°.   ◀

**DO EXERCISE 7.**

 ## Trigonometric Functions of Angles or Rotations

Many applied problems in trigonometry involve the use of angles that are not acute. Thus we need to extend the domains of the trigonometric functions defined in Section 6.1 to angles, or rotations, of any size. To do this, we first consider a right triangle with one vertex at the origin of a coordinate system and one vertex on the positive *x*-axis. The other vertex is at *P*, a point on the circle whose center is at the origin and whose radius *r* is the length of the hypotenuse of the triangle. This triangle is a **reference triangle** for angle θ, which is in standard position.

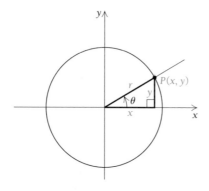

6. Find the complement and the supplement of 85°12′38″.

7. Find the complement and the supplement of 34.89°.

Note that three of the trigonometric functions of angle $\theta$ are defined as follows:

$$\sin \theta = \frac{\text{side opposite } \theta}{\text{hypotenuse}} = \frac{y}{r},$$

$$\cos \theta = \frac{\text{side adjacent to } \theta}{\text{hypotenuse}} = \frac{x}{r}$$

$$\tan \theta = \frac{\text{side opposite } \theta}{\text{side adjacent to } \theta} = \frac{y}{x}.$$

Since $x$ and $y$ are coordinates of the point $P$, we can also define these functions as follows:

$$\sin \theta = \frac{y\text{-coordinate}}{\text{radius}},$$

$$\cos \theta = \frac{x\text{-coordinate}}{\text{radius}},$$

$$\tan \theta = \frac{y\text{-coordinate}}{x\text{-coordinate}}.$$

We will use these definitions for functions of angles of any measure. The following figure shows angles whose terminal sides lie in quadrants II, III, and IV.

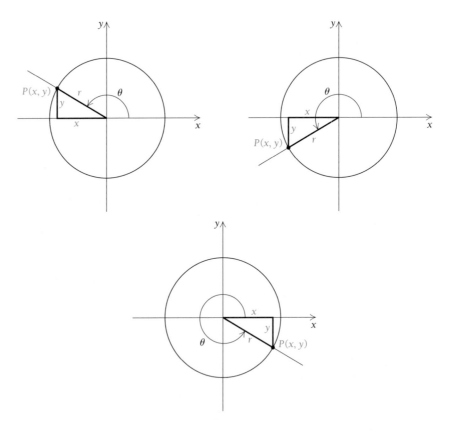

The point $P$, which is a point other than the vertex on the terminal side of the angle, can be anywhere on a circle of radius $r$. Its coordinates may be positive, negative, or zero, depending on the quadrant in which the terminal side lies. The length of the radius, which is also the hypotenuse of what we call a *reference triangle,* is always considered positive. The angle is always measured from the *positive* half of the x-axis. Regardless of the location of point $P$, we have

$$\sin \theta = \frac{y}{r}, \quad \cos \theta = \frac{x}{r}, \quad \text{and} \quad \tan \theta = \frac{y}{x}.$$

In quadrants I and II, $y$ and $r$ are both positive, so $\sin \theta$ is positive. In quadrants III and IV, $y$ is negative (and $r$ is positive), so $\sin \theta$ is negative.

---

**DEFINITION**          **Trigonometric Functions**

Suppose that $P(x, y)$ is any point on the terminal side of any angle $\theta$ in standard position, and $r$ is the radius, or distance, from the origin to $P(x, y)$. Then the trigonometric functions are defined as follows.

$$\sin \theta = \frac{\text{second coordinate}}{\text{radius}} = \frac{y}{r} \qquad \csc \theta = \frac{\text{radius}}{\text{second coordinate}} = \frac{r}{y}$$

$$\cos \theta = \frac{\text{first coordinate}}{\text{radius}} = \frac{x}{r} \qquad \sec \theta = \frac{\text{radius}}{\text{first coordinate}} = \frac{r}{x}$$

$$\tan \theta = \frac{\text{second coordinate}}{\text{first coordinate}} = \frac{y}{x} \qquad \cot \theta = \frac{\text{first coordinate}}{\text{second coordinate}} = \frac{x}{y}$$

---

Given any point $P(x, y)$ other than the vertex on the terminal side of an angle $\theta$ in standard position, we can find the length $r$ using the distance formula:

$$r = \sqrt{(x - 0)^2 + (y - 0)^2}$$
$$= \sqrt{x^2 + y^2}.$$

**Example 9**   Find the six trigonometric function values for the angle shown.

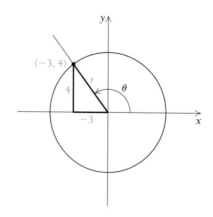

**Solution**   We first determine $r$, the distance from the origin to the point $(-3, 4)$:

$$r = \sqrt{x^2 + y^2}$$
$$= \sqrt{(-3)^2 + 4^2} \qquad \text{Substituting } -3 \text{ for } x \text{ and } 4 \text{ for } y$$
$$= \sqrt{9 + 16}$$
$$= \sqrt{25}, \quad \text{or } 5.$$

8. Find the six trigonometric function values for the angle shown.

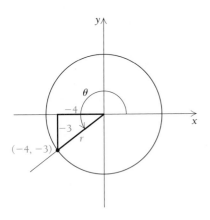

Using the definitions of the trigonometric functions, we can now find the function values for $\theta$. We substitute $-3$ for $x$, $4$ for $y$, and $5$ for $r$:

$$\sin \theta = \frac{y}{r} = \frac{4}{5} = 0.8, \qquad \csc \theta = \frac{r}{y} = \frac{5}{4} = 1.25,$$

$$\cos \theta = \frac{x}{r} = \frac{-3}{5} = -0.6, \qquad \sec \theta = \frac{r}{x} = \frac{5}{-3} \approx -1.67,$$

$$\tan \theta = \frac{y}{x} = \frac{4}{-3} \approx -1.33, \qquad \cot \theta = \frac{x}{y} = \frac{-3}{4} = -0.75. \qquad \blacktriangleleft$$

**Example 10**   Find the six trigonometric function values for the angle shown.

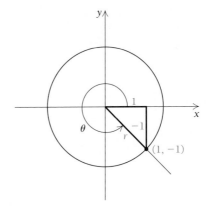

**Solution**   We first determine $r$, the distance from the origin to the point $(1, -1)$:

$$r = \sqrt{x^2 + y^2}$$
$$= \sqrt{1^2 + (-1)^2} \qquad \text{Substituting 1 for } x \text{ and } -1 \text{ for } y$$
$$= \sqrt{1 + 1}$$
$$= \sqrt{2}.$$

Substituting 1 for $x$, $-1$ for $y$, and $\sqrt{2}$ for $r$, we find that the trigonometric function values of $\theta$ are

$$\sin \theta = \frac{y}{r} = \frac{-1}{\sqrt{2}} = -\frac{\sqrt{2}}{2}, \qquad \csc \theta = \frac{r}{y} = \frac{\sqrt{2}}{-1} = -\sqrt{2},$$

$$\cos \theta = \frac{x}{r} = \frac{1}{\sqrt{2}} = \frac{\sqrt{2}}{2}, \qquad \sec \theta = \frac{r}{x} = \frac{\sqrt{2}}{1} = \sqrt{2},$$

$$\tan \theta = \frac{y}{x} = \frac{-1}{1} = -1, \qquad \cot \theta = \frac{x}{y} = \frac{1}{-1} = -1. \qquad \blacktriangleleft$$

9. Find the six trigonometric function values for the angle shown.

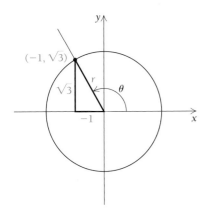

**DO EXERCISES 8 AND 9.**

Any point other than the origin on the terminal side of an angle can be used to determine the trigonometric function values of that angle. The function values are the same regardless of which point is used. We illustrate this in the following example.

**Example 11**   The terminal side of angle $\theta$ in standard position lies on the line $x + 2y = 0$. If the terminal side is in quadrant II, find $\sin \theta$, $\cos \theta$, and $\tan \theta$.

**Solution**   First we draw the graph of $x + 2y = 0$ and determine a second quadrant solution of the equation.

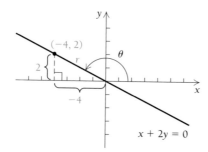

Using $(-4, 2)$, we then determine $r$:

$$r = \sqrt{(-4)^2 + 2^2} = \sqrt{20} = 2\sqrt{5}.$$

Then using

$$x = -4, \quad y = 2, \quad \text{and} \quad r = 2\sqrt{5},$$

we find that

$$\sin \theta = \frac{2}{2\sqrt{5}} = \frac{\sqrt{5}}{5}, \qquad \cos \theta = \frac{-4}{2\sqrt{5}} = -\frac{2\sqrt{5}}{5},$$

and

$$\tan \theta = \frac{2}{-4} = -\frac{1}{2}.$$

Would we get the same function values for $\theta$ if we had chosen a different point on the terminal side? The point $(-8, 4)$ is also a solution of the equation that is in the second quadrant and on the terminal side of $\theta$.

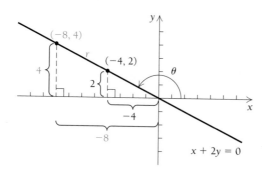

We determine $r$ and find the function values:

$$r = \sqrt{(-8)^2 + 4^2} = \sqrt{80} = 4\sqrt{5}.$$

Then

$$\sin \theta = \frac{4}{4\sqrt{5}} = \frac{\sqrt{5}}{5}, \qquad \cos \theta = \frac{-8}{4\sqrt{5}} = -\frac{2\sqrt{5}}{5},$$

$$\tan \theta = \frac{4}{-8} = -\frac{1}{2}.$$

10. Suppose that $\theta$ is an angle whose terminal side lies in quadrant III and has equation $y - 2x = 0$. Find $\sin \theta$, $\cos \theta$, and $\tan \theta$.

11. Complete the following table with the appropriate signs of the trigonometric functions of $\theta$ when the terminal side of angle $\theta$ lies in each of the four quadrants.

|  | I | II | III | IV |
|---|---|---|---|---|
| $\sin \theta$ |  |  |  |  |
| $\cos \theta$ |  |  |  |  |
| $\tan \theta$ |  |  |  |  |
| $\cot \theta$ |  |  |  |  |
| $\sec \theta$ |  |  |  |  |
| $\csc \theta$ |  |  |  |  |

Give the signs of the six trigonometric function values for each of the following rotations.

12. $-30°$

13. $160°$

14. $343°$

We see that the function values are the same as what we had obtained earlier.   ◀

Any point other than the origin on the terminal side of an angle can be used to determine the trigonometric function values.

> The trigonometric function values of $\theta$ depend only on the angle, not on the choice of the point on the terminal side.

**DO EXERCISE 10.** _____

## Signs of the Functions

Function values of the generalized trigonometric functions can be positive, negative, or zero, depending on where the terminal side of the angle lies. In the first quadrant, all function values are positive because $x$, $y$, and $r$ are all positive. In the second quadrant, first coordinates are negative and second coordinates are positive.

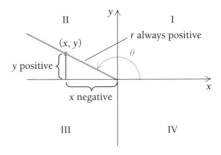

Thus if $\theta$ terminates in quadrant II,

$$\sin \theta = \frac{y}{r} \text{ is positive, because } y \text{ and } r \text{ are positive;}$$

$$\cos \theta = \frac{x}{r} \text{ is negative, because } x \text{ is negative and } r \text{ is positive;}$$

$$\tan \theta = \frac{y}{x} \text{ is negative, because } y \text{ is positive and } x \text{ is negative;}$$

and so on.

**DO EXERCISE 11.** _____

**Example 12**   Determine the signs of the six trigonometric function values for a rotation of $225°$.

Solution   We have

$$180° < 225° < 270°, \quad \text{so } P(x, y) \text{ is in the third quadrant.}$$

The tangent and the cotangent are positive, and the other four function values are negative.   ◀

**DO EXERCISES 12–14.** _____

## Terminal Side on an Axis

Now let us suppose that the terminal side of an angle falls on one of the axes. In that case, one of the coordinates is zero. The definitions of the functions still apply, but in some cases, functions will not be defined because a denominator will be 0.

**Example 13**   Find the trigonometric function values for 180°.

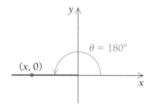

**Solution**   Let $(x, 0)$ represent any point, other than the vertex, on the terminal side of a 180° angle in standard position. We note that the first coordinate is negative, that the second coordinate is 0, and that $x$ and $r$ have the same absolute value ($r$ being always positive). Thus we have

$$\sin 180° = \frac{y}{r} = \frac{0}{r} = 0,$$

$$\cos 180° = \frac{x}{r} = -1, \qquad \text{Since } |x| = |r|, \text{ but } x \text{ and } r \text{ have opposite signs}$$

$$\tan 180° = \frac{y}{x} = \frac{0}{x} = 0,$$

$$\cot 180° = \frac{x}{y} = \frac{x}{0}, \qquad \text{Thus, } \cot 180° \text{ is undefined.}$$

$$\sec 180° = \frac{r}{x} = -1, \qquad \text{The reciprocal of } \cos 180°$$

$$\csc 180° = \frac{r}{y} = \frac{r}{0}. \qquad \text{Thus, } \csc 180° \text{ is undefined.} \qquad \blacktriangleleft$$

**DO EXERCISE 15.** _____

 ## Function Values for Any Angle

We are now able to determine trigonometric function values for many other angles. If the terminal side of an angle falls on one of the axes, the function values are 0, 1, −1, or undefined. Thus we can determine function values for any multiple of 90°. We can also determine the function values for any angle whose terminal side makes a 30°, 45°, or 60° angle with the $x$-axis.

Consider, for example, an angle of 150°. The terminal side makes a 30° angle with the $x$-axis, since 180° − 150° = 30°. As the figure below shows, $\triangle ONR$ is congruent to $\triangle ON'R'$; the ratios of the sides of the two triangles are the same. Thus the trigonometric function values are the same except perhaps for sign. We could determine the function values directly from $\triangle ONR$, but this is not necessary. If we remember that in quadrant II the sine is positive and the cosine and the tangent are negative, we can simply use the values of 30°, prefixing the appropriate sign. The $\triangle ONR$ is

15. Find the six trigonometric function values for 270°.

16. Find the trigonometric function values for 120°.

called a **reference triangle** and its acute angle, $\angle NOR$, is called a **reference angle.** The reference angle for a rotation, or angle, is the acute angle formed by the terminal side and the x-axis.

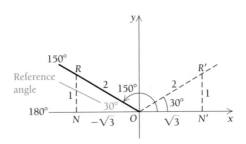

**Example 14**  Find the trigonometric function values for 225°.

Solution  We draw a figure showing the terminal side of a 225° angle. The reference angle is 225° − 180°, or 45°.

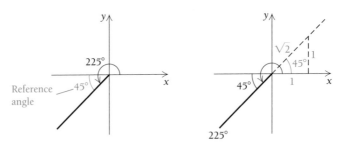

We recall from Section 6.1 that sin 45° = $\sqrt{2}/2$, cos 45° = $\sqrt{2}/2$, and tan 45° = 1. We also note that in the third quadrant, the sine and the cosine are negative and the tangent is positive. We can easily determine the other three function values—the cosecant, the secant, and the cotangent—by remembering that they are the respective reciprocals of the sine, cosine, and tangent function values. Thus we have

$$\sin 225° = -\frac{\sqrt{2}}{2}, \qquad \csc 225° = -\frac{2}{\sqrt{2}}, \quad \text{or} -\sqrt{2},$$

$$\cos 225° = -\frac{\sqrt{2}}{2}, \qquad \sec 225° = -\frac{2}{\sqrt{2}}, \quad \text{or} -\sqrt{2},$$

$$\tan 225° = 1, \qquad \cot 225° = 1.$$

◄

**DO EXERCISE 16.** _____

**Example 15**  Find the trigonometric function values for 660°.

Solution  We find the multiple of 180° nearest 660°:

$$180° \times 2 = 360°$$
$$180° \times 3 = 540°$$

and ⟵ 660°

$$180° \times 4 = 720°.$$

The nearest multiple is 720°. The difference between 720° and 660° is 60°. This gives us the reference angle.

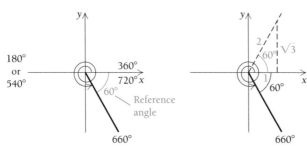

**17.** Find the trigonometric function values for 570°.

We recall that sin 60° = $\sqrt{3}/2$, cos 60° = 1/2, and tan 60° = $\sqrt{3}$. In the fourth quadrant, the cosine is positive and the sine and tangent are negative. Thus we have

$$\sin 660° = -\frac{\sqrt{3}}{2}, \qquad \csc 660° = -\frac{2}{\sqrt{3}}, \quad \text{or } -\frac{2\sqrt{3}}{3},$$

$$\cos 660° = \frac{1}{2}, \qquad \sec 660° = \frac{2}{1} = 2,$$

$$\tan 660° = -\sqrt{3}, \qquad \cot 660° = -\frac{1}{\sqrt{3}}, \quad \text{or } -\frac{\sqrt{3}}{3}.$$

**DO EXERCISE 17.** _____

We can use the same procedure for negative rotations.

**Example 16**   Find the sine, the cosine, and the tangent of $-1050°$.

**Solution**   We find the multiple of $-180°$ nearest $-1050°$:

$$-180° \times 5 = -900° \quad \text{and} \quad -180° \times 6 = -1080°.$$

The nearest multiple is $-1080°$. The difference between $-1050°$ and $-1080°$ is 30°. This gives us the reference angle.

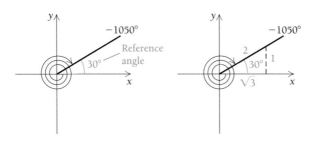

**18.** Find the trigonometric function values for $-945°$.

All trigonometric function values are positive in the first quadrant, and sin 30° = 1/2, cos 30° = $\sqrt{3}/2$, and tan 30° = $\sqrt{3}/3$. Thus we have

$$\sin(-1050°) = \frac{1}{2}, \qquad \cos(-1050°) = \frac{\sqrt{3}}{2},$$

and

$$\tan(-1050°) = \frac{\sqrt{3}}{3}.$$

**DO EXERCISE 18.** _____

**19.** An airplane flies 150 km from an airport in a direction of 120°. How far east of the airport is the plane then? How far south?

This text presents two methods of giving direction, or bearing. We discussed the first type in Example 5 of Section 6.2. The second type is in Example 17 below.

*Bearing: Second type.*    In aerial navigation, directions are given in degrees clockwise from north. Thus, east is 90°, south is 180°, and west is 270°. Several aerial directions or bearings are given below.

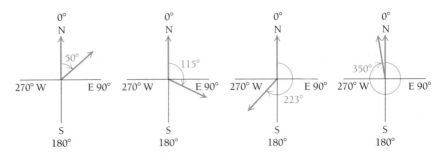

**Example 17**    An airplane leaves an airport and travels for 100 mi in a direction of 300°. How far north of the airport is the plane then? How far west?

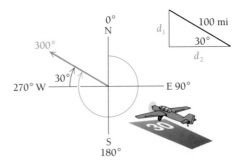

**Solution**    The direction of flight is shown in the figure. In the triangle, $d_1$ is the northerly distance and $d_2$ is the westerly distance. Then

$$\frac{d_1}{100} = \sin 30° \quad \text{and} \quad \frac{d_2}{100} = \cos 30°.$$

Thus, to the nearest mile, we have

$$d_1 = 100 \sin 30° = 50 \text{ mi};$$
$$d_2 = 100 \cos 30° = 87 \text{ mi}.$$

The plane is 50 mi north and 87 mi west of the airport.

**DO EXERCISE 19.** ⎯⎯⎯⎯⎯⎯⎯⎯⎯⎯⎯⎯⎯⎯

When the terminal side of an angle does not fall on one of the axes or make a 30°, 45°, or 60° angle with the *x*-axis, we can find the function values using tables or a calculator. With a calculator, we can find the trigonometric function values of *any* angle without using a reference angle.

**Example 18**    Find cos 823°.

**Solution**    We enter 823 and then press the $\boxed{\text{COS}}$ key. We find cos 823° ≈ −0.2250.

**DO EXERCISE 20.** _____

**Example 19**   Find csc (−389°).

Solution   The cosecant function value is the reciprocal of the sine function value. We enter −389, press the SIN key, and then press the reciprocal key:

$$\csc(-389°) = \frac{1}{\sin(-389°)} \approx -2.063.$$

◀

**DO EXERCISE 21.** _____

**Examples**   Find the specified function value.

20.  cos 112.7° ≈ −0.3859
21.  tan 421°28′ ≈ tan 421.47° ≈ 1.839
22.  sin (−180°34′) ≈ sin (−180.57°) ≈ 0.0099
23.  sec 2400.15° ≈ −2.009      $\sec 2400.15° = \dfrac{1}{\cos 2400.15°}$
24.  csc (−560°17′) ≈ csc (−560.28°) ≈ 2.885

◀

**DO EXERCISES 22–27.** _____

In many applications, we have a trigonometric function value and want to find the measure of an angle. To do so, we use a calculator or a table in reverse.

**Example 25**   Given that sin θ = 0.2812, find θ between 90° and 180°.

Solution   We first find the reference angle. We enter 0.2812 and press the SIN⁻¹ key. The reference angle is approximately 16.33°.

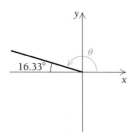

We find the angle θ by subtracting 16.33° from 180°:

180° − 16.33° = 163.67°.

Thus, θ ≈ 163.67°.

◀

**Example 26**   Given that tan θ = −6.2051, find θ between 270° and 360°.

Solution   We find the reference angle, ignoring the fact that tan θ is negative. We enter 6.2051 and press the TAN⁻¹ key. The reference angle is approximately 80.85°.

---

20.  Find sin 229°.

21.  Find cot (−1007°).

Find the specified function value.

22.  sin 243.8°

23.  cot 515°16′

24.  sin (−270°41′)

25.  csc 1086.22°

26.  cos (−93°)

27.  tan 175°20′

Find $\theta$ in the interval indicated.

**28.** $\cos \theta = 0.5712$, $(270°, 360°)$

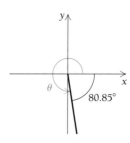

We find the angle $\theta$ by subtracting $80.85°$ from $360°$:

$$360° - 80.85° = 279.15°.$$

Thus, $\theta \approx 279.15°$.    ◀

**DO EXERCISES 28–31.** _____

## The Six Trigonometric Functions Related

The six trigonometric functions are related in a number of interesting ways. When we know one of the function values for an angle, we can find the other five if we know the quadrant in which the terminal side lies. The idea is to sketch a triangle in the appropriate quadrant, use the Pythagorean theorem as needed to find the lengths of its sides, and then read off the ratios of the sides.

**29.** $\tan \theta = -2.4778$, $(90°, 180°)$

**Example 27**    Given that $\tan \theta = -\frac{2}{3}$ and $\theta$ is in the second quadrant, find the other function values.

**Solution**    We first sketch a second-quadrant triangle. Since $\tan \theta = -\frac{2}{3}$, we make the legs of lengths 2 and 3. The hypotenuse must then have length $\sqrt{13}$. Now we can read off the appropriate ratios:

**30.** $\sin \theta = -0.2363$, $(180°, 270°)$

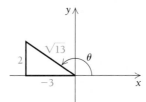

**31.** $\sec \theta = -1.2020$, $(180°, 270°)$

$$\sin \theta = \frac{2}{\sqrt{13}}, \qquad \csc \theta = \frac{\sqrt{13}}{2},$$

$$\cos \theta = -\frac{3}{\sqrt{13}}, \qquad \sec \theta = -\frac{\sqrt{13}}{3},$$

$$\tan \theta = -\frac{2}{3}, \qquad \cot \theta = -\frac{3}{2}.$$

◀

**Example 28**    Given that $\cos \theta = -\frac{3}{4}$ and $\theta$ is in the third quadrant, find $\sin \theta$ and $\tan \theta$.

**Solution**    We first sketch a third-quadrant triangle. Since $\cos \theta = -\frac{3}{4}$, we make the hypotenuse 4 and the leg on the $x$-axis 3. The other leg must have length $\sqrt{7}$. Now we can read off the appropriate ratios:

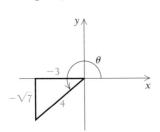

$$\sin \theta = -\frac{\sqrt{7}}{4},$$

$$\tan \theta = \frac{\sqrt{7}}{3}.$$

**DO EXERCISES 32 AND 33.**

**32.** Given that $\cot \theta = -3$ and the terminal side is in quadrant II, find the other function values.

**33.** Given that $\cos \theta = \frac{3}{4}$ and the terminal side is in quadrant IV, find the other function values.

---

## ● EXERCISE SET 6.3

**A**    For angles of the following measures, state in which quadrant the terminal side lies.

**1.** 34°      **2.** 320°      **3.** −120°

**4.** 175°     **5.** −210°     **6.** −345°

**7.** 800°     **8.** 1075°     **9.** −460°

**10.** 537°    **11.** −189°    **12.** −912°

Find two positive angles and two negative angles that are co-terminal with the given angle.

**13.** 58°                    **14.** 320°

**15.** −120°                  **16.** −215°

Find the complement and the supplement.

**17.** 57°23′                 **18.** 47°38′

**19.** 73°45′11″              **20.** 12°03′14″

**21.** 67.31°                 **22.** 13.68°

**23.** 11.2344°               **24.** 85.06312°

**B**    Find the six trigonometric function values for the angle θ.

**25.**                        **26.**

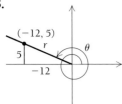            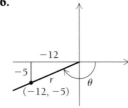

**27.**                        **28.**

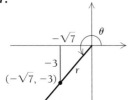            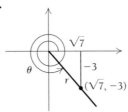

The terminal side of angle θ in standard position lies on the given line in the given quadrant. Find sin θ, cos θ, and tan θ.

**29.** $2x + 3y = 0$; quadrant IV

**30.** $4x + y = 0$; quadrant II

**31.** $5x - 4y = 0$; quadrant I

**32.** $y = 0.8x$; quadrant III

Find the signs of the six trigonometric function values for the given rotation.

**33.** 319°                   **34.** −57°

**35.** −620°                  **36.** 194°

**37.** −215°                  **38.** 290°

**39.** 91°                    **40.** −272°

**C**, **D**    Find the following, if they exist.

**41.** sin 360°               **42.** cos 180°

**43.** tan 240°               **44.** cot 240°

**45.** sec 315°               **46.** csc 315°

**47.** tan 90°                **48.** cot (−270°)

**49.** sin 150°               **50.** cos 150°

**51.** sec (−210°)            **52.** csc (−210°)

**53.** sec (−90°)             **54.** csc 720°

**55.** csc 225°               **56.** sin 225°

**57.** cot 570°               **58.** tan 570°

**59.** sec 300°               **60.** csc 300°

**61.** csc 270°               **62.** sin (−450°)

**63.** tan 480°               **64.** cot 855°

**65.** sin (−45°)             **66.** cos (−120°)

**67.** cot (−225°)            **68.** sec (−225°)

**69.** cos (−540°)            **70.** sec 180°

**71.** sin 1050°              **72.** cos 675°

**73.** tan (−135°)            **74.** sin (−135°)

**75.** sec 1125°              **76.** csc 1125°

**77.** sin 495°

**78.** cos 495°

**79.** tan 330°

**80.** cot 330°

**81.** sec (−855°)

**82.** csc (−840°)

**83.** cos 5220°

**84.** sin 7560°

Use a calculator in Exercises 85–88, but do not use the trigonometric function keys.

**85.** Given that

$$\sin 41° = 0.6561,$$
$$\cos 41° = 0.7547,$$
$$\tan 41° = 0.8693,$$

find the trigonometric function values for 319°.

**86.** Given that

$$\sin 27° = 0.4540,$$
$$\cos 27° = 0.8910,$$
$$\tan 27° = 0.5095,$$

find the trigonometric function values for 333°.

**87.** Given that

$$\sin 65° = 0.9063,$$
$$\cos 65° = 0.4226,$$
$$\tan 65° = 2.1445,$$

find the trigonometric function values for 115°.

**88.** Given that

$$\sin 35° = 0.5736,$$
$$\cos 35° = 0.8192,$$
$$\tan 35° = 0.7002,$$

find the trigonometric function values for 215°.

**89.** An airplane travels at 120 km/h for 2 hr in a direction of 243° from Chicago. At the end of this time, how far south of Chicago is the plane?

**90.** An airplane travels at 150 km/h for 2 hr in a direction of 138° from Omaha. At the end of this time, how far southeast of Omaha is the plane?

Find the following function values.

**91.** tan 295°14′

**92.** cos 230°53′

**93.** sec 146.9°

**94.** sin 98.4°

**95.** sin 756°25′

**96.** cot 820°40′

**97.** cos (−1000.85°)

**98.** tan (−1086.15°)

**99.** cot (−16°37′)

**100.** csc (−13°28′)

**101.** sin 3824°

**102.** cos 5417°

Find θ in the interval indicated.

**103.** sin θ = −0.9956, (270°, 360°)

**104.** sin θ = 0.4313, (90°, 180°)

**105.** cos θ = −0.9388, (180°, 270°)

**106.** cos θ = −0.0990, (90°, 180°)

**107.** tan θ = 0.2460, (180°, 270°)

**108.** tan θ = −3.0541, (270°, 360°)

**109.** sec θ = −1.0485, (90°, 180°)

**110.** csc θ = 1.0480, (0°, 90°)

In Exercises 111–116, a function value and a quadrant are given. Find the other five function values.

**111.** $\sin \theta = -\frac{1}{3}$, III

**112.** $\sin \theta = -\frac{1}{5}$, IV

**113.** $\cos \theta = \frac{3}{5}$, IV

**114.** $\cos \theta = -\frac{4}{5}$, II

**115.** $\cot \theta = -2$, IV

**116.** $\tan \theta = 5$, III

● SYNTHESIS

**117.** The valve cap on a bicycle wheel is 24.5 in. from the center of the wheel. From the position shown, the wheel starts rolling. After the wheel has turned 390°, how far above the ground is the valve cap? Assume that the outer radius of the tire is 26 in.

**118.** The seats of a ferris wheel are 35 ft from the center of the wheel. When you board the wheel, you are 5 ft above the ground. After you have rotated through an angle of 765°, how far above the ground are you?

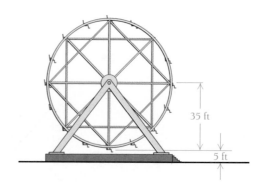

# 6.4

# RADIANS, ARC LENGTH, AND ANGULAR SPEED

Another useful unit of angle measure is called a *radian*. To introduce radian measure and to further develop the trigonometric, or circular, functions in Section 6.5, we use a circle centered at the origin with a radius of length 1. Such a circle is called a **unit circle.** It has equation $u^2 + v^2 = 1$. We use $u$ and $v$ for the variables here because we wish to reserve $x$ and $y$ for later purposes.

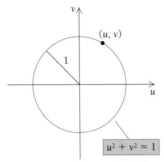

**OBJECTIVES**

You should be able to:

**A**   Given a real number that is an integral multiple of halves, thirds, fourths, or sixths of $\pi$, find on the unit circle the point determined by it. Given a point on the unit circle that is an integral multiple of fourths, sixths, eighths, or twelfths of the distance around the circle, find the real numbers between $-2\pi$ and $2\pi$ that determine that point.

**B**   Given the coordinates of a point on the unit circle, find its reflection across the $u$-axis, the $v$-axis, and the origin.

**C**   Convert between radian measure and degree measure.

**D**   Find the length of an arc of a circle, given the measure of its central angle and the length of a radius. Also, find the measure of a central angle of a circle, given the length of its arc and the length of a radius.

**E**   Convert between linear and angular speed.

**F**   Find total distance or total angle, when speed, radius, and time are given.

## A   Distances on the Unit Circle

The circumference of a circle of radius $r$ is $2\pi r$. Thus for the unit circle, where $r = 1$, the circumference is $2\pi$. If a point starts at $A$ and travels counterclockwise around the circle, it will travel a distance of $2\pi$. If it travels halfway around the circle, it will travel a distance of $\frac{1}{2} \cdot 2\pi$, or $\pi$.

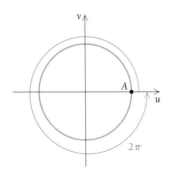

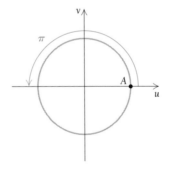

**Example 1**   How far will a point travel if it goes $\frac{1}{12}$ of the way around the unit circle?

**Solution**   The distance will be $\frac{1}{12}$ of the total distance around the circle, or $\frac{1}{12} \cdot 2\pi$, which is $\pi/6$.

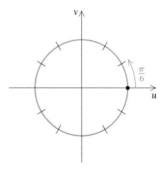

1. How far will a point travel if it goes:

   a) $\frac{1}{4}$ of the way around the unit circle?

   b) $\frac{3}{8}$ of the way around the unit circle?

   c) $\frac{3}{4}$ of the way around the unit circle?

2. How far will a point move on the unit circle, counterclockwise, going from point $A$ to:

   a) point $M$?

   b) point $N$?

   c) point $P$?

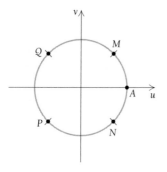

3. How far will a point move on the unit circle, counterclockwise, in going from point $A$ to:

   a) point $F$?

   b) point $H$?

   c) point $J$?

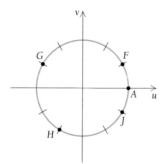

**DO EXERCISE 1.** _____

If a point travels $\frac{1}{8}$ of the way around the circle, it will travel a distance of $\frac{1}{8} \cdot 2\pi$, or $\pi/4$. Note that the stopping point is $\frac{1}{4}$ of the way from $A$ to $C$ and $\frac{1}{2}$ of the way from $A$ to $B$.

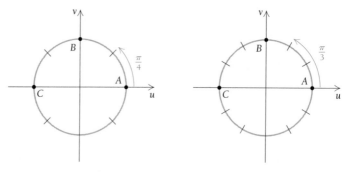

If a point travels $\frac{1}{6}$ of the way around the circle ($\frac{2}{12}$ of the way, as shown), it will travel a distance of $\frac{1}{6} \cdot 2\pi$, or $\pi/3$. Note that the stopping point is $\frac{1}{3}$ of the way from $A$ to $C$ and $\frac{2}{3}$ of the way from $A$ to $B$.

**Example 2**   How far will a point move on the unit circle, counterclockwise, going from point $A$ to point $Q$?

**Solution**   If the point went to $C$, it would move a distance of $\pi$. It goes $\frac{3}{4}$ of that distance to get to $Q$. Hence it goes a distance of $3\pi/4$ to get to $Q$.

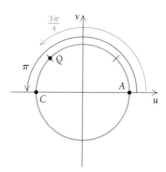

**Example 3**   How far will a point move on the unit circle, counterclockwise, going from point $A$ to point $G$?

**Solution**   If the point went to $C$, it would move a distance of $\pi$. It goes $\frac{5}{6}$ of that distance to $G$. Hence it goes a distance of $5\pi/6$ to get to $G$.

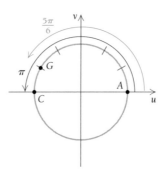

**DO EXERCISES 2 AND 3.** _____

A point may travel completely around the circle and then continue. For example, if it goes around once and then continues $\frac{1}{4}$ of the way around, it will have traveled a distance of $2\pi + \frac{1}{4} \cdot 2\pi$, or $5\pi/2$.

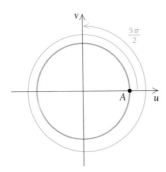

Any positive number thus determines a point on the unit circle. For the number 35, for example, we start at $A$ and travel counterclockwise a distance of 35. The point at which we stop is the point "determined" by the number 35.

**Example 4**   On the unit circle, mark the point determined by $5\pi/4$.

**Solution**   From $A$ to $C$, the distance is $\pi$, or $(\frac{4}{4})\pi$, so we need to go another distance of $\pi/4$. That is, think of $5\pi/4$ as $\pi + \pi/4$.

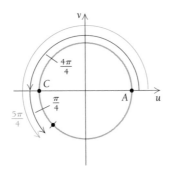

**DO EXERCISES 4 AND 5.**

Negative numbers also determine points on the unit circle. For a negative number, we move *clockwise* around the circle. Points for $-\pi/4$ and $-3\pi/2$ are shown. Any negative number thus determines a point on the unit circle. The number 0 determines the point $A$.

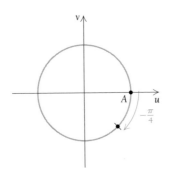
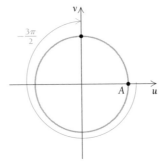

4. On this unit circle, mark the points determined by each of the following.

   a) $\dfrac{\pi}{4}$     b) $\dfrac{7\pi}{4}$

   c) $\dfrac{9\pi}{4}$     d) $\dfrac{13\pi}{4}$

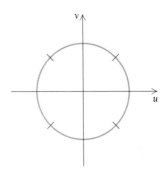

5. On this unit circle, mark the points determined by each of the following.

   a) $\dfrac{\pi}{6}$     b) $\dfrac{7\pi}{6}$

   c) $\dfrac{13\pi}{6}$     d) $\dfrac{25\pi}{6}$

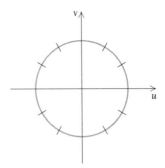

6. On this unit circle, mark the points determined by each of the following.

a)  $-\dfrac{\pi}{2}$     b)  $-\dfrac{3\pi}{4}$

c)  $-\dfrac{7\pi}{4}$     d)  $-2\pi$

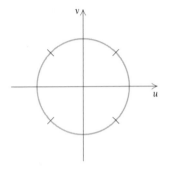

---

> There is a point on the unit circle for every real number.

**Example 5**   On the unit circle, mark the point determined by $-7\pi/6$.

**Solution**   From $A$ to $C$, the distance is $\pi$, or $\frac{6}{6}\pi$, clockwise so we need to go another distance of $\pi/6$, clockwise.

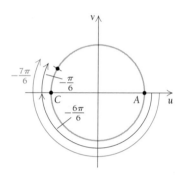

**DO EXERCISE 6.** _____

## ▶ **Reflections**

A unit circle is symmetric with respect to the $u$-axis, the $v$-axis, and the origin. We can use the coordinates of one point to find coordinates of reflections.

**Example 6**   The point $E$ on the unit circle, as shown below, has coordinates $(\sqrt{2}/2, \sqrt{2}/2)$. Find the coordinates of points $F$, $G$, and $H$, which are the reflections of $E$.

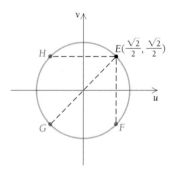

**Solution**

a)  Point $F$ is the reflection of point $E$ across the $u$-axis. Hence its coordinates are $(\sqrt{2}/2, -\sqrt{2}/2)$.

b)  Point $G$ is the reflection of point $E$ across the origin. Hence its coordinates are $(-\sqrt{2}/2, -\sqrt{2}/2)$.

c)  Point $H$ is the reflection of point $E$ across the $v$-axis. Hence its coordinates are $(-\sqrt{2}/2, \sqrt{2}/2)$.

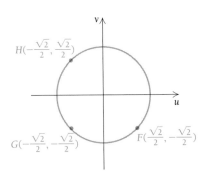

**DO EXERCISE 7.**

**Example 7**   The point $(1/2, \sqrt{3}/2)$ is on the unit circle. Find its reflection across (a) the $u$-axis, (b) the $v$-axis, and (c) the origin.

Solution

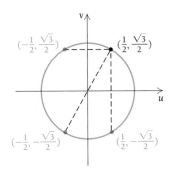

a)  The reflection across the $u$-axis is $(1/2, -\sqrt{3}/2)$.
b)  The reflection across the $v$-axis is $(-1/2, \sqrt{3}/2)$.
c)  The reflection across the origin is $(-1/2, -\sqrt{3}/2)$.

By symmetry, all these points are on the circle.

**DO EXERCISES 8 AND 9.**

 **Radian Measure**

Degree measure is a common unit of angle measure in many everyday applications and in such fields as navigation and surveying. But in many scientific fields and in mathematics and calculus, there is another commonly used unit of measure called the *radian*.

Consider a circle with its center at the origin and radius of length 1 (a *unit circle*). Suppose we measure an arc, moving counterclockwise, of length 1, and mark a point $T$ on the circle. If we draw a ray from the origin through $T$, we have formed an angle. The measure of that angle is 1 **radian.** The word radian comes from the word *radius*. Thus, measuring 1 radius along the circumference of the circle determines an angle whose measure is 1 *radian*. One radian is about $57.3°$.

**7.** The point $M$, as shown, has coordinates $(\sqrt{3}/2, 1/2)$. Find the coordinates of the points $N$, $P$, and $R$, which are reflections of $M$.

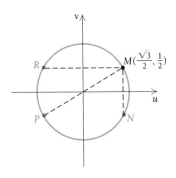

**8.** The point $(\frac{3}{5}, -\frac{4}{5})$ is on the unit circle. Find the coordinates of its reflection across:

a)  the $u$-axis;

b)  the $v$-axis;

c)  the origin.

d)  Are these points on the circle? Why?

**9.** The point $(-\sqrt{35}/6, -1/6)$ is on the unit circle. Find the coordinates of its reflection across:

a)  the $u$-axis;

b)  the $v$-axis;

c)  the origin.

d)  Are these points on the circle? Why?

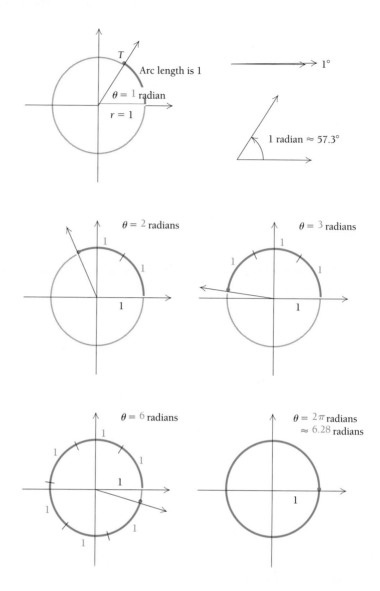

Angles that measure 2 radians and 3 radians are also shown above. When we make a complete (counterclockwise) revolution, the terminal side coincides with the initial side on the positive *x*-axis. We then have an angle whose measure is $2\pi$ radians, or about 6.28 radians, which is the circumference of the circle:

$$2\pi r = 2\pi(1) = 2\pi.$$

Thus a rotation of 360° (1 revolution) has a measure of $2\pi$ radians. Half of a revolution is a rotation of 180°, or $\pi$ radians. A quarter revolution is a rotation of 90°, or $\pi/2$ radians, and so on.

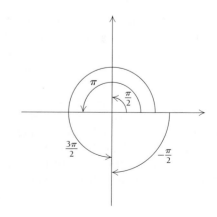

**DO EXERCISE 10.** _____

To convert between degrees and radians, we first note that

$360° = 2\pi$ radians.

It follows that

$$180° = \pi \text{ radians.}$$

To make conversions, we use the notion of "multiplying by one" (see Section 1.9), noting the following:

$$\frac{\pi \text{ radians}}{180°} = \frac{180°}{\pi \text{ radians}} = 1.$$

When a rotation is given in radians, the word "radians" is optional and is most often omitted. Thus, if no unit is given for a rotation, the rotation is understood to be in radians.

**Example 8**   Convert 60° to radians.

**Solution**   We have

$$60° = 60° \cdot \frac{\pi \text{ radians}}{180°}$$

$$= \frac{60°}{180°} \pi \text{ radians}$$

$$= \frac{\pi}{3} \text{ radians,} \quad \text{or} \quad \frac{\pi}{3}.$$

We find that $\pi/3$ radians is about 1.047 radians.

◀

**Example 9**   Convert $\frac{3\pi}{4}$ radians to degrees.

**Solution**

$$\frac{3\pi}{4} \text{ radians} = \frac{3\pi}{4} \text{ radians} \cdot \frac{180°}{\pi \text{ radians}}$$

$$= \frac{3\pi}{4\pi} \cdot 180°$$

$$= \frac{3}{4} \cdot 180°$$

$$= 135°$$

◀

**Example 10**   Convert 1 radian to degrees.

10. In which quadrant does the terminal side of each angle lie?

a) $\dfrac{5\pi}{4}$

b) $\dfrac{17\pi}{8}$

c) $-\dfrac{\pi}{15}$

d) $37.3\pi$

11. Convert to radian measure. Leave answers in terms of $\pi$.

a) $225°$

b) $315°$

c) $-720°$

12. Convert to radian measure. Do not leave answers in terms of $\pi$ (use 3.14 for $\pi$).

a) $72\frac{1}{2}°$ (a safe working angle for ladders)

b) $300°$

c) $-315°$

13. Convert to degree measure.

a) $\dfrac{4\pi}{3}$

b) $\dfrac{5\pi}{2}$

c) $-\dfrac{4\pi}{5}$

**Solution**

$$1 \text{ radian} = 1 \text{ radian} \cdot \frac{180°}{\pi \text{ radians}}$$

$$= \frac{180°}{\pi}$$

$$\approx 57.3°$$

◄

**DO EXERCISES 11–13.**

## ▷ Arc Length and Central Angles

Radian measure can be determined using a circle other than a unit circle. In the following figure, a unit circle is shown along with another circle. The angle shown is a **central angle** of both circles; hence the arcs that it intercepts have their lengths in the same ratio as the radii of the circles. The radii of the circles are $r$ and 1. The corresponding arc lengths are $s$ and $s_1$.

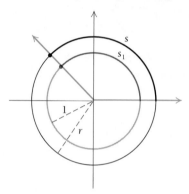

Thus we have the proportion

$$\frac{s}{s_1} = \frac{r}{1}, \quad \text{or} \quad \frac{s}{r} = \frac{s_1}{1}.$$

Now $s_1$ is the radian measure of the rotation in question. It is more common to use a Greek letter, such as $\theta$, for the measure of an angle or rotation. We commonly use the letter $s$ for arc length. Adopting this convention, we rewrite the proportion above as $s/r = \theta$. In any circle, arc length, central angle, and length of radius are related in this fashion. Or, in general, the following is true.

---

**DEFINITION**        **Radian Measure**

The *radian measure* $\theta$ of a rotation is the ratio of the distance $s$ traveled by a point at a radius $r$ from the center of rotation, to the length of the radius $r$:

$$\theta = \frac{s}{r}.$$

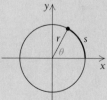

**Example 11**   Find the measure of a rotation in radians where a point 2 m from the center of rotation travels 4 m.

**Solution**

$$\theta = \frac{s}{r} = \frac{4 \text{ m}}{2 \text{ m}} = 2 \qquad \text{The unit is understood to be radians.}$$   ◀

**Example 12**   Find the length of an arc of a circle of 5-cm radius associated with a central angle of $\pi/3$ radians.

**Solution**   We have

$$\theta = \frac{s}{r}, \quad \text{or} \quad s = r\theta.$$

Therefore, $s = 5 \cdot \pi/3$ cm, or about 5.23 cm.   ◀

A look at Examples 11 and 12 will show why the word radian is most often omitted. In Example 11, we have the division 4 m/2 m, which simplifies to the number 2, since m/m = 1. From this point of view, it would seem preferable to omit the word radians. In Example 12, had we used the word radians all the way through, our answer would have come out to be 5.23 cm-radians. It is a distance we seek; hence we know the unit should be centimeters. Thus we must omit the word radians. Since a measure in radians is simply a number, it is usually preferable to omit the word radians.

---
**CAUTION!**   In using the formula $\theta = s/r$, you must make sure that $\theta$ is in radians and that $s$ and $r$ are expressed in the same unit.

---

**DO EXERCISES 14–16.**

   **Angular Speed**

**Linear speed** is defined to be distance traveled per unit of time. Similarly, **angular speed** is defined to be amount of rotation per unit of time. For example, we might speak of the angular speed of a wheel as 150 revolutions per minute or the angular speed of the earth as $2\pi$ radians per day. The Greek letter $\omega$ (omega) is generally used for angular speed. Thus angular speed is defined as

$$\omega = \frac{\theta}{t}.$$

**Relating Linear and Angular Speed**

For many applications, it is important to know a relationship between *angular speed* and *linear speed*. For example, we might wish to find the linear speed of a point on the earth, knowing its angular speed. Or, we might wish to know the linear speed of an earth satellite, knowing its angular speed. To develop the relationship we seek, we recall the relation between angle and distance: $\theta = s/r$. This is equivalent to

$$s = r\theta.$$

14. Find the length of an arc of a circle with 10-cm radius associated with a central angle of measure $11\pi/6$.

15. Find the radian measure of a rotation where a point 2.5 cm from the center of rotation travels 15 cm.

16. Find the radian measure of a rotation where a point 24 in. from the center of rotation travels 3 ft.

We divide by time, $t$, to obtain

$$\frac{s}{t} = r\frac{\theta}{t}$$

$$\downarrow \qquad \downarrow$$
$$v \qquad \omega$$

Now $s/t$ is linear speed $v$, and $\theta/t$ is angular speed $\omega$. Thus we have the relation we seek.

---

**DEFINITION**                      **Linear Speed**

The *linear speed* $v$ of a point a distance $r$ from the center of rotation is given by

$v = r\omega$,

where $\omega$ is the *angular speed* in radians per unit time.

---

In deriving this formula, we used the equation $s = r\theta$, in which the units for $s$ and $r$ must be the same and $\theta$ must be in radians. So, for our new formula $v = r\omega$, the units of distance for $v$ and $r$ must be the same, $\omega$ must be in radians per unit of time, and the units of time must be the same for $v$ and $\omega$.

**Example 13**   An earth satellite in circular orbit 1200 km high makes one complete revolution every 90 min. What is its linear speed? Use 6400 km for the length of a radius of the earth.

1200 km

**Solution**   We will use the formula

$v = r\omega$;

thus we will need to know $r$ and $\omega$:

$r = 6400 \text{ km} + 1200 \text{ km}$          Radius of earth plus height of satellite

$= 7600 \text{ km},$

$\omega = \dfrac{\theta}{t} = \dfrac{2\pi \text{ radians}}{90 \text{ min}} = \dfrac{\pi}{45 \text{ min}}.$          We have, as usual, omitted the word radians.

Now, using $v = r\omega$, we have

$$v = 7600 \text{ km} \cdot \frac{\pi}{45 \text{ min}} = \frac{7600\pi}{45} \cdot \frac{\text{km}}{\text{min}}$$

$$\approx 531 \frac{\text{km}}{\text{min}}.*$$

◀

---

\* It may be appropriate at this point to review Section 1.9 on handling units.

**DO EXERCISE 17.** _____

**Example 14**   An anchor is being hoisted at a rate of 2 ft/sec, the chain being wound around a capstan with a 1.8-yd diameter. What is the angular speed of the capstan?

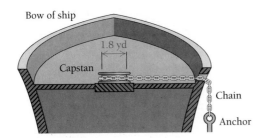

Bow of ship

1.8 yd

Capstan

Chain

Anchor

**Solution**   We will use the formula $v = r\omega$ in the form $\omega = v/r$, taking care to use the proper units. Since $v$ is given in feet per second, we need $r$ in feet. Then $\omega$ will be in radians per second:

$$r = \frac{1.8}{2}\,\text{yd} \cdot \frac{3\,\text{ft}}{\text{yd}} = 2.7\,\text{ft},$$

$$\omega = \frac{v}{r} = \frac{2}{2.7} \approx 0.741\,\text{radian/sec}.$$ ◀

> **CAUTION!**   In applying the formula $v = r\omega$, we must be sure that the distance units for $v$ and $r$ are the same and that $\omega$ is in radians per unit of time. The units of time must be the same for $v$ and $\omega$.

**DO EXERCISES 18–20.** _____

## Total Distance and Total Angle

The formulas $\theta = s/r$ and $v = r\omega$ can be used in combination to find distances and angles in various situations involving rotational motion.

**Example 15**   A car is traveling at a speed of 45 mph. Its tires have a 20-in. radius. Find the angle through which a wheel turns in 5 sec.

**Solution**   Recall that $\omega = \theta/t$, or $\theta = \omega t$. Thus we can find $\theta$ if we know $\omega$ and $t$. To find $\omega$, we use $v = r\omega$. For convenience, we will first convert 45 mph to ft/sec:

$$v = 45\,\frac{\text{mi}}{\text{hr}} \cdot \frac{1\,\text{hr}}{60\,\text{min}} \cdot \frac{1\,\text{min}}{60\,\text{sec}} \cdot \frac{5280\,\text{ft}}{1\,\text{mi}}$$

$$= 66\,\frac{\text{ft}}{\text{sec}}.$$

17. A wheel with 12-cm diameter is rotating at a speed of 10 revolutions per second. What is the velocity of a point on the rim?

18. In using $v = r\omega$, if $v$ is given in centimeters per second, what must be the units for $r$ and $\omega$?

19. In using $v = r\omega$, if $\omega$ is given in radians per year and $r$ is in kilometers, what must be the units for $v$?

20. The old oaken bucket is being raised at a rate of 3 ft/sec. The radius of the drum is 5 in. What is the angular speed of the handle?

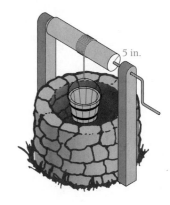

5 in.

**21.** The diameter of a wheel of a car is 30 in. When the car is traveling at a speed of 60 mph (88 ft/sec), how many revolutions does the wheel make in 1 sec?

Now $r = 20$ in. We will convert to ft, since $v$ is in ft/sec:

$$r = 20 \text{ in.} \cdot \frac{1 \text{ ft}}{12 \text{ in.}} = \frac{20}{12} \text{ ft} = \frac{5}{3} \text{ ft}.$$

Using $v = r\omega$, we have

$$66 \frac{\text{ft}}{\text{sec}} = \frac{5}{3} \text{ ft} \cdot \omega,$$

so

$$\omega = 39.6 \frac{\text{radians}}{\text{sec}}.$$

Then

$$\theta = \omega t = 39.6 \frac{\text{radians}}{\text{sec}} \cdot 5 \text{ sec} = 198 \text{ radians}.$$

**DO EXERCISE 21.**

● **EXERCISE SET** **6.4**

**A**   For each of Exercises 1–6, sketch a unit circle and mark the points determined by the given real numbers.

**1. a)** $\dfrac{\pi}{4}$   **b)** $\dfrac{3\pi}{2}$   **c)** $\dfrac{3\pi}{4}$

**d)** $\pi$   **e)** $\dfrac{11\pi}{4}$   **f)** $\dfrac{17\pi}{4}$

**2. a)** $\dfrac{\pi}{2}$   **b)** $\dfrac{5\pi}{4}$   **c)** $2\pi$

**d)** $\dfrac{9\pi}{4}$   **e)** $\dfrac{13\pi}{4}$   **f)** $\dfrac{23\pi}{4}$

**3. a)** $\dfrac{\pi}{6}$   **b)** $\dfrac{2\pi}{3}$   **c)** $\dfrac{7\pi}{6}$

**d)** $\dfrac{10\pi}{6}$   **e)** $\dfrac{14\pi}{6}$   **f)** $\dfrac{23\pi}{4}$

**4. a)** $\dfrac{\pi}{3}$   **b)** $\dfrac{5\pi}{6}$   **c)** $\dfrac{11\pi}{6}$

**d)** $\dfrac{13\pi}{6}$   **e)** $\dfrac{23\pi}{6}$   **f)** $\dfrac{33\pi}{6}$

**5. a)** $-\dfrac{\pi}{2}$   **b)** $-\dfrac{3\pi}{4}$   **c)** $-\dfrac{5\pi}{6}$

**d)** $-\dfrac{5\pi}{2}$   **e)** $-\dfrac{17\pi}{6}$   **f)** $-\dfrac{9\pi}{4}$

**6. a)** $-\dfrac{3\pi}{2}$   **b)** $-\dfrac{\pi}{3}$   **c)** $-\dfrac{7\pi}{6}$

**d)** $-\dfrac{13\pi}{6}$   **e)** $-\dfrac{19\pi}{6}$   **f)** $-\dfrac{11\pi}{4}$

Find two real numbers between $-2\pi$ and $2\pi$ that determine each of the points on the unit circle.

**7.**

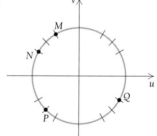

**8.**

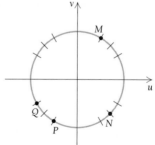

**B**   The following points are on the unit circle. Find the coordinates of their reflections across (a) the $u$-axis, (b) the $v$-axis, and (c) the origin.

**9.** $\left(-\dfrac{\sqrt{2}}{2}, \dfrac{\sqrt{2}}{2}\right)$   **10.** $\left(-\dfrac{3}{4}, \dfrac{\sqrt{7}}{4}\right)$

**11.** $\left(\dfrac{2}{3}, \dfrac{\sqrt{5}}{3}\right)$     **12.** $\left(-\dfrac{\sqrt{3}}{2}, -\dfrac{1}{2}\right)$

**13.** The number $\pi/6$ determines a point on the unit circle with coordinates $(\sqrt{3}/2, 1/2)$. What are the coordinates of the point determined by $-\pi/6$?

**14.** The number $\pi/3$ determines a point on the unit circle with coordinates $(1/2, \sqrt{3}/2)$. What are the coordinates of the point determined by $-\pi/3$?

**15.** A number $\alpha$ determines a point on the unit circle with coordinates $(3/4, -\sqrt{7}/4)$. What are the coordinates of the point determined by $-\alpha$?

**16.** A number $\beta$ determines a point on the unit circle with coordinates $(-2/3, \sqrt{5}/3)$. What are the coordinates of the point determined by $-\beta$?

 Convert to radian measure. Leave answers in terms of $\pi$.

**17.** 30°             **18.** 15°
**19.** 60°             **20.** 200°
**21.** 75°             **22.** 300°
**23.** 37.71°          **24.** 12.73°
**25.** 214.6°          **26.** 73.87°

Convert to radian measure. Do not leave answers in terms of $\pi$.

**27.** 120°            **28.** 240°
**29.** 320°            **30.** 75°
**31.** 200°            **32.** 300°
**33.** 117.8°          **34.** 231.2°
**35.** 1.354°          **36.** 327.9°

Convert to degree measure.

**37.** 1 radian        **38.** 2 radians
**39.** $8\pi$          **40.** $-12\pi$
**41.** $\dfrac{3\pi}{4}$   **42.** $\dfrac{5\pi}{4}$
**43.**  1.303     **44.** ▦ 2.347
**45.** ▦ $0.7532\pi$   **46.** ▦ $-1.205\pi$

**47.** Certain positive angles are marked here in degrees. Find the corresponding radian measures.

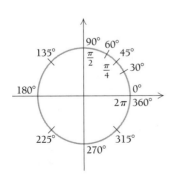

**48.** Certain negative angles are marked here in degrees. Find the corresponding radian measures.

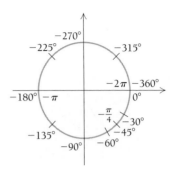

**D**

**49.** In a circle with 120-cm radius, an arc 132 cm long subtends a central angle of how many radians? how many degrees, to the nearest degree?

**50.** In a circle with 200-cm radius, an arc 65 cm long subtends a central angle of how many radians? how many degrees, to the nearest degree?

**51.** Through how many radians does the minute hand of a clock rotate in 50 min?

**52.** A wheel on a car has a 14-in. radius. Through what angle (in radians) does the wheel turn while the car travels 1 mi?

**53.** In a circle with 10-m radius, how long is an arc associated with a central angle of 1.6 radians?

**54.** In a circle with 5-m radius, how long is an arc associated with a central angle of 2.1 radians?

**E**

**55.** A flywheel with 15-cm diameter is rotating at a rate of 7 radians/sec. What is the linear speed of a point on its rim, in centimeters per minute?

**56.** A wheel with 30-cm radius is rotating at a rate of 3 radians/sec. What is the linear speed of a point on its rim, in meters per minute?

**57.** A $33\frac{1}{3}$-rpm record has a radius of 15 cm. What is the linear velocity of a point on the rim, in centimeters per second?

**58.** A 45-rpm record has a radius of 8.7 cm. What is the linear velocity of a point on the rim, in centimeters per second?

**59.** The earth has a 4000-mi radius and rotates one revolution every 24 hr. What is the linear speed of a point on the equator, in miles per hour?

**60.** The earth is 93,000,000 miles from the sun and traverses its orbit, which is nearly circular, every 365.25 days. What is the linear velocity of the earth in its orbit, in miles per hour?

**61.** A wheel has a 32-cm diameter. The speed of a point on its rim is 11 m/s. What is its angular speed?

**62.** A horse on a merry-go-round is 7 m from the center and travels at a speed of 10 km/h. What is its angular speed?

**63. *Determining the speed of a river.*** A water wheel has a 10-ft radius. To get a good approximation of the speed of the river, you count the revolutions of the wheel and find that it makes 14 revolutions per minute (rpm). What is the speed of the river?

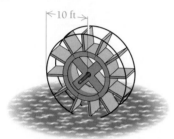

**64. *Determining the speed of a river.*** A water wheel has a 10-ft radius. To get a good approximation of the speed of the river, you count the revolutions of the wheel and find that it makes 16 rpm. What is the speed of the river?

**65.** The wheels of a bicycle have a 24-in. diameter. When the bicycle is being ridden so that the wheels make 12 rpm, how far will the bike travel in 1 min?

**66.** The wheels of a car have a 15-in. radius. When the car is being driven so that the wheels make 10 revolutions per second, how far will the car travel in 1 min?

**67.** A car is traveling at a speed of 30 mph. Its wheels have a 14-in. radius. Find the angle through which a wheel rotates in 10 sec.

**68.** A car is traveling at a speed of 40 mph. Its wheels have a 15-in. radius. Find the angle through which a wheel rotates in 12 sec.

● **SYNTHESIS** _____

**69.** For each of the points given in Exercise 7, find a real number between $2\pi$ and $4\pi$.

**70.** For each of the points given in Exercise 8, find a real number between $-2\pi$ and $-4\pi$.

**71.** A point on the unit circle has $u$-coordinate $-1/3$. What is its $v$-coordinate?

**72.** A point on the unit circle has $v$-coordinate $-\sqrt{21}/5$. What is its $u$-coordinate?

**73.** The **grad** is a unit of angle measure similar to a degree. A right angle has a measure of 100 grads. Convert each of the following to grads.

  **a)** $48°$            **b)** $153°$
  **c)** $\pi/8$ radians     **d)** $5\pi/7$ radians

**74.** A **mil** is a unit of angle measure. A right angle has a measure of 1600 mils. Convert each of the following to degrees, minutes, and seconds.

  **a)** 100 mils         **b)** 350 mils

**75.** On the earth, one degree of latitude is how many kilometers? how many miles? (Assume that the radius of the earth is 6400 km, or 4000 mi, approximately.)

**76.** One minute of latitude on the earth is equivalent to one *nautical mile*. Find the circumference and the radius of the earth in nautical miles.

**77.** An astronaut on the moon observes the earth, about 240,000 mi away. The diameter of the earth is about 8000 mi. Find the angle $\alpha$.

**78.** The circumference of the earth was computed by Eratosthenes (276–195 B.C.). He knew the distance from Aswan to Alexandria to be about 500 mi. From each town he observed the sun at noon, finding the angular difference to be 7°12′. Do Eratosthenes' calculation.

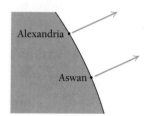

**79.** Two pulleys, 50 cm and 30 cm in diameter, respectively, are connected by a belt. The larger pulley makes 12 revolutions per minute. Find the angular speed of the smaller pulley, in radians per second.

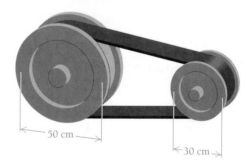

**80.** One gear wheel turns another, the teeth being on the rims. The wheels have 40-cm and 50-cm radii, and the smaller wheel rotates at 20 rpm. Find the angular speed of the larger wheel, in radians per second.

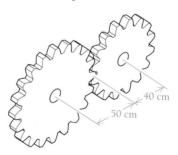

**81.** An airplane engine is idling at 800 rpm. When the throttle is opened, it takes 4.3 sec for the speed to come up to

2500 rpm. What was the angular acceleration:

**a)** in revolutions per minute per second?
**b)** in radians per second per second?

**82.** The linear speed of an airplane, flying at low level over the sea, is 175 knots (nautical miles per hour). It accelerates to 325 knots in 12 sec.

**a)** What was its linear acceleration in knots per second?
**b)** What was its angular acceleration in radians per second per second? (*Hint:* One nautical mile is equivalent to one minute of latitude.)

● CHALLENGE _____

**83.** What is the angle between the hands of a clock at 7:45?

**84.** At what time between noon and 1:00 P.M. are the hands of a clock perpendicular?

**85.** The diameter of the earth at the equator has been measured to be 7926.4 statute miles. Use this and the result of Exercise 76 to find the number of statute miles in a nautical mile.

**86.** To find the distance between two points on the earth when their latitude and longitude are known, we can use a plane right triangle for an excellent approximation if the points are not too far apart. Point *A* is at latitude 38°27′30″ N, longitude 82°57′15″ W; and point *B* is at latitude 38°28′45″ N, longitude 82°56′30″ W. Find the distance from *A* to *B* in nautical miles (one minute of latitude is one nautical mile).

# 6.5

# GRAPHS OF BASIC CIRCULAR FUNCTIONS

The domains of the trigonometric functions, thus far, have been sets of angles or rotations measured in a real number of degree units. We can also consider the domains to be sets of real numbers, or radians. Let us again consider radian measure and the unit circle.

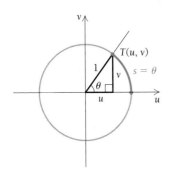

$$\cos \theta = \frac{u}{1} = u,$$

$$\sin \theta = \frac{v}{1} = v$$

We defined radian measure as $\theta = s/r$. When $r = 1$, $\theta = s/1$, or $\theta = s$. Thus the arc length *s* on the unit circle is the same as the radian measure of the angle $\theta$. In the figure above, $T(u, v)$ is the point where the terminal side of the angle with radian measure *s* intersects the unit circle. We can now extend our definitions of the trigonometric functions using domains of real numbers, or radians. In the definitions, *s* can be considered as the radian measure of an angle or as the measure of an arc length as a real number on the unit circle. In either consideration, *s* is a real number. To each real number *s*, there corresponds an arc length *s* on the unit circle. Trigonometric functions with a domain of real numbers are called **circular functions.**

**OBJECTIVES**

You should be able to:

**A** Find trigonometric function values of any angle given in radian measure.

**B** Sketch a graph of the sine function. State the properties of the sine function and state or complete the following identities:

$$\sin s = \sin(s + 2k\pi),$$
$$k \text{ any integer,}$$
$$\sin(-s) = -\sin s,$$
$$\sin(s \pm \pi) = -\sin s,$$
$$\sin(\pi - s) = \sin s.$$

**C** Sketch a graph of the cosine function. State the properties of the cosine function and state or complete the following identities:

$$\cos s = \cos(s + 2k\pi),$$
$$k \text{ any integer,}$$
$$\cos(-s) = \cos s,$$
$$\cos(s \pm \pi) = -\cos s,$$
$$\cos(\pi - s) = -\cos s.$$

**D** Sketch graphs of the tangent, the cotangent, the secant, and the cosecant functions, describe properties of the functions, and find function values (when they exist) for multiples of $\pi/6$, $\pi/4$, and $\pi/3$.

**E** State and prove basic and Pythagorean identities and use them to find function values given a function value of *s* and the quadrant in which *s* lies.

---

**DEFINITION**              **Basic Circular Functions**

For a real number $s$ that determines a point $(u, v)$ on a unit circle:

$$\sin s = v = \text{second coordinate}, \qquad \csc s = \frac{1}{v} \ (v \neq 0) = \frac{1}{\text{second coordinate}},$$

$$\cos s = u = \text{first coordinate}, \qquad \sec s = \frac{1}{u} \ (u \neq 0) = \frac{1}{\text{first coordinate}},$$

$$\tan s = \frac{v}{u} = \frac{\text{second coordinate}}{\text{first coordinate}}, \qquad \cot s = \frac{u}{v} \ (v \neq 0) = \frac{\text{first coordinate}}{\text{second coordinate}}.$$

---

As a result of these definitions, we can consider the trigonometric functions using a domain of real numbers without referring to angles. There are many applications in calculus that use the circular functions without referring to angles.

##  Finding Function Values

The process we use in finding trigonometric function values will not change if we are using radians. If an angle, or rotation, is given in radians, we can change to degrees to determine the function values. For example, $\sin \pi/6$ is the same as $\sin 30°$. The diagrams below show unit circles marked in both radians and degrees. Just as with degrees, there are circular function values that need to be memorized.

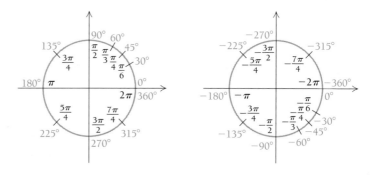

**Examples**   Find the following function values.

1.  $\cos \dfrac{\pi}{3}$

    From the diagram above, we have $\pi/3 = 60°$. Thus,

    $$\cos \frac{\pi}{3} = \cos 60° = \frac{1}{2}.$$

2.  $\tan\left(-\dfrac{5\pi}{4}\right)$

    Since $-5\pi/4 = -225°$, we have

    $$\tan\left(-\frac{5\pi}{4}\right) = \tan\left(225°\right).$$

The reference angle, or number, for $-5\pi/4$ is $\pi/4$, or $45°$. We also recall that the tangent value is negative in the second quadrant. Thus we have

$$\tan\left(-\frac{5\pi}{4}\right) = -1.$$

3. $\sin 9\pi$

We have

$$\sin 9\pi = \sin(9 \cdot 180°), \quad \text{or} \quad \sin 1620°.$$

The terminal side of a rotation of $9\pi$ radians, or $1620°$, lies on the negative $x$-axis. Thus,

$$\sin 9\pi = 0.$$ ◀

**DO EXERCISES 1–5.** _____

Using a calculator, we can find circular function values directly without first converting to degrees. Most scientific calculators have both degrees and radian modes. When directly finding function values of radian measures, or real numbers, we must set the calculator in radian mode.

**Examples**   Find the following function values using a calculator.

4. $\cos\dfrac{2\pi}{5}$

Using the $\boxed{\pi}$ key, we get $\dfrac{2\pi}{5} \approx 1.2566$. Then, remembering to use the radian mode, we press the $\boxed{\text{COS}}$ key. We then find that

$$\cos\frac{2\pi}{5} \approx \cos 1.2566 \approx 0.3090.$$

5. $\sin 13$

We enter 13 and press the $\boxed{\text{SIN}}$ key. The calculator must be in radian mode. We find $\sin 13 \approx 0.4202$.

6. $\tan\left(-\dfrac{\pi}{7}\right)$

Using the $\boxed{\pi}$ key, we find that $-\pi/7 \approx -0.4488$. Then we press the $\boxed{\text{TAN}}$ key. The result is

$$\tan\left(-\frac{\pi}{7}\right) \approx \tan(-0.4488) \approx -0.4816.$$

7. $\csc 429$

The cosecant function value can be found by taking the reciprocal of the sine function value. Remembering to use the radian mode, we enter 429 and press the $\boxed{\text{SIN}}$ key to get

$$\sin 429 \approx 0.9851.$$

We then press the reciprocal key $\boxed{1/\text{x}}$:

$$\csc 429 \approx 1.0151.$$ ◀

**DO EXERCISES 6–9.** _____

Find the function values.

1. $\cos\dfrac{\pi}{4}$

2. $\tan\dfrac{\pi}{6}$

3. $\sin\left(-\dfrac{2\pi}{3}\right)$

4. $\cos(-12\pi)$

5. $\tan\dfrac{5\pi}{2}$

Find the function values using a calculator.

6. $\tan\dfrac{\pi}{9}$

7. $\sin 21$

8. $\cot 1026$

9. $\cos\left(-\dfrac{3\pi}{11}\right)$

10. Construct a graph of the sine function. From the unit circle, transfer vertical distances with a compass.

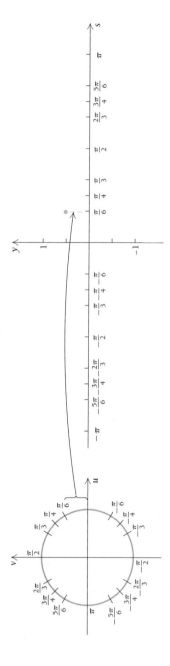

## Properties of the Sine Function

Let us consider making a graph of the sine function, making a table of values, plotting points, and connecting them. It is helpful to first draw a unit circle and label a few points with coordinates.

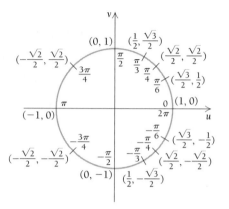

Since $\sin s = v =$ second coordinate, a table of function values can be made directly from the circle.

| $s$ | 0 | $\dfrac{\pi}{6}$ | $\dfrac{\pi}{4}$ | $\dfrac{\pi}{3}$ | $\dfrac{\pi}{2}$ | $\dfrac{3\pi}{4}$ | $\pi$ | $\dfrac{5\pi}{4}$ | $\dfrac{3\pi}{2}$ | $2\pi$ |
|---|---|---|---|---|---|---|---|---|---|---|
| $\sin s$ | 0 | $\dfrac{1}{2}$ | $\dfrac{\sqrt{2}}{2}$ | $\dfrac{\sqrt{3}}{2}$ | 1 | $\dfrac{\sqrt{2}}{2}$ | 0 | $-\dfrac{\sqrt{2}}{2}$ | $-1$ | 0 |

| $s$ | $-\dfrac{\pi}{6}$ | $-\dfrac{\pi}{4}$ | $-\dfrac{\pi}{3}$ | $-\dfrac{\pi}{2}$ | $-\dfrac{3\pi}{4}$ | $-\pi$ | $-\dfrac{5\pi}{4}$ | $-\dfrac{3\pi}{2}$ | $-2\pi$ |
|---|---|---|---|---|---|---|---|---|---|
| $\sin s$ | $-\dfrac{1}{2}$ | $-\dfrac{\sqrt{2}}{2}$ | $-\dfrac{\sqrt{3}}{2}$ | $-1$ | $-\dfrac{\sqrt{2}}{2}$ | 0 | $\dfrac{\sqrt{2}}{2}$ | 1 | 0 |

We find decimal approximations of the expressions involving square roots. Then we plot points, look for patterns, and connect them with a curve. The graph is as follows.

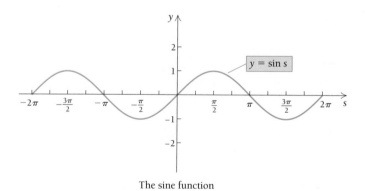

The sine function

The second way to construct a graph is by considering a unit circle. Then from the unit circle, we transfer vertical distances with a compass. You are asked to draw such a graph in Margin Exercise 10.

**DO EXERCISES 10–13. (EXERCISE 10 IS ON THE PRECEDING PAGE.)**

Note in the graph of the sine function that function values increase from 0 at 0 to 1 at $\pi/2$, then decrease to 0 at $\pi$, decrease further to $-1$ at $3\pi/2$, and then increase to 0 at $2\pi$. A similar pattern follows for negative inputs.

Certain functions with a repeating pattern are called **periodic.** The sine function is an example. The function values of the sine function repeat themselves every $2\pi$ units. In other words, for any $s$, we have $\sin(s + 2\pi) = \sin s$. To see this another way, think of the part of the graph between 0 and $2\pi$ and note that the rest of the graph consists of copies of it. If we translate the graph $2\pi$ units to the left or right, the original graph will be obtained. We say that the sine function has a *period* of $2\pi$.

---

**DEFINITION**        **Periodic Function**

A function $f$ is said to be *periodic* if these exists a positive constant $p$ such that

$$f(x + p) = f(x)$$

for all $x$ in the domain of $f$. The smallest positive number $p$ is called the *period* of the function.

---

The period $p$ can be thought of as the length of the shortest recurring interval.

We can use the unit circle to verify some properties of the sine function. Let us first verify that the period is $2\pi$. Consider any real number $s$ and the point $T$ that it determines on a unit circle. Now we increase $s$ by $2\pi$. The point for $s + 2\pi$ is again the point $T$.

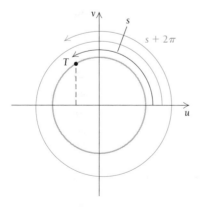

Hence for any real number $s$,

$$\sin(s + 2\pi) = \sin s.$$

There is no number smaller than $2\pi$ for which this is true. Thus the period of the sine function is $2\pi$.

11. Does the sine function appear to be continuous?

12. What is the domain of the sine function?

13. What is the range of the sine function?

14. A real number $s$ determines the point $T$, as shown.

a) Find the point for $s + \pi$. What are its coordinates?

b) Find the point for $s - \pi$. What are its coordinates?

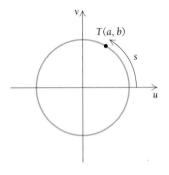

## Identities Involving the Sine Function

The statement $\sin s = \sin (s + 2\pi)$ holds for any real number $s$. Equations that hold for all meaningful replacements for the variable are known as **identities**. Thus we have shown that $\sin s = \sin (s + 2\pi)$ is an identity. Actually, we know that $\sin s = \sin [s + k(2\pi)]$, or

$$\sin s = \sin (s + 2k\pi)$$

for any integer $k$.

For example, if $k = 1$, we have $\sin s = \sin (s + 2\pi)$. For $k = 2$, we have $\sin s = \sin (s + 4\pi)$; the point for $s + 4\pi$ is found by going around the unit circle from $s$, two full circuits. For $k = -5$, we have $\sin s = \sin (s - 10\pi)$, corresponding to 5 full circuits, *but in the clockwise direction*.

Consider any real number $s$ and its opposite, $-s$. These numbers determine points $T$ and $T_1$ on a unit circle that are symmetric with respect to the $u$-axis.

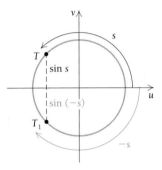

Because their second coordinates are opposites of each other, we know that for any number $s$,

$$\sin (-s) = -\sin s.$$

We have shown another identity as well as showing that the sine function is *odd*.

**DO EXERCISE 14.**

Consider any real number $s$ and the point $T$ that it determines on a unit circle. Then $s + \pi$ and $s - \pi$ both determine a point $T_1$ symmetric to $T$ with respect to the origin.

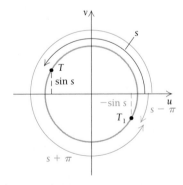

Thus the sine of $s + \pi$ and the sine of $s - \pi$ are the same, and each is the opposite of $\sin s$. This gives us another important identity. For any real number $s$,

$$\sin(s + \pi) = -\sin s$$

and

$$\sin(s - \pi) = -\sin s.$$

Let us now add to our list a property concerning $\sin(\pi - s)$. Consider any real number $s$ and the point $T$ that it determines on a unit circle. Then $\pi - s$ determines a point $T_1$ symmetric to $T$ with respect to the $v$-axis.

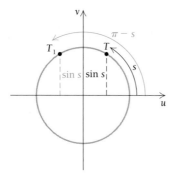

Because their second coordinates are the same, we know that for any number $s$,

$$\sin(\pi - s) = \sin s.$$

We have proven the following identities concerning the sine function.

---

### Identities for the Sine Function

$$\sin s = \sin(s + 2k\pi), k \text{ any integer}$$
$$\sin(-s) = -\sin s$$
$$\sin(s \pm \pi) = -\sin s$$
$$\sin(\pi - s) = \sin s$$

---

The **amplitude** of a periodic function is defined to be one-half the difference between its maximum and minimum function values. It is always positive. We can see from either the graph or a unit circle that the maximum value of the sine function is 1, whereas the minimum value is $-1$. Thus,

**The amplitude of the sine function** $= \frac{1}{2}[1 - (-1)] = 1$.

The following is a summary of properties of the sine function.

---

### Properties of the Sine Function

1. The sine function is periodic, with period $2\pi$.
2. The sine function is odd. Thus it is symmetric with respect to the origin, and $\sin(-s) = -\sin s$ for all real numbers $s$.
3. It is continuous everywhere.
4. The domain is the set of all real numbers. The range is the interval $[-1, 1]$, that is, the set of all real numbers from $-1$ to 1.
5. The amplitude of the sine function is 1.

15. Construct a graph of the cosine function. From the unit circle, transfer horizontal distances with a compass.

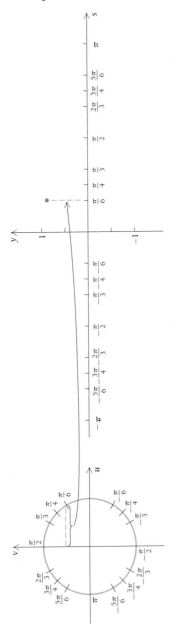

## Properties of the Cosine Function

Let us consider making a graph of the cosine function. There are again two ways in which we can make such a graph. The first is to take the function values that we have found in the preceding examples and margin exercises, make a table of values, plot points, look for patterns, and complete the graph. Below is a table of function values.

| $s$ | 0 | $\dfrac{\pi}{6}$ | $\dfrac{\pi}{4}$ | $\dfrac{\pi}{3}$ | $\dfrac{\pi}{2}$ | $\dfrac{3\pi}{4}$ | $\pi$ | $\dfrac{5\pi}{4}$ | $\dfrac{3\pi}{2}$ | $2\pi$ |
|---|---|---|---|---|---|---|---|---|---|---|
| $\cos s$ | 1 | $\dfrac{\sqrt{3}}{2}$ | $\dfrac{\sqrt{2}}{2}$ | $\dfrac{1}{2}$ | 0 | $-\dfrac{\sqrt{2}}{2}$ | $-1$ | $-\dfrac{\sqrt{2}}{2}$ | 0 | 1 |

| $s$ | $-\dfrac{\pi}{6}$ | $-\dfrac{\pi}{4}$ | $-\dfrac{\pi}{3}$ | $-\dfrac{\pi}{2}$ | $-\dfrac{3\pi}{4}$ | $-\pi$ | $-\dfrac{5\pi}{4}$ | $-\dfrac{3\pi}{2}$ | $-2\pi$ |
|---|---|---|---|---|---|---|---|---|---|
| $\cos s$ | $\dfrac{\sqrt{3}}{2}$ | $\dfrac{\sqrt{2}}{2}$ | $\dfrac{1}{2}$ | 0 | $-\dfrac{\sqrt{2}}{2}$ | $-1$ | $-\dfrac{\sqrt{2}}{2}$ | 0 | 1 |

We find decimal approximations of the expressions involving square roots. Then we plot points, look for patterns, and connect them with a curve. The graph is as follows.

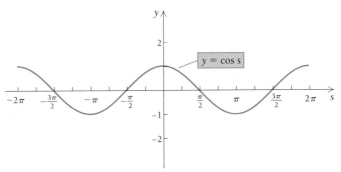

The cosine function

The second way to construct a graph is by considering a unit circle. Then from the unit circle, we transfer horizontal distances with a compass. You are asked to draw such a graph in Margin Exercise 15.

**DO EXERCISES 15–18. (EXERCISES 16–18 ARE ON THE FOLLOWING PAGE.)** _____

Note in the graph of the cosine function that function values start at 1 when $s = 0$, and decrease to 0 at $\pi/2$. They decrease further to $-1$ at $\pi$, and then begin to increase, reaching the value 1 at $2\pi$. A similar pattern follows for negative inputs. Moreover, the function is periodic, with period $2\pi$.

We can use the unit circle to verify some properties of the cosine function. Let us verify that the period is $2\pi$. Consider any real number $s$ and the point $T$ that it determines on a unit circle. Now we increase $s$ by $2\pi$. The point for $s + 2\pi$ is again the point $T$.

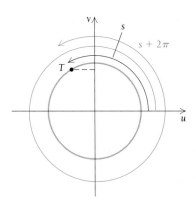

Hence for any real number $s$,

$$\cos(s + 2\pi) = \cos s.$$

There is no number smaller than $2\pi$ for which this is true. Thus the period of the cosine function is $2\pi$.

## Identities Involving the Cosine Function

The unit circle can be used to verify that

$$\cos s = \cos(s + 2k\pi)$$

holds for any real number $s$ and any integer $k$.

Consider any real number $s$ and its opposite, $-s$. These numbers determine points $T$ and $T_1$ symmetric with respect to the $u$-axis.

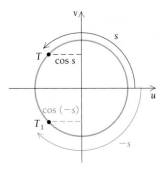

Because $T$ and $T_1$ have the same first coordinates, we know that for any number $s$,

$$\cos(-s) = \cos s.$$

We have proved another identity as well as showing that the cosine function is *even*.

Consider any real number $s$ and the point $T$ that it determines on a unit circle. Then $s + \pi$ and $s - \pi$ both determine a point $T_1$ symmetric to $T$ with respect to the origin.

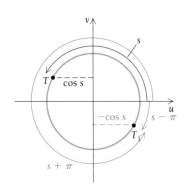

16. Does the cosine function appear to be continuous?

17. What is the domain of the cosine function?

18. What is the range of the cosine function?

Thus the cosine of $s + \pi$ and the cosine of $s - \pi$ are the same, and each is the opposite of cos $s$. This gives us another important identity and property. For any real number $s$,

$$\cos (s + \pi) = -\cos s$$

and

$$\cos (s - \pi) = -\cos s.$$

Let us consider cos $(\pi - s)$ as well as any real number $s$ and the point $T$ that it determines on a unit circle. Then $\pi - s$ determines a point $T_1$ symmetric to $T$ with respect to the $v$-axis.

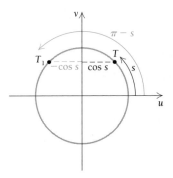

Because the first coordinates of $T$ and $T_1$ are opposites, we know that for any number $s$,

$$\cos (\pi - s) = -\cos s.$$

---

**Identities for the Cosine Function**

$$\cos s = \cos (s + 2k\pi), k \text{ any integer}$$
$$\cos (-s) = \cos s$$
$$\cos (s \pm \pi) = -\cos s$$
$$\cos (\pi - s) = -\cos s$$

---

The following is a summary of the properties of the cosine function.

---

**Properties of the Cosine Function**

1. The cosine function is periodic, with period $2\pi$.
2. The cosine function is even. Thus it is symmetric with respect to the $y$-axis, and cos $(-s) = \cos s$ for all real numbers $s$.
3. It is continuous everywhere.
4. The domain is the set of all real numbers. The range is the interval $[-1, 1]$, that is, the set of all real numbers from $-1$ to $1$.
5. The amplitude of the cosine function is $1$.

---

For your understanding, we have used the letter $s$ for arc length and have avoided the letters $x$ and $y$, which usually represent first and second coordinates. In fact, we can represent the arc length on a unit circle by any variable, such as $s, t, x,$ or $\theta$ (the Greek letter "theta"). Each arc length determines a point that can be labeled with an ordered pair. **The first coor-**

dinate of that ordered pair is the *cosine* of the arc length, and the second coordinate is the *sine* of the arc length. This is illustrated in the following figures.

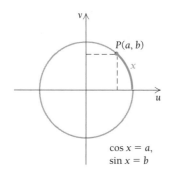

$\cos x = a,$
$\sin x = b$

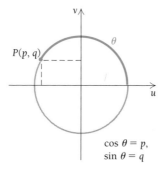

$\cos \theta = p,$
$\sin \theta = q$

The identities we have developed hold no matter what symbols are used for variables. For example, $\cos(-s) = \cos s$, $\cos(-x) = \cos x$, $\cos(-\theta) = \cos \theta$, and $\cos(-y) = \cos y$.

**DO EXERCISE 19.**

 **Properties of the Tangent, Cotangent, Secant, and Cosecant Functions**

Four of the circular functions can be defined in terms of only the sine and the cosine functions. The resulting identities are as follows:

$$\tan s = \frac{\sin s}{\cos s} \qquad \csc s = \frac{1}{\sin s} \qquad \cot s = \frac{\cos s}{\sin s} \qquad \sec s = \frac{1}{\cos s}$$

These identities are very useful in finding function values.

**DO EXERCISES 20 AND 21.**

Let us now find some function values.

**Example 8** Find $\tan 0$, $\cot 0$, $\sec 0$, and $\csc 0$.

**Solution** We note that 0 determines a point with coordinates $(1, 0)$ on the unit circle. Then

$$\tan 0 = \frac{\sin 0}{\cos 0} = \frac{0}{1} = 0;$$

$$\cot 0 = \frac{\cos 0}{\sin 0} = \frac{1}{0}, \text{ which is not defined;}$$

$$\sec 0 = \frac{1}{\cos 0} = \frac{1}{1} = 1;$$

$$\csc 0 = \frac{1}{\sin 0} = \frac{1}{0}, \text{ which is not defined.}$$

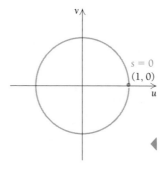

19. Complete each of the following in reference to the figure.

$\cos y = $ _____

$\sin y = $ _____

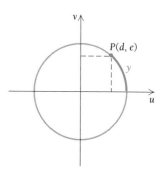

Using the definitions of the basic circular functions, verify the identity.

20. $\tan s = \dfrac{\sin s}{\cos s}$

21. $\cot s = \dfrac{\cos s}{\sin s}$

**22.** Find $\tan \pi$, $\cot \pi$, $\sec \pi$, and $\csc \pi$.

**DO EXERCISE 22.**

**Example 9**   Find $\tan \dfrac{\pi}{4}$, $\cot \dfrac{\pi}{4}$, $\sec \dfrac{\pi}{4}$, and $\csc \dfrac{\pi}{4}$.

**Solution**   The number $\pi/4$ determines a point with coordinates $(\sqrt{2}/2, \sqrt{2}/2)$ on the unit circle.

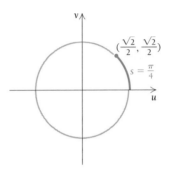

Then

$$\tan \frac{\pi}{4} = \frac{\sin \dfrac{\pi}{4}}{\cos \dfrac{\pi}{4}} = \frac{\dfrac{\sqrt{2}}{2}}{\dfrac{\sqrt{2}}{2}} = 1;$$

$$\cot \frac{\pi}{4} = \frac{\cos \dfrac{\pi}{4}}{\sin \dfrac{\pi}{4}} = \frac{\dfrac{\sqrt{2}}{2}}{\dfrac{\sqrt{2}}{2}} = 1;$$

$$\sec \frac{\pi}{4} = \frac{1}{\cos \dfrac{\pi}{4}} = \frac{1}{\dfrac{\sqrt{2}}{2}}$$

$$= \frac{2}{\sqrt{2}} = \frac{2\sqrt{2}}{2} = \sqrt{2};$$

$$\csc \frac{\pi}{4} = \frac{1}{\sin \dfrac{\pi}{4}} = \frac{1}{\dfrac{\sqrt{2}}{2}}$$

$$= \frac{2}{\sqrt{2}} = \sqrt{2}.$$

**23.** Find $\tan \dfrac{3\pi}{4}$, $\cot \dfrac{3\pi}{4}$, $\sec \dfrac{3\pi}{4}$, and $\csc \dfrac{3\pi}{4}$.

**DO EXERCISE 23.**

**Example 10**   Find $\tan \dfrac{\pi}{3}$, $\cot \dfrac{\pi}{3}$, $\sec \dfrac{\pi}{3}$, and $\csc \dfrac{\pi}{3}$.

**Solution**   The number $\pi/3$ determines a point with coordinates $(1/2, \sqrt{3}/2)$

on the unit circle.

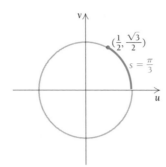

Then

$$\tan \frac{\pi}{3} = \frac{\sin \dfrac{\pi}{3}}{\cos \dfrac{\pi}{3}} = \frac{\dfrac{\sqrt{3}}{2}}{\dfrac{1}{2}} = \sqrt{3};$$

$$\cot \frac{\pi}{3} = \frac{\cos \dfrac{\pi}{3}}{\sin \dfrac{\pi}{3}} = \frac{\dfrac{1}{2}}{\dfrac{\sqrt{3}}{2}} = \frac{1}{\sqrt{3}} = \frac{\sqrt{3}}{3};$$

$$\sec \frac{\pi}{3} = \frac{1}{\cos \dfrac{\pi}{3}} = \frac{1}{\dfrac{1}{2}} = 2;$$

$$\csc \frac{\pi}{3} = \frac{1}{\sin \dfrac{\pi}{3}} = \frac{1}{\dfrac{\sqrt{3}}{2}} = \frac{2}{\sqrt{3}} = \frac{2\sqrt{3}}{3}.$$

◀

**DO EXERCISES 24–26.**

## The Tangent Function

Note the following:

$$\tan \frac{\pi}{2} = \frac{\sin \dfrac{\pi}{2}}{\cos \dfrac{\pi}{2}} = \frac{1}{0},$$

which is undefined, since division by 0 is not possible. The tangent function is undefined for any number whose cosine is 0. Thus it is undefined for $(\pi/2) + k\pi$, where $k$ is any integer. In the first quadrant, the function values are positive. In the second quadrant, however, the sine and cosine values have opposite signs, so the tangent values are negative. Tangent values are also negative in the fourth quadrant, and they are positive in the third quadrant.

A graph of the tangent function is shown below. It can be constructed using the values we have found in the preceding examples and margin exercises and by computing other values as needed. Note that the function value is 0 when $x = 0$, and the values increase as $x$ increases toward $\pi/2$. As we approach $\pi/2$, the denominator becomes very small, so the tangent

**24.** Find $\tan \dfrac{\pi}{6}$, $\cot \dfrac{\pi}{6}$, $\sec \dfrac{\pi}{6}$, and $\csc \dfrac{\pi}{6}$.

**25.** Find $\tan \left( -\dfrac{\pi}{3} \right)$, $\cot \left( -\dfrac{\pi}{3} \right)$, $\sec \left( -\dfrac{\pi}{3} \right)$, and $\csc \left( -\dfrac{\pi}{3} \right)$.

**26.** Find $\tan \dfrac{\pi}{2}$, $\tan \left( -\dfrac{\pi}{2} \right)$, and $\tan \left( -\dfrac{3\pi}{2} \right)$.

27. Find three real numbers other than $\pi/2$ at which the tangent function is undefined.

values become very large. In fact, they increase without bound. The dashed vertical line $x = \pi/2$ is called a **vertical asymptote.** The graph gets closer and closer to the asymptote as $x$ gets closer to $\pi/2$, but it never crosses the line. Each of the vertical dashed lines is an asymptote.

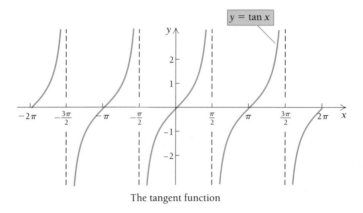

The tangent function

**DO EXERCISES 27–32.**

28. Graph the tangent function.

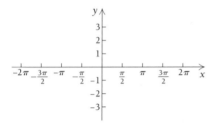

The following is a summary of the properties of the tangent function.

---

**Properties of the Tangent Function**

1. The tangent function is periodic, with period $\pi$.
2. The domain of the tangent function is the set of all real numbers except $(\pi/2) + k\pi$, where $k$ is any integer. Tan $x$ is not defined where $\cos x = 0$.
3. The range of the tangent function is the set of all real numbers.
4. The tangent function is continuous except where it is not defined.

---

29. What is the period of the tangent function?

## The Cotangent Function

The graph of the cotangent function is shown below. The cotangent of a number is found by dividing its cosine by its sine. The cotangent function is undefined for any number whose sine is 0. Thus it is undefined for $k\pi$, where $k$ is any integer.

30. What is the domain of the tangent function?

31. What is the range of the tangent function?

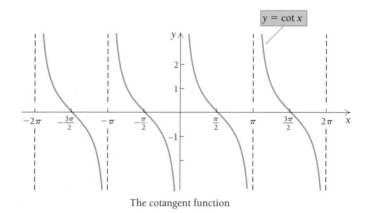

The cotangent function

32. Does the tangent function appear to be even? Is it odd?

**DO EXERCISES 33–39.**

The following is a summary of the properties of the cotangent function.

---

### Properties of the Cotangent Function

1. The cotangent function is periodic, with period $\pi$.
2. The domain of the cotangent function is the set of all real numbers except $k\pi$, where $k$ is any integer. Cot $x$ is not defined where $\sin x = 0$.
3. The range of the cotangent function is the set of all real numbers.
4. The cotangent function is continuous except where it is not defined.

---

## The Secant Function

The secant and cosine functions are reciprocal functions. The graph of the secant function can be constructed by finding the reciprocal of the values of the cosine function. Thus the functions will be positive together and negative together. The secant function is not defined for those numbers $x$ in which $\cos x = 0$. Thus it is undefined for $(\pi/2) + k\pi$, where $k$ is any integer. A graph of the secant function follows. The cosine function is shown for reference by the dashed curve.

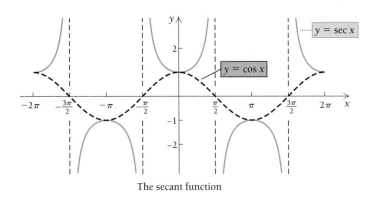

The secant function

The following is a summary of the properties of the secant function.

---

### Properties of the Secant Function

1. The secant function is periodic, with period $2\pi$.
2. The domain of the secant function is the set of all real numbers except $(\pi/2) - k\pi$, where $k$ is any integer. Sec $x$ is not defined where $\cos x = 0$.
3. The range of the secant function consists of all real numbers 1 and greater, in addition to all real numbers $-1$ and less.
4. The secant function is continuous wherever it is defined.

---

33. Find three real numbers at which the cotangent function is undefined.

34. Graph the cotangent function.

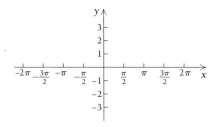

35. What is the period of the cotangent function?

36. What is the domain of the cotangent function?

37. What is the range of the cotangent function?

38. In which quadrants are the function values of the cotangent function positive? negative?

39. Does the cotangent function appear to be even? Is it odd?

40. a) Find three numbers at which the secant function is undefined.
    b) Find three numbers at which the cosecant function is undefined.

41. Graph the secant function.

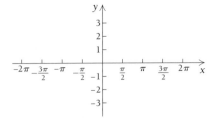

42. Graph the cosecant function.

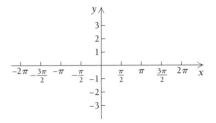

43. What is the period of the secant function? the cosecant function?

44. What is the domain of the secant function? the cosecant function?

45. What is the range of the secant function? the cosecant function?

46. In which quadrants are the function values of the secant function positive? negative?

47. In which quadrants are the function values of the cosecant function positive? negative?

48. Does the secant function appear to be even? odd?

49. Does the cosecant function appear to be even? odd?

## The Cosecant Function

The cosecant and sine functions are reciprocal functions. The graph of the cosecant function can be constructed by finding the reciprocal of the values of the sine function. Thus the functions will be positive together and negative together. The cosecant function is not defined for those numbers $x$ for which $\sin x = 0$. Thus it is undefined for $k\pi$, where $k$ is any integer. A graph of the cosecant function follows. The sine function is shown for reference by the dashed curve.

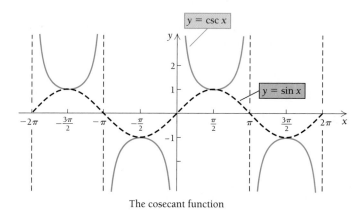

The cosecant function

The following is a summary of the properties of the cosecant function.

---

### Properties of the Cosecant Function

1. The cosecant function is periodic, with period $2\pi$.
2. The domain of the cosecant function is the set of all real numbers except $k\pi$, where $k$ is any integer. Csc $x$ is not defined where $\sin x = 0$.
3. The range of the cosecant function consists of all real numbers 1 and greater, in addition to all real numbers $-1$ and less.
4. The cosecant function is continuous wherever it is defined.

---

**DO EXERCISES 40–49.** _____

## Signs of the Functions

Suppose that a real number $s$ determines a point on the unit circle in the first quadrant. Since both coordinates are positive there, the function values for all the circular functions will be positive. In the second quadrant, the first coordinate is negative and the second coordinate is positive. Thus the sine function has positive values and the cosine function has negative values. The secant, being the reciprocal of the cosine, also has negative values, and the cosecant, being the reciprocal of the sine, has positive values. The tangent and the cotangent both have negative values in the second quadrant.

**DO EXERCISE 50.**

**50.** What are the signs of the six circular functions in each of the four quadrants?

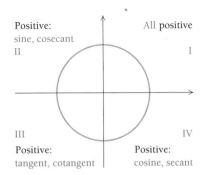

Positive: sine, cosecant II

All positive I

III Positive: tangent, cotangent

IV Positive: cosine, secant

The diagram shows the quadrants in which the circular function values are positive. It need not be memorized, because the signs can be readily determined by referring to a unit circle.

 **TECHNOLOGY CONNECTION**

A grapher can be used to graph any trigonometric function that can be written in the form $y = f(x)$. Before this can be done, however, there are a few preparations.

First, make sure your grapher is in RADIAN mode. There are two common systems for measuring angles, *radians* and *degrees*. So far, all the work we have done in this chapter has used radian mode; degree mode will be introduced later. All graphers allow you to select a mode. (A third system, *grads*, may also be available on some graphers. We will not use this system.)

Next, the appropriate range values should be set. Some graphers have standard or default settings that are set automatically whenever a trig function is involved. Most of the time, however, you must set the values for the viewing box. The most common values for the $x$-axis are $-2\pi$ and $+2\pi$ (or $-6.28$ and $+6.28$ if your grapher doesn't accept the symbol $\pi$). On the $y$-axis, values slightly larger than the amplitude are generally used so that the entire graph can be viewed.

Finally, if the function includes sec, csc, or cot, it must be rewritten in terms of sin, cos, and tan (these are the only trig keys that are available). For this step, use the following

formulas:

$$\sec x = \frac{1}{\cos x},$$

$$\csc x = \frac{1}{\sin x},$$

$$\cot x = \frac{1}{\tan x} = \frac{\cos x}{\sin x}.$$

Once these preliminaries have been tended to, you are ready to draw the graph. All you need to do is enter the function as you would any other, except that you will use one or more of the trig function keys. In addition, you may need to use parentheses around the part of the trig function that follows *sin* or *cos*. For example, $y = \sin x$ may need to be entered as $Y = SIN(X)$.

Graph each of the following. Use a viewing box of $[-2\pi, 2\pi] \times [-2, 2]$, and make sure your grapher is in radian mode.

**TC 1.** $y = \cos^2 x \sin x$      **TC 2.** $y = (\cos x + 1)(\tan x)$

**TC 3.** $y = \csc x + \cot x$      **TC 4.** $y = \sec x \cot x$

**51.** Prove the identity $\tan s = \dfrac{1}{\cot s}$.

**52.** Prove the identity $\cos s = \dfrac{1}{\sec s}$.

▶ ## The Functions Interrelated

The definitions of the four trigonometric functions introduced in this section give us eight basic identities.

---

**Basic Identities**

$$\sin s = \frac{1}{\csc s} \qquad \csc s = \frac{1}{\sin s} \qquad \tan s = \frac{\sin s}{\cos s}$$

$$\cos s = \frac{1}{\sec s} \qquad \sec s = \frac{1}{\cos s} \qquad \cot s = \frac{\cos s}{\sin s}$$

$$\tan s = \frac{1}{\cot s} \qquad \cot s = \frac{1}{\tan s}$$

---

Some of the above identities have not been proved, but they are straightforward.

**Example 11**    Prove the identity $\cot s = \dfrac{1}{\tan s}$.

*Proof*

$$\cot s = \frac{\cos s}{\sin s} = \frac{1}{\dfrac{\sin s}{\cos s}} = \frac{1}{\tan s}$$  ◀

**DO EXERCISE 51.**

---

**Example 12**    Prove the identity $\sin s = \dfrac{1}{\csc s}$.

*Proof*

$$\sin s = \frac{1}{\dfrac{1}{\sin s}} = \frac{1}{\csc s}$$  ◀

**DO EXERCISE 52.**

---

### The Pythagorean Identities

Here we consider three other identities that are fundamental to a study of trigonometry. They are called the **Pythagorean identities.** Recall that the equation of a unit circle in a $uv$-plane is

$$u^2 + v^2 = 1.$$

For any point on a unit circle, the coordinates $u$ and $v$ satisfy this equation. Suppose that a real number $s$ determines a point $T$ on a unit circle, with coordinates $(u, v)$, or $(\cos s, \sin s)$. Then $u = \cos s$ and $v = \sin s$.

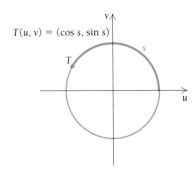

53. Prove the identity $1 + \tan^2 s = \sec^2 s$.

Substituting cos $s$ for $u$ and sin $s$ for $v$ in the equation of the unit circle gives us the identity

$$(\cos s)^2 + (\sin s)^2 = 1,$$

which can be expressed as

$$\sin^2 s + \cos^2 s = 1.$$

It is conventional in trigonometry to use the notation $\sin^2 s$ rather than $(\sin s)^2$—there are fewer symbols to write. This identity relates the sine and the cosine of any real number $s$. It is an important Pythagorean identity. We now develop another. We divide on both sides of the preceding identity by $\sin^2 s$:

$$\frac{\sin^2 s}{\sin^2 s} + \frac{\cos^2 s}{\sin^2 s} = \frac{1}{\sin^2 s}.$$

Simplifying, we get

$$1 + \cot^2 s = \csc^2 s.$$

This equation is true for any replacement of $s$ by a real number for which $\sin^2 s \neq 0$, since we divided by $\sin^2 s$. But the numbers for which $\sin^2 s = 0$ (or $\sin s = 0$) are exactly the ones for which the cotangent and the cosecant functions are undefined. Hence our new equation holds for all numbers $s$ for which cot $s$ and csc $s$ are defined, and is thus an identity.

**DO EXERCISE 53.**

The third Pythagorean identity, obtained by dividing on both sides of the first Pythagorean identity by $\cos^2 s$, is

$$1 + \tan^2 s = \sec^2 s.$$

---

### The Pythagorean Identities

$$\sin^2 s + \cos^2 s = 1$$
$$1 + \cot^2 s = \csc^2 s$$
$$1 + \tan^2 s = \sec^2 s$$

---

The Pythagorean identities should be memorized. Certain variations of them, although most useful, need not be memorized, because they are so easily obtained from these three.

**54.** Derive identities that give $\sin^2 x$ and $\sin x$ in terms of $\cos x$.

**Example 13** Derive identities that give $\cos^2 x$ and $\cos x$ in terms of $\sin x$.

Solution We begin with $\sin^2 x + \cos^2 x = 1$, and solve it for $\cos^2 x$. We have

$$\cos^2 x = 1 - \sin^2 x.$$

This gives us $\cos^2 x$ in terms of $\sin x$. Then taking square roots, we have

$$\cos x = \pm\sqrt{1 - \sin^2 x},$$

with the understanding that the sign must be determined by the quadrant in which the point determined by $x$ lies on the unit circle. ◄

**DO EXERCISE 54.**

Using identities, we can find the rest of the trigonometric function values from one value if we know the quadrant in which $s$ lies on a unit circle.

**Example 14** Given that $\cos s = \frac{1}{5}$ and that $s$ is in the fourth quadrant, find the other function values.

Solution From Margin Exercise 54, we know that

$$\sin s = \pm\sqrt{1 - \cos^2 s},$$

where the sign of the radical depends on the quadrant in which $s$ lies. Since $s$ is in the fourth quadrant, the sine function is negative. We can then solve for $\sin s$ by substituting $\frac{1}{5}$ for $\cos s$ in the preceding identity:

$$\sin s = -\sqrt{1 - \cos^2 s} = -\sqrt{1 - \left(\frac{1}{5}\right)^2} = -\sqrt{1 - \frac{1}{25}}$$

$$= -\sqrt{\frac{24}{25}} = -\frac{\sqrt{24}}{5} = -\frac{2\sqrt{6}}{5}.$$

**55.** Given that $\sin s = -\frac{3}{8}$ and that $s$ is in the third quadrant, find the other function values.

Now that we have $\sin s$ and $\cos s$, we can find the other function values:

$$\tan s = \frac{\sin s}{\cos s} = \frac{-\frac{2\sqrt{6}}{5}}{\frac{1}{5}} = -2\sqrt{6};$$

$$\cot s = \frac{1}{\tan s} = \frac{1}{-2\sqrt{6}} = -\frac{\sqrt{6}}{12};$$

$$\sec s = \frac{1}{\cos s} = \frac{1}{\frac{1}{5}} = 5;$$

$$\csc s = \frac{1}{\sin s} = \frac{1}{-\frac{2\sqrt{6}}{5}} = -\frac{5}{2\sqrt{6}} = -\frac{5\sqrt{6}}{12}.$$ ◄

**DO EXERCISE 55.**

**Example 15** Given that $\cot s = 2$ and that $s$ is in the third quadrant, find the other function values.

Solution   Solving the identity $1 + \cot^2 s = \csc^2 s$ for $\csc s$, we get

$$\csc s = \pm\sqrt{1 + \cot^2 s},$$

where the sign of the radical depends on the quadrant in which $s$ lies. Since $s$ is in the third quadrant, $\csc s = 1/\sin s$, and the sine function is negative in the third quadrant, it follows that the cosecant function is negative in the third quadrant. Thus,

$$\csc s = -\sqrt{1 + \cot^2 s} = -\sqrt{1 + 2^2} = -\sqrt{5}.$$

From the identity $\sin s = 1/\csc s$, we can find $\sin s$ as follows:

$$\sin s = \frac{1}{\csc s} = \frac{1}{-\sqrt{5}} = -\frac{\sqrt{5}}{5}.$$

From Example 13, we know that

$$\cos s = \pm\sqrt{1 - \sin^2 s}.$$

Since the cosine function is negative in the third quadrant, we can find $\cos s$ as follows:

$$\cos s = -\sqrt{1 - \sin^2 s} = -\sqrt{1 - \left(-\frac{\sqrt{5}}{5}\right)^2}$$

$$= -\sqrt{1 - \frac{5}{25}} = -\sqrt{\frac{4}{5}} = -\frac{2\sqrt{5}}{5}.$$

Finally, we know the secant and the tangent as the reciprocals of the cosine and the cotangent, respectively:

$$\sec s = \frac{1}{\cos s} = \frac{1}{-\dfrac{2\sqrt{5}}{5}} = -\frac{5}{2\sqrt{5}} = -\frac{\sqrt{5}}{2},$$

$$\tan s = \frac{1}{\cot s} = \frac{1}{2}.$$

**DO EXERCISE 56.**

**56.** Given that $\tan s = -3$ and that $s$ is in the second quadrant, find the other function values.

● **EXERCISE SET** **6.5**

A   Find the function values, if they exist, using your knowledge of degrees.

**1.** $\sin \pi$

**2.** $\sin(-3\pi)$

**3.** $\tan \dfrac{3\pi}{4}$

**4.** $\tan \dfrac{5\pi}{6}$

**5.** $\cos\left(-\dfrac{\pi}{3}\right)$

**6.** $\cos \dfrac{7\pi}{6}$

**7.** $\sin \dfrac{\pi}{2}$

**8.** $\cos \dfrac{2\pi}{3}$

**9.** $\cos(-11\pi)$

**10.** $\sin 10\pi$

**11.** $\tan \dfrac{11\pi}{4}$

**12.** $\tan\left(-\dfrac{11\pi}{4}\right)$

**13.** $\sin\left(-\dfrac{9\pi}{4}\right)$

**14.** $\cos\left(-\dfrac{\pi}{2}\right)$

**15.** $\tan \dfrac{4\pi}{3}$

**16.** $\tan \dfrac{\pi}{3}$

Find the function values, if they exist, using a calculator.

**17.** $\sin 37$

**18.** $\cos 810$

**19.** $\cos(-10)$

**20.** $\sin(-4)$

**21.** $\tan 5\pi$

**22.** $\cot 28\pi$

**23.** $\sec\left(-\dfrac{\pi}{5}\right)$

**24.** $\csc\left(-\dfrac{7\pi}{5}\right)$

**25.** $\cot 1000$

**26.** $\tan 13$

**27.** $\sec \dfrac{10\pi}{7}$

**28.** $\cot \dfrac{2\pi}{7}$

**29.** $\cos(-13\pi)$

**30.** $\sin(-10\pi)$

**31.** $\tan 2.5$

**32.** $\cot 5.2$

**B, C**    Complete.

**33.** $\cos(-x) = $ _____

**34.** $\sin(-x) = $ _____

**35.** $\sin(x + \pi) = $ _____

**36.** $\sin(x - \pi) = $ _____

**37.** $\cos(\pi - x) = $ _____

**38.** $\sin(\pi - x) = $ _____

**39.** $\cos(x + 2k\pi) = $ _____

**40.** $\sin(x + 2k\pi) = $ _____

**41.** $\cos(x - \pi) = $ _____

**42.** $\cos(x + \pi) = $ _____

**43. a)** Sketch a graph of $y = \sin x$.
   **b)** By reflecting the graph in part (a), sketch a graph of $y = \sin(-x)$.
   **c)** By reflecting the graph in part (a), sketch a graph of $y = -\sin x$.
   **d)** How do the graphs in parts (b) and (c) compare?

**44. a)** Sketch a graph of $y = \cos x$.
   **b)** By reflecting the graph in part (a), sketch a graph of $y = \cos(-x)$.
   **c)** By reflecting the graph in part (a), sketch a graph of $y = -\cos x$.
   **d)** How do the graphs in parts (a) and (b) compare?

**45. a)** Sketch a graph of $y = \sin x$.
   **b)** By translating, sketch a graph of $y = \sin(x + \pi)$.
   **c)** By reflecting the graph of part (a), sketch a graph of $y = -\sin x$.
   **d)** How do the graphs of parts (b) and (c) compare?

**46. a)** Sketch a graph of $y = \sin x$.
   **b)** By translating, sketch a graph of $y = \sin(x - \pi)$.
   **c)** By reflecting the graph of part (a), sketch a graph of $y = -\sin x$.
   **d)** How do the graphs of parts (b) and (c) compare?

**47. a)** Sketch a graph of $y = \cos x$.
   **b)** By translating, sketch a graph of $y = \cos(x + \pi)$.
   **c)** By reflecting the graph of part (a), sketch a graph of $y = -\cos x$.
   **d)** How do the graphs of parts (b) and (c) compare?

**48. a)** Sketch a graph of $y = \cos x$.
   **b)** By translating, sketch a graph of $y = \cos(x - \pi)$.
   **c)** By reflecting the graph of part (a), sketch a graph of $y = -\cos x$.
   **d)** How do the graphs of parts (b) and (c) compare?

**D**    Find the function values.

**49.** $\cot \dfrac{\pi}{4}$

**50.** $\tan\left(-\dfrac{\pi}{4}\right)$

**51.** $\tan \dfrac{\pi}{6}$

**52.** $\cot \dfrac{5\pi}{6}$

**53.** $\sec \dfrac{\pi}{4}$

**54.** $\csc \dfrac{3\pi}{4}$

**55.** $\tan \dfrac{3\pi}{2}$

**56.** $\cot \pi$

**57.** $\tan \dfrac{2\pi}{3}$

**58.** $\cot\left(-\dfrac{2\pi}{3}\right)$

**59.** $\sec\left(-\dfrac{7\pi}{4}\right)$

**60.** $\csc\left(-\dfrac{4\pi}{3}\right)$

**61.** $\tan \dfrac{5\pi}{6}$

**62.** $\cot \dfrac{7\pi}{6}$

**63.** Of the six circular functions, which are even?

**64.** Of the six circular functions, which are odd?

**65.** Which of the six circular functions have period $2\pi$?

**66.** Which of the six circular functions have period $\pi$?

**67.** In which quadrants is the tangent function positive? negative?

**68.** In which quadrants is the cotangent function positive? negative?

**69.** In which quadrants is the secant function positive? negative?

**70.** In which quadrants is the cosecant function positive? negative?

**71.** Complete this table of approximate function values. Do not use trigonometric function keys. Round to five decimal places.

|      | $\pi/16$ | $\pi/8$ | $\pi/6$ |
| --- | --- | --- | --- |
| sin | 0.19509 | 0.38268 | |
| cos | 0.98079 | 0.92388 | |
| tan | | | |
| cot | | | |
| sec | | | |
| csc | | | |

|      | $\pi/4$ | $3\pi/8$ | $7\pi/16$ |
| --- | --- | --- | --- |
| sin | | 0.92388 | 0.98079 |
| cos | | 0.38268 | 0.19509 |
| tan | | | |
| cot | | | |
| sec | | | |
| csc | | | |

**72.** Complete this table of approximate function values. Do not use trigonometric function keys. Round to five decimal places.

|      | $-\pi/16$ | $-\pi/8$ | $-\pi/6$ |
|------|-----------|----------|----------|
| sin  | $-0.19509$ | $-0.38268$ | $-0.50000$ |
| cos  | $0.98079$ | $0.92388$ | $0.86603$ |
| tan  |           |          |          |
| cot  |           |          |          |
| sec  |           |          |          |
| csc  |           |          |          |

|      | $-\pi/4$ | $-\pi/3$ |
|------|----------|----------|
| sin  | $-0.70711$ | $-0.86603$ |
| cos  | $0.70711$ | $0.50000$ |
| tan  |          |          |
| cot  |          |          |
| sec  |          |          |
| csc  |          |          |

Find the other function values, given each set of conditions.

**73.** $\cos s = \frac{1}{3}$, $s$ in the first quadrant

**74.** $\sin s = \frac{2}{3}$, $s$ in the first quadrant

**75.** $\tan s = 3$, $s$ in the third quadrant

**76.** $\cot s = 4$, $s$ in the third quadrant

**77.** $\sec s = -\frac{5}{3}$, $s$ in the second quadrant

**78.** $\csc s = -\frac{5}{4}$, $s$ in the fourth quadrant

**79.** $\sin s = -\frac{2}{5}$, $s$ in the third quadrant

**80.** $\cos s = -\frac{1}{6}$, $s$ in the third quadrant

● SYNTHESIS _____

**81.** For which numbers is:

  **a)** $\sin x = 1$?          **b)** $\sin x = -1$?

**82.** For which numbers is:

  **a)** $\cos x = 1$?          **b)** $\cos x = -1$?

**83.** Solve for $x$: $\sin x = 0$.

**84.** Solve for $x$: $\cos x = 0$.

**85.** Find $f \circ g$ and $g \circ f$, where $f(x) = x^2 + 2x$ and $g(x) = \cos x$.

**86.** Show that the sine function does not have the *linearity* property—that is, it is not true that

$$\sin (x + y) = \sin x + \sin y$$

for all numbers $x$ and $y$.

**87.** Graph: $y = 3 \sin x$.

**88.** Graph: $y = \sin x + \cos x$.

What are the domain, the range, the period, and the amplitude?

**89.** $y = \sin^2 x$          **90.** $y = |\cos x| + 1$

What is the domain?

**91.** $f(x) = \sqrt{\cos x}$          **92.** $g(x) = \dfrac{1}{\sin x}$

**93.** $f(x) = \dfrac{\sin x}{\cos x}$          **94.** $y = \log (\sin x)$

**95.** Consider $(\sin x)/x$, when $x$ is between 0 and $\pi/2$. What can you say about this function? As $x$ approaches 0, this function approaches a limit. What is it?

**96.** Solve $\sin x < \cos x$ in the interval $[-\pi, \pi]$.

**97.** Verify the identity $\sec (x - \pi) = -\sec x$ graphically.

**98.** Verify the identity $\tan (x + \pi) = \tan x$ graphically.

**99.** Describe how the graphs of the tangent and the cotangent functions are related.

**100.** Describe how the graphs of the secant and the cosecant functions are related.

**101.** Which pairs of circular functions have the same zeros? (A "zero" of a function is an input that produces an output of 0.)

**102.** Describe how the asymptotes of the tangent, cotangent, secant, and cosecant functions are related to the inputs that produce outputs of 0.

Graph.

**103.** $f(x) = |\tan x|$

**104.** $f(x) = |\sin x|$

**105.** $g(x) = \sin |x|$

**106.** $f(x) = |\cos x|$

**107.** One of the motivations for developing trigonometry with a unit circle is that you can actually "see" $\sin \theta$ and $\cos \theta$ on the circle. Note in the figure that $AP = \sin \theta$ and $OA = \cos \theta$. It turns out that you can also "see" the other four trigonometric functions. Prove each of the following.

  **a)** $BD = \tan \theta$          **b)** $OD = \sec \theta$
  **c)** $OE = \csc \theta$          **d)** $CE = \cot \theta$

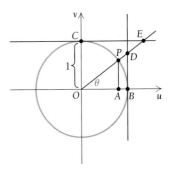

● CHALLENGE _____

Solve graphically.

**108.** $\sin x > \csc x$          **109.** $\cos x \leq \sec x$

1. Sketch a graph of $y = \cos x - 2$.

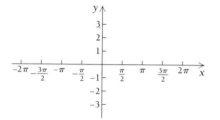

# 6.6

# GRAPHS OF TRANSFORMED SINE AND COSINE FUNCTIONS

## A, B Variations of Basic Graphs

In Sections 6.2 and 6.3, we graphed the six circular, or trigonometric, functions. It might be helpful to review those graphs and have their shapes well in mind because in this section, we will consider variations of these graphs. In particular, we are interested in graphs in the form

$$y = A \sin (Cx - D) + B$$

and

$$y = A \cos (Cx - D) + B,$$

where $A$, $B$, $C$, and $D$ are constants, some of which may be 0. These constants have the effect of translating, stretching, reflecting, or shrinking the basic graphs. (It might be helpful also to review Section 3.8.) Let us first examine the effect of each constant individually. Then we will consider the combined effect of more than one of the constants.

We first consider the effect of the constant $B$.

**Example 1** Sketch a graph of $y = \sin x + 3$.

Solution The graph of $y = \sin x + 3$ is a vertical translation of the graph of $y = \sin x$ up 3 units. One way to sketch the graph is to first consider $y = \sin x$ on an interval of length $2\pi$, say, $[0, 2\pi]$. The zeros of the function and the maximum and minimum values can be considered key points. These are

$$(0, 0), \quad \left(\frac{\pi}{2}, 1\right), \quad (\pi, 0), \quad \left(\frac{3\pi}{2}, -1\right), \quad (2\pi, 0).$$

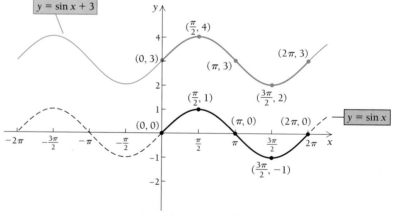

These key points are transformed 3 units up to obtain the key points of the graph of $y = \sin x + 3$. These are

$$(0, 3), \quad \left(\frac{\pi}{2}, 4\right), \quad (\pi, 3), \quad \left(\frac{3\pi}{2}, 2\right), \quad (2\pi, 3).$$

The graph of $y = \sin x + 3$ can be sketched over the interval $[0, 2\pi]$ and extended to obtain the rest of the graph by repeating the graph over intervals of length $2\pi$.

**DO EXERCISE 1.**

Next we consider the effect of the constant $A$. (Also see the Technology Connection at the bottom of p. 432).

**Example 2**   Sketch a graph of $y = 2 \sin x$. What is the amplitude?

*Solution*   The constant 2 in $y = 2 \sin x$ has the effect of stretching the graph of $y = \sin x$ vertically by a factor of 2 units. The function values of $y = \sin x$ are such that $-1 \le \sin x \le 1$, and $|\sin x| \le 1$. The function values of $y = 2 \sin x$ are such that $-2 \le 2 \sin x \le 2$, or $|2 \sin x| \le 2$. The maximum value of $y = 2 \sin x$ is 2, and the minimum value is $-2$. Thus the amplitude is $\frac{1}{2}[2 - (-2)]$, or 2.

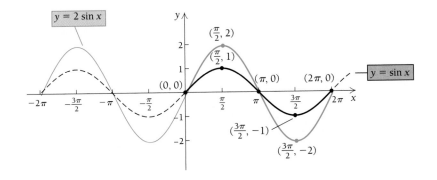

We draw the graph of $y = \sin x$ and consider its key points

$$(0, 0), \quad \left(\frac{\pi}{2}, 1\right), \quad (\pi, 0), \quad \left(\frac{3\pi}{2}, -1\right), \quad (2\pi, 0)$$

over the interval $[0, 2\pi]$.

We then multiply the second coordinates by 2 to obtain the key points of $y = 2 \sin x$. These are

$$(0, 0), \quad \left(\frac{\pi}{2}, 2\right), \quad (\pi, 0), \quad \left(\frac{3\pi}{2}, -2\right), \quad (2\pi, 0).$$

We plot these points and sketch the graph over the interval $[0, 2\pi]$. Then we repeat this part of the graph over other intervals of length $2\pi$. ◀

**DO EXERCISE 2.**

If the absolute value of $A$ is greater than 1, then there will be a vertical stretching. If the absolute value of $A$ is less than 1, then there will be a vertical shrinking.

---

| **DEFINITION** | **Amplitude** |
| --- | --- |

The *amplitude* of $y = A \sin (Cx - D) + B$ is $|A|$.

---

2. Sketch a graph of $y = \frac{2}{3} \cos x$. What is the amplitude?

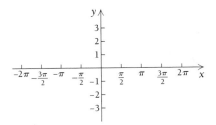

**3.** Sketch a graph of $y = -2 \cos x$. What is the amplitude?

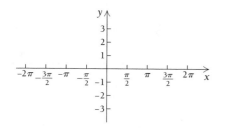

If the constant $A$ in $y = A \sin x$ is negative, there will also be a reflection across the $x$-axis.

**Example 3** Sketch a graph of $y = -\frac{1}{2} \sin x$.

**Solution** The amplitude of the graph is $|-\frac{1}{2}|$, or $\frac{1}{2}$. The graph of $y = -\frac{1}{2} \sin x$ is a vertical shrinking and a reflection of the graph of $y = \sin x$. In graphing, the key points of $y = \sin x$,

$$(0, 0), \quad \left(\frac{\pi}{2}, 1\right), \quad (\pi, 0), \quad \left(\frac{3\pi}{2}, -1\right), \quad (2\pi, 0),$$

are transformed to

$$(0, 0), \quad \left(\frac{\pi}{2}, -\frac{1}{2}\right), \quad (\pi, 0), \quad \left(\frac{3\pi}{2}, \frac{1}{2}\right), \quad (2\pi, 0).$$

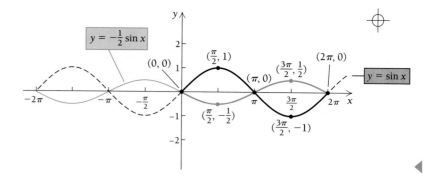

**DO EXERCISE 3.**

Now we consider the effect of the constant $C$ in $y = A \sin (Cx - D) + B$. (Also see the Technology Connection at the top of p. 433.)

**Example 4** Sketch a graph of $y = \sin 2x$. What is the period?

**Solution** Each of the graphs in Examples 1–3, as well as the graph of $y = \sin x$, has period $2\pi$. The constant $C$ has the effect of changing the period. Recall from Section 3.8 that the graph of $y = f(2x)$ is obtained from the graph of $y = f(x)$ by shrinking the graph horizontally. The new graph is obtained by dividing the first coordinate of each ordered-pair solution of $y = f(x)$ by 2. The key points of $y = \sin x$ are

$$(0, 0), \quad \left(\frac{\pi}{2}, 1\right), \quad (\pi, 0), \quad \left(\frac{3\pi}{2}, -1\right), \quad (2\pi, 0).$$

These are transformed to the key points of $y = \sin 2x$, which are

$$(0, 0), \quad \left(\frac{\pi}{4}, 1\right), \quad \left(\frac{\pi}{2}, 0\right), \quad \left(\frac{3\pi}{4}, -1\right), \quad (\pi, 0).$$

We plot these key points and sketch in the graph over the shortened interval $[0, \pi]$, which is of length $\pi$. Then we repeat the graph over other intervals of length $\pi$.

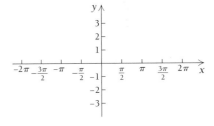

**DEFINITION**                    **Period**

The *period* of the graph of $y = A \sin(Cx - D) + B$ is $\left|\dfrac{2\pi}{C}\right|$.

**DO EXERCISES 4 AND 5.**

Now we examine the effect of the constant $D$ in $y = A \sin(Cx - D) + B$. (Also see the Technology Connection in the middle of p. 433.)

**Example 5**   Sketch a graph of $y = \sin\left(x - \dfrac{\pi}{2}\right)$.

**Solution**   The graph of $y = \sin(x - \pi/2)$ is a translation of the graph of $y = \sin x$ to the right $\pi/2$ units. The key points of $y = \sin x$,

$$(0, 0), \quad \left(\frac{\pi}{2}, 1\right), \quad (\pi, 0), \quad \left(\frac{3\pi}{2}, -1\right), \quad (2\pi, 0),$$

are transformed by adding $\pi/2$ to each of the first coordinates to obtain the following key points of $y = \sin(x - \pi/2)$:

$$\left(\frac{\pi}{2}, 0\right), \quad (\pi, 1), \quad \left(\frac{3\pi}{2}, 0\right), \quad (2\pi, -1), \quad \left(\frac{5\pi}{2}, 0\right).$$

We plot these key points and sketch the curve over the interval $[\pi/2, 5\pi/2]$. Then we repeat the graph over other intervals of length $2\pi$.

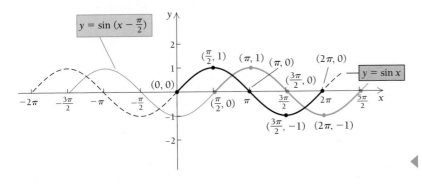

**DO EXERCISE 6.**

4. Sketch a graph of $y = \cos \frac{1}{2}x$. What is the period?

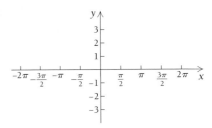

5. Sketch a graph of $y = \sin(-2x)$. What is the period?

6. Sketch a graph of

$$y = \cos\left(x + \frac{\pi}{2}\right).$$

**TECHNOLOGY CONNECTION**

Just as we did in Section 3.8, we can use a grapher to determine the effect of each of the constants in equations of the form

$$f(x) = A \sin (Cx - D) + B$$

or

$$f(x) = A \cos (Cx - D) + B.$$

The effect of each constant will be investigated first individually, and then in combination.

Again, as we did earlier, we will use simple *base functions* in our investigations. These base functions are shown here:

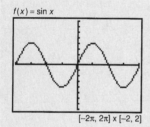

$f(x) = \sin x$

$[-2\pi, 2\pi] \times [-2, 2]$

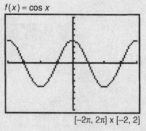

$f(x) = \cos x$

$[-2\pi, 2\pi] \times [-2, 2]$

Start by graphing one of the base functions. Use radian mode and a $[-2\pi, 2\pi] \times [-5, 5]$ viewing box. Next, add 3 to the function, forming the new function $f(x) + 3$. (Thus if your base function is $\sin x$, your new function is $f(x) = \sin x + 3$.) Finally, subtract 2 from the base function, forming $f(x) - 2$.

Try repeating this process with the other base function. What is the rule that describes the difference between the graph of $f(x)$ and the graph of $f(x) + B$?

Predict what each of the following graphs will look like. Then check your prediction using a grapher.

**TC 1.** $f(x) = \sin x + 4$        **TC 2.** $f(x) = \cos x - 2$

**TC 3.** $f(x) = \cos x + 3.5$      **TC 4.** $f(x) = \sin x - 0.6$

 **TECHNOLOGY CONNECTION**

Beginning with the same base function that you used in the last investigation, draw its graph on your grapher. Next, multiply the function by 3 and draw the graph of this new function, $3 \cdot f(x)$. How does it compare to the graph of $f(x)$?

Now replace the 3 with $\frac{1}{4}$, and draw the graph of $\frac{1}{4} \cdot f(x)$. How does it compare with the graph of $f(x)$?

Repeat this three-part process with the other base function.

What rule predicts the shape of the graph that results from multiplying a function by a constant $A$ that is greater than 1? What about the multiplication of a function by a constant $A$ that is between 0 and 1?

Beginning with the graph of your original base function, try multiplying the function by the negative constant $-3$.

How does its graph compare with the graph of $f(x)$? What about multiplying by the negative constant $-\frac{1}{4}$?

After you have tried multiplying the other base function by negative constants, state a rule that predicts the result of multiplying a function by a negative constant $A$.

Predict the shape of each of the following graphs. Then use your grapher to check your answer.

**TC 5.** $f(x) = 5 \sin x$

**TC 6.** $f(x) = -5 \sin x$

**TC 7.** $f(x) = \frac{1}{2} \cos x$

**TC 8.** $f(x) = -\frac{1}{2} \cos x$

**TECHNOLOGY CONNECTION**

Beginning with the graph of the base function you've been using, graph the function that results when the variable $x$ is multiplied by 3, forming $f(3x)$. (If your base function is $\sin x$, the new function will be $\sin 3x$.) Note the change in the graph. Now replace the 3 with $\frac{1}{2}$, forming $f(\frac{1}{2}x)$. How is its graph different from the graph of the base function? Now try these steps with the other base function.

Start with your base function again, but this time multiply by *negative* constants. Graph $f(-3x)$ and $f(-\frac{1}{2}x)$. What effect does this negative sign have?

What is the general rule that describes the effect of the

constant $C$ in a function $f(Cx)$? (Think of both positive *and* negative values of $C$.)

Predict the shape of each of the following graphs. Then use your grapher to check your prediction. Use radian mode, with a viewing box of $[-2\pi, 2\pi] \times [-2, 2]$.

**TC 9.** $f(x) = \sin 4x$

**TC 10.** $f(x) = \sin(-4x)$

**TC 11.** $f(x) = \cos(-\frac{1}{4}x)$

**TC 12.** $f(x) = \cos 2.5x$

 **TECHNOLOGY CONNECTION**

Once again, start with the graph of your favorite base function and your grapher in radian mode. This time, subtract $\pi/2$ from $x$, forming $f(x - \pi/2)$. How does this graph compare with the graph of the base function? Replace the $\pi/2$ with $-\pi/3$, and draw the graph of $f(x + \pi/3)$.

If your function is in degrees, the values of the constants added to, or subtracted from, $x$ will probably be significantly different. For example, instead of the $\pi/2$ and $-\pi/3$ used above, you would use $90°$ and $-60°$, respectively. Change your grapher to degree mode and then graph the same base function you used above. However, this time, the second function will be $f(x - 90°)$ and the third function will be $f(x + 60°)$. The resulting graphs should be identical to the graphs you found at the start of this investigation.

Try this with the other base function, in both radian and degree modes. Can you state a rule that predicts the effect of subtracting (or adding) a constant $D$ from $x$ in a trigonometric function? That is, how is the graph of $f(x - D)$ different from the graph of $f(x)$?

For each of the following functions, predict the shape of the graph. If the constant includes a $\pi$, you can assume that all angles are measured in radians; otherwise, assume that the angles are in degrees.

**TC 13.** $f(x) = \cos(x - \pi/4)$

**TC 14.** $f(x) = \cos(x + 135°)$

**TC 15.** $f(x) = \sin(x + 3\pi/2)$

**TC 16.** $f(x) = \sin(x - 45°)$

 **TECHNOLOGY CONNECTION**

Now, we can put everything together, predicting the shape of functions such as

$$f(x) = A \sin(Cx - D) + B$$

or

$$f(x) = A \cos(Cx - D) + B.$$

In the last four investigations, we have examined the effect of each of the constants $A$ through $D$ individually. By breaking the function down step by step, you will be able to predict the shape of functions that include any combination of these constants. If you have a grapher available, you can check each step of the solution in Examples 6–8 and Margin Exercises 7–9. (You should be able to go through each of

these steps *without* a grapher. Use the grapher only to check your work or to deal with cases in which the constants are "difficult" to work with.)

For each of the following functions, predict the shape of its graph by following the steps illustrated in Examples 6–8. Write a description of the effect of each transformation *before* you graph it and then use your grapher to check your prediction.

**TC 17.** $f(x) = 12.5 \cos(5x - 2.5\pi) - 3$

**TC 18.** $f(x) = -20 \sin(-3x - 2\pi) + 1.5$

**TC 19.** $f(x) = 0.25 \cos(4x + \pi/3) + 3.5$

Now we consider combined transformations of graphs. (Also see the Technology Connection at the bottom of p. 425.)

**Example 6** Sketch a graph of $y = \sin(2x - \pi)$. What is the period?

**Solution** The graph of

$$y = \sin(2x - \pi)$$

is the same as the graph of

$$y = \sin[2(x - \pi/2)].$$

The $\pi/2$ translates the graph of $y = \sin 2x$ to the right $\pi/2$ units. (See Example 5.) The 2 factored out of the parentheses shrinks the period by half, making the period $|2\pi/2|$, or $\pi$.

Thus to form this graph, we first graph $y = \sin 2x$, as in Example 4. The key points of $y = \sin 2x$ are

$$(0,0), \quad \left(\frac{\pi}{4}, 1\right), \quad \left(\frac{\pi}{2}, 0\right), \quad \left(\frac{3\pi}{4}, -1\right), \quad (\pi, 0).$$

We use them to obtain the key points of $y = \sin[2(x - \pi/2)]$, which are obtained, in a manner similar to Example 5, by adding $\pi/2$ to each first coordinate:

$$\left(\frac{\pi}{2}, 0\right), \quad \left(\frac{3\pi}{4}, 1\right), \quad (\pi, 0), \quad \left(\frac{5\pi}{4}, -1\right), \quad \left(\frac{3\pi}{2}, 0\right).$$

We draw one period of the graph over the interval $[\pi/2, 3\pi/2]$, which has length $\pi$. Then we repeat the graph over other intervals of length $\pi$.

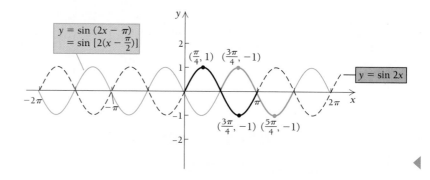

---

**DEFINITION**  **Phase Shift**

For the graph of

$$y = A\sin(Cx - D) + B = A\sin\left[C\left(x - \frac{D}{C}\right)\right] + B,$$

the quantity $D/C$, called the *phase shift*, translates the graph to the right if $D/C$ is positive and to the left if $D/C$ is negative. Be sure to do the horizontal stretching or shrinking based on the constant $C$ *before* the translation based on the phase shift $D/C$.

In Example 6, the phase shift is $\pi/2$. This means that the graph of $y = \sin 2x$ has been shifted $\pi/2$ units to the right.

**DO EXERCISE 7.**

Let us summarize the effects of the constants. We carry out the procedures in the *order* listed.

(1) Stretches the graph horizontally if $|C| < 1$. Shrinks the graph horizontally if $|C| > 1$. If $C < 0$, the graph is also reflected across the y-axis. The *period* is $|2\pi/C|$.

(2) Stretches the graph vertically if $|A| > 1$. Shrinks the graph vertically if $|A| < 1$. If $A < 0$, the graph is also reflected across the x-axis. The *amplitude* of the graph is $|A|$.

$$y = A \sin(Cx - D) + B = A \sin\left[C\left(x - \frac{D}{C}\right)\right] + B$$

(3) Translates the graph $|D/C|$ units to the right if $D/C > 0$ or $|D/C|$ units to the left if $D/C < 0$. The *phase shift* is $D/C$.

(4) Translates the graph $|B|$ units up if $B > 0$ or $|B|$ units down if $B < 0$.

Similar statements hold for

$$y = A \cos(Cx - D) + B = A \cos\left[C\left(x - \frac{D}{C}\right)\right] + B.$$

**Example 7**   Sketch a graph of $y = 3 \sin(2x + \pi/2)$. Find the amplitude, the period, and the phase shift.

**Solution**   We first note that

$$y = 3 \sin\left(2x + \frac{\pi}{2}\right) = 3 \sin\left[2\left(x - \left(-\frac{\pi}{4}\right)\right)\right].$$

7. Sketch a graph of

$$y = \cos(2x + \pi).$$

What is the period? the phase shift?

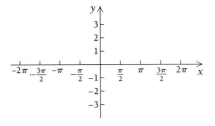

8. Sketch a graph of

$$y = 3\cos\left(2x - \frac{\pi}{2}\right).$$

What is the amplitude? the period? the phase shift?

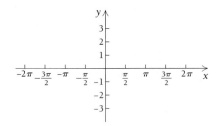

Then

$$\text{Amplitude} = |A| = |3| = 3,$$

$$\text{Period} = \left|\frac{2\pi}{C}\right| = \left|\frac{2\pi}{2}\right| = \pi,$$

$$\text{Phase shift} = \frac{D}{C} = -\frac{\pi}{4}.$$

To create the final graph, we sketch graphs of each of the following equations in sequence:

1. $y = \sin 2x$,
2. $y = 3\sin 2x$,
3. $y = 3\sin\left[2\left(x - \left(-\frac{\pi}{4}\right)\right)\right].$

Start with $y = \sin x$.

1. Then graph $y = \sin 2x$. (See Example 4.)

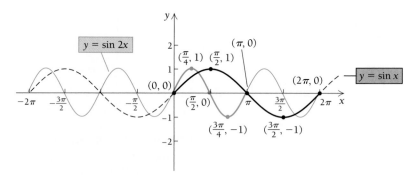

2. Next graph $y = 3\sin 2x$ by stretching the graph vertically by a factor of 3.

3. Then graph $y = 3\sin\left[2(x - (-\pi/4))\right]$ by translating the graph of $y = 3\sin 2x$ to the left $\pi/4$ units.

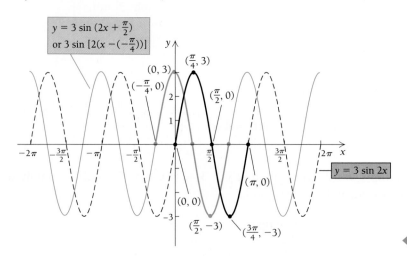

**DO EXERCISE 8.**

**Example 8**   Sketch a graph of $y = 3\cos(2\pi x) - 1$. Find the amplitude, the period, and the phase shift.

**Solution**   First we note the following:

Amplitude $= |A| = |3| = 3$,

Period $= \left|\dfrac{2\pi}{C}\right| = \left|\dfrac{2\pi}{2\pi}\right| = |1| = 1$,

Phase shift $= \dfrac{D}{C} = \dfrac{0}{2\pi} = 0$.

There is no phase shift in this problem since the constant $D = 0$. Thus there is no horizontal translation.

To create the final graph, we graph each of the following equations in sequence:

1.  $y = \cos(2\pi x)$,
2.  $y = 3\cos(2\pi x)$,
3.  $y = 3\cos(2\pi x) - 1$.

Start with $y = \cos x$.

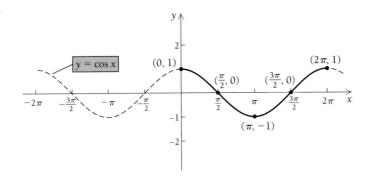

1.  Then graph $y = \cos(2\pi x)$. The period is 1 and the graph is shrunk horizontally.

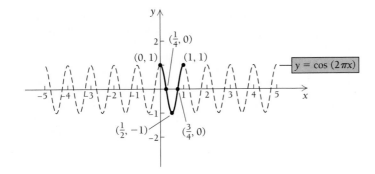

2.  Next graph $y = 3\cos(2\pi x)$. The amplitude is 3. The graph is stretched vertically by a factor of 3.

9. Sketch a graph of

$$y = -2 \sin(\pi x) + 2.$$

What is the amplitude? the period? the phase shift?

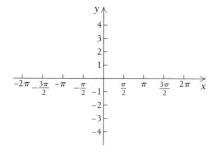

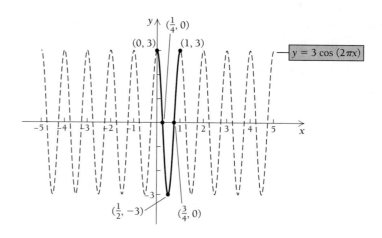

3. Then graph $y = 3 \cos(2\pi x) - 1$. The graph is translated down 1 unit.

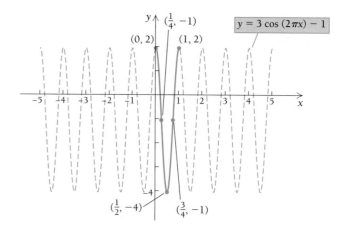

**DO EXERCISE 9.**

The oscilloscope shown here is an electronic device that draws graphs like those in the preceding examples. These graphs are often called **sinusoidal.** By manipulating the controls, we can change such things as the amplitude, the period, and the phase shift. The oscilloscope has many applications, and the trigonometric functions play a major role in many of them.

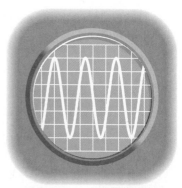

An oscilloscope display, showing a sinusoidal (sine-shaped) curve.

## Graphs of Sums: Addition of Ordinates

A function that is a sum of two functions can often be graphed by a method called **addition of ordinates.** We graph the two functions separately and then add the second coordinates (called **ordinates**) graphically. A compass may be helpful.

**Example 9**   Graph: $y = 2 \sin x + \sin 2x$.

**Solution**   We graph $y = 2 \sin x$ and $y = \sin 2x$ using the same set of axes.

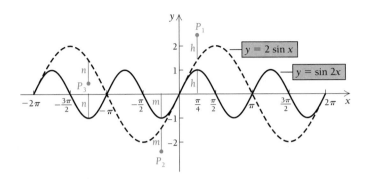

Now we graphically add some ordinates to obtain points on the graph we seek. At $x = \pi/4$, we transfer the distance $h$, which is the value of $\sin 2x$, up to add it to the value of $2 \sin x$. Point $P_1$ is on the graph we seek. At $x = -\pi/4$, we do a similar thing, but the distance $m$ is negative. Point $P_2$ is on our graph. At $x = -5\pi/4$, the distance $n$ is negative, so we in effect subtract it from the value of $2 \sin x$. Point $P_3$ is on the graph we seek. We continue to plot points in this fashion and then connect them to get the desired graph, shown here.

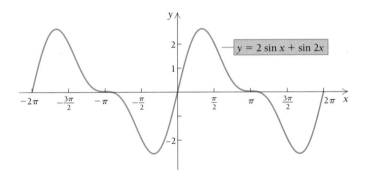

**DO EXERCISE 10.**

A "sawtooth" function, such as the one shown on this oscilloscope, has numerous applications—for example, in television circuits. This sawtooth function can be approximated extremely well by adding, electronically, several sine and cosine functions. The graph in Example 9 could be a first step in such an approximation.

An oscilloscope displaying a sawtooth function. The function has been synthesized using sines and cosines.

10. By addition of ordinates, sketch a graph of $y = \cos x + \sin 2x$.

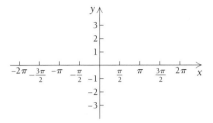

**TECHNOLOGY CONNECTION**

A grapher can easily graph any functions shown in this section. It is especially useful when a function is made up of more than one trigonometric function. The only step is to select the appropriate viewing box.

Use a grapher to graph each of the following functions.

**TC 20.**   $f(x) = 5 \cos x + \cos 5x$

**TC 21.**   $f(x) = \cos 2x - \sin \frac{1}{2}x$

**TC 22.**   $f(x) = 4 \cos x + \sin 3x$

## ● EXERCISE SET 6.6

**A , B**    Use graph paper to sketch the graph of the function. Determine the amplitude, the period, and the phase shift.

**1.** $y = \sin x + 2$

**2.** $y = \cos x - 3$

**3.** $y = \dfrac{1}{2}\sin x$

**4.** $y = \dfrac{1}{3}\cos x$

**5.** $y = 2\cos x$

**6.** $y = 3\cos x$

**7.** $y = -\dfrac{1}{2}\cos x$

**8.** $y = -2\sin x$

**9.** $y = \cos 2x$

**10.** $y = \cos 3x$

**11.** $y = \cos(-2x)$

**12.** $y = \cos(-3x)$

**13.** $y = \sin\dfrac{1}{2}x$

**14.** $y = \cos\dfrac{1}{3}x$

**15.** $y = \sin\left(-\dfrac{1}{2}x\right)$

**16.** $y = \cos\left(-\dfrac{1}{3}x\right)$

**17.** $y = \cos(2x - \pi)$

**18.** $y = \sin(2x + \pi)$

**19.** $y = 2\cos\left(\dfrac{1}{2}x - \dfrac{\pi}{2}\right)$

**20.** $y = 4\sin\left(\dfrac{1}{4}x + \dfrac{\pi}{8}\right)$

**21.** $y = -3\cos(4x - \pi)$

**22.** $y = -3\sin\left(2x + \dfrac{\pi}{2}\right)$

**23.** $y = 2 + 3\cos(\pi x - 3)$

**24.** $y = 5 - 2\cos\left(\dfrac{\pi}{2}x + \dfrac{\pi}{2}\right)$

**B**    Determine the amplitude, the period, and the phase shift.

**25.** $y = 3\cos\left(3x - \dfrac{\pi}{2}\right)$

**26.** $y = 4\sin\left(4x - \dfrac{\pi}{3}\right)$

**27.** $y = -5\cos\left(4x + \dfrac{\pi}{3}\right)$

**28.** $y = -4\sin\left(5x + \dfrac{\pi}{2}\right)$

**29.** $y = \dfrac{1}{2}\sin(2\pi x + \pi)$

**30.** $y = -\dfrac{1}{4}\cos(\pi x - 4)$

**C**    Graph.

**31.** $y = 2\cos x + \cos 2x$

**32.** $y = 3\cos x + \cos 3x$

**33.** $y = \sin x + \cos 2x$

**34.** $y = 2\sin x + \cos 2x$

**35.** $y = \sin x - \cos x$

**36.** $y = 3\cos x - \sin x$

**37.** $y = 3\cos x + \sin 2x$

**38.** $y = 3\sin x - \cos 2x$

● **SYNTHESIS** _____

**39. *Temperature during an illness.*** The temperature $T$ of a patient during a 12-day illness is given by

$$T(t) = 101.6° + 3°\sin\left(\dfrac{\pi}{8}t\right).$$

**a)** Sketch a graph of the function over the interval [0, 12].
**b)** What are the maximum and the minimum temperatures during the illness?

**40. *Periodic sales.*** A company in a northern climate has sales of skis as given by

$$S(t) = 10\left(1 - \cos\dfrac{\pi}{6}t\right),$$

where $t$ = time, in months ($t = 0$ corresponds to July 1), and $S(t)$ is in thousands of dollars.

**a)** Sketch a graph of the function over a 12-month interval [0, 12].
**b)** What is the period of the function?
**c)** What is the minimum amount of sales and when does it occur?
**d)** What is the maximum amount of sales and when does it occur?

**41. *Satellite location.*** A satellite circles the earth in such a way that it is $y$ miles from the equator (north or south, height not considered) $t$ minutes after its launch, where

$$y = 3000\left[\cos\dfrac{\pi}{45}(t - 10)\right].$$

**a)** Sketch a graph of the function.
**b)** What are the amplitude, the period, and the phase shift?

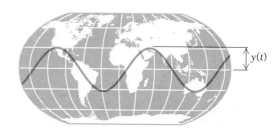

**42. *Water wave.*** The cross-section of a water wave is given by

$$y = 3\sin\left(\dfrac{\pi}{4}x + \dfrac{\pi}{4}\right),$$

where $y$ is the vertical height of the water wave and $x$ is the distance from the origin to the wave.

**a)** Sketch a graph of the function.
**b)** What are the amplitude, the period, and the phase shift?

**43.** Find a sine function that has amplitude 8.6, period 5, and phase shift $-11$.

**44.** Find a cosine function that has amplitude 6.7, period $\pi$, and phase shift $\frac{4}{7}$.

**45.** Find a cosine function that has amplitude 16, period $\pi/3$, and phase shift $-2/\pi$.

**46.** Find a sine function that has amplitude 34, period $\pi/5$, and phase shift $-8/\pi$.

Graph.

**47.** $y = |4\sin x|$

**48.** $y = |\cos 3x|$

● CHALLENGE _____

Graph using addition of ordinates. You might want to use a calculator to find some function values to plot.

**49.** $y = x + \sin x$     **50.** $y = \cos x - x$

☰ TECHNOLOGY CONNECTION _____

Use a grapher to graph the function.

**51.** $y = \cos 2x + 2x$     **52.** $y = \cos 3x + \sin 3x$

**53.** $y = 4\cos 2x - 2\sin x$     **54.** $y = 7.5\cos x + \sin 2x$

# 6.7
# TRANSFORMED GRAPHS OF OTHER TRIGONOMETRIC FUNCTIONS

> **OBJECTIVES**
>
> You should be able to:
>
> ◢ Sketch graphs that are transformations of the tangent, cotangent, secant, and cosecant functions.
>
> ◢ Sketch graphs of functions found by multiplying trigonometric functions by other functions.

## ◢ Transformations of Tangent, Cotangent, Secant, and Cosecant

The transformational techniques that we learned in Section 6.6 for graphing the sine and cosine can also be applied to the other trigonometric functions.

**Example 1**   Sketch a graph of $y = \dfrac{1}{2}\tan\left(x - \dfrac{\pi}{4}\right)$.

**Solution**   The graph of $y = \tan x$ is shown below. Recall that it has period $\pi$ and is defined for all real numbers except $\pi/2 + k\pi$.

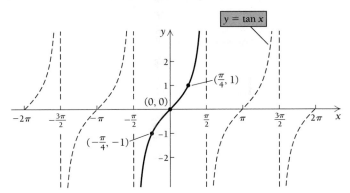

Let us consider the following ordered pairs as key points in the interval $(-\pi/2, \pi/2)$. Note that the function is not defined at the endpoints, so we indicate the following as key points,

$$\left(-\frac{\pi}{2}, \text{undefined}\right), \quad \left(-\frac{\pi}{4}, -1\right), \quad (0,0), \quad \left(\frac{\pi}{4}, 1\right), \quad \left(\frac{\pi}{2}, \text{undefined}\right),$$

keeping in mind that "undefined" means that the first coordinate is not in the domain.

There is no amplitude since the function has no maximum or minimum value. Nevertheless, the constant $\frac{1}{2}$ shrinks each function value of $y = \tan x$ by a factor of $\frac{1}{2}$. Thus the graph of $y = \frac{1}{2}\tan x$ is obtained from the graph of $y = \tan x$ by a vertical shrinking. The key points are transformed by multiplying each second coordinate by $\frac{1}{2}$:

$$\left(-\frac{\pi}{2}, \text{undefined}\right), \quad \left(-\frac{\pi}{4}, -\frac{1}{2}\right), \quad (0,0), \quad \left(\frac{\pi}{4}, \frac{1}{2}\right), \quad \left(\frac{\pi}{2}, \text{undefined}\right).$$

1. Sketch a graph of

$$y = 2 \cot\left(x + \frac{\pi}{4}\right).$$

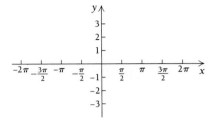

We plot these key points and sketch the curve over the interval $(-\pi/2, \pi/2)$. We can use a calculator to find other function values, if desired. Then we repeat the graph over other intervals of length $\pi$.

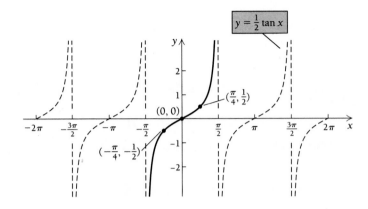

Next, we obtain the graph of $y = \frac{1}{2}\tan(x - \pi/4)$ by translating the graph of $y = \frac{1}{2}\tan x$ to the right $\pi/4$ units.

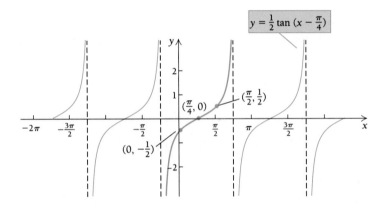

**DO EXERCISE 1.**

**Example 2** Sketch a graph of $y = 3\sec\left(x + \dfrac{\pi}{2}\right)$.

Solution The graph of $y = \sec x$ is shown below. Recall that it has period $2\pi$ and is defined for all real numbers except $\pi/2 + k\pi$.

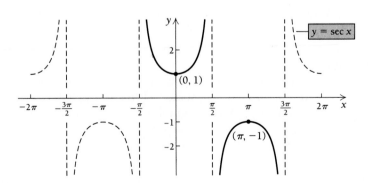

Let us consider the following ordered pairs as key points in the interval $(-\pi/2, 3\pi/2)$. Note that the function is not defined at the endpoints or at $\pi/2$, so we indicate the following as key points,

$$\left(-\frac{\pi}{2}, \text{undefined}\right), \quad (0, 1), \quad \left(\frac{\pi}{2}, \text{undefined}\right),$$

$$(\pi, -1), \quad \left(\frac{3\pi}{2}, \text{undefined}\right),$$

keeping in mind that "undefined" means that the first coordinate is not in the domain.

There is no amplitude since the function has no maximum or minimum value. Nevertheless, the constant 3 stretches each function value of $y = \sec x$ by a factor of 3. Thus the graph of $y = 3 \sec x$ is obtained from the graph of $y = \sec x$ by a vertical stretching. The key points are transformed by multiplying each second coordinate by 3:

$$\left(-\frac{\pi}{2}, \text{undefined}\right), \quad (0, 3), \quad \left(\frac{\pi}{2}, \text{undefined}\right),$$

$$(\pi, -3), \quad \left(\frac{3\pi}{2}, \text{undefined}\right).$$

 **TECHNOLOGY CONNECTION**

All the work that we did in the last section concerning the graphs of functions of the form

$$f(x) = A \sin (Cx - D) + B$$

or

$$f(x) = A \cos (Cx - D) + B$$

also applies to similar functions involving tan, sec, csc, and cot. In order to use what we learned, we need to review what the base functions look like.

The effects of the constants $A$, $B$, $C$, and $D$ are exactly the same on these four base functions as they are on the functions $\sin x$ and $\cos x$. Therefore, you can use the same steps used in the analysis of $\sin x$ and $\cos x$ functions to predict the shapes of these graphs. Finally, you can use your grapher to verify your predictions or to handle functions that include difficult values.

Predict the shapes of each of the following functions, using the steps learned in Section 6.6. Then check your predictions using your grapher. (These functions are in radians.)

**TC 1.** $f(x) = 5 \tan (2x - 3\pi) + 6$

**TC 2.** $f(x) = -4 \csc (-x - \pi/2) + 3.5$

**TC 3.** $f(x) = 2.5 \sec (-3x + 2\pi) - 4.5$

**TC 4.** $f(x) = 0.4 \cot (8x + 2\pi) - 1.2$

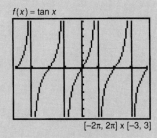

$f(x) = \tan x$

$[-2\pi, 2\pi] \times [-3, 3]$

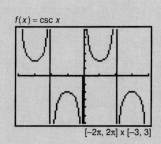

$f(x) = \csc x$

$[-2\pi, 2\pi] \times [-3, 3]$

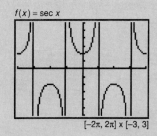

$f(x) = \sec x$

$[-2\pi, 2\pi] \times [-3, 3]$

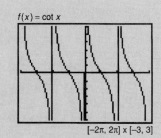

$f(x) = \cot x$

$[-2\pi, 2\pi] \times [-3, 3]$

2. Sketch a graph of

$$y = -\frac{2}{3} \csc\left(x - \frac{\pi}{2}\right).$$

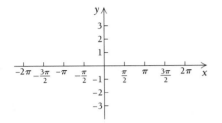

We plot these key points and sketch the curve over the interval $(-\pi/2, 3\pi/2)$. We can use a calculator to find other function values, if desired. Then we repeat the graph over other intervals of length $2\pi$.

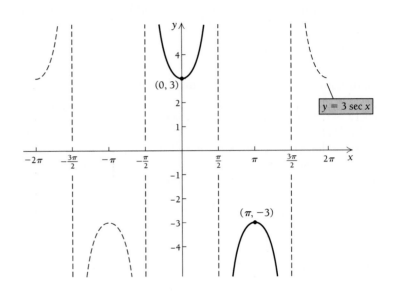

Next, we obtain the graph of $y = 3 \sec [x - (-\pi/2)]$ by translating the graph of $y = 3 \sec x$ to the left $\pi/2$ units.

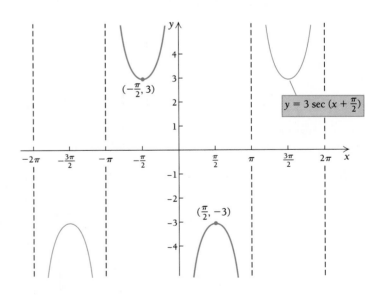

**DO EXERCISE 2.**

## ▶ Multiplication of Ordinates

Suppose that a weight is attached to a spring and the spring is stretched and put into motion. The weight oscillates up and down. If we could assume falsely that the weight will bob up and down forever, then its height $h$ after time $t$ in seconds might be approximated by a function like

$$h(t) = 5 + 2\sin(6\pi t).$$

Over a short time period, this might be a valid model, but experience tells us that eventually the spring will come to rest. A more appropriate model is provided by the following example, which illustrates **damped oscillation.**

**Example 3**    Sketch a graph of $f(x) = e^{-x/2}\sin x$.

**Solution**    The function $f$ is the product of two functions $g$ and $h$, where

$$g(x) = e^{-x/2} \quad \text{and} \quad h(x) = \sin x.$$

Thus, to find function values, we can **multiply ordinates.** Let us do more analysis before graphing. Note that for any real number $x$,

$$-1 \le \sin x \le 1.$$

Recall from Chapter 5 that all values of the exponential function are positive. Thus we can multiply by $e^{-x/2}$ and obtain the inequality

$$-e^{-x/2} \le e^{-x/2}\sin x \le e^{-x/2}.$$

The direction of the inequality symbols does not change since $e^{-x/2} > 0$. This also tells us that the original function crosses the $x$-axis only at values for which $\sin x = 0$. These are the numbers $k\pi$, for any integer $k$.

The inequality tells us that the function $f$ is constrained between the graphs of $y = -e^{-x/2}$ and $y = e^{-x/2}$. We start by graphing these functions using dashed lines. Since we also know that $f(x) = 0$ when $x = k\pi$, $k$ an integer, we mark these points on the graph. Then we use a calculator and compute other function values. The graph is as follows.

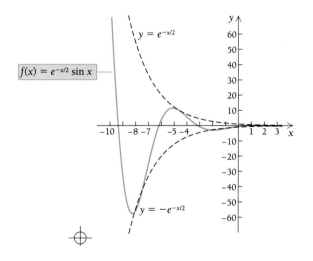

**DO EXERCISE 3.**

3. Sketch a graph of

$$f(x) = \frac{1}{2}x^2\sin x.$$

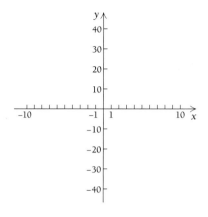

**TECHNOLOGY CONNECTION**

The following functions are very simple to graph using a grapher. In fact, the constants involved can be any real number—they are all equally easy. (You may need to determine how your grapher handles exponential functions of the form $e^{ax}$.)

Use a grapher to draw the graphs of each of these functions.

**TC 5.** $f(x) = e^{-0.6x}\cos x$

**TC 6.** $f(x) = 0.2x^2\sin x$

**TC 7.** $f(x) = 4^{-x}\cos x$

● **EXERCISE SET** **6.7**

A    Sketch a graph of each of the following.

**1.** $y = -\tan x$      **2.** $y = -\cot x$

**3.** $y = -\csc x$      **4.** $y = -\sec x$

**5.** $y = \sec(-x)$      **6.** $y = \csc(-x)$

**7.** $y = \cot(-x)$      **8.** $y = \tan(-x)$

**9.** $y = -2 + \cot x$      **10.** $y = -3 + \sec x$

**11.** $y = -\dfrac{3}{2}\csc x$      **12.** $y = 0.25 \cot x + 3$

**13.** $y = \cot 2x$      **14.** $y = 2\tan\dfrac{1}{2}x$

**15.** $y = \sec\left(-\dfrac{1}{2}x\right)$      **16.** $y = \csc\left(-\dfrac{1}{3}x\right)$

**17.** $y = 2\sec(x - \pi)$      **18.** $y = 3\csc(x + \pi)$

**19.** $y = 4\tan\left(\dfrac{1}{4}x + \dfrac{\pi}{8}\right)$      **20.** $y = 2\sec\left(\dfrac{1}{2}x - \dfrac{\pi}{2}\right)$

**21.** $y = -3\cot\left(2x + \dfrac{\pi}{2}\right)$      **22.** $y = -3\tan(4x - \pi)$

**23.** $y = 4\sec(2x - \pi)$      **24.** $y = 2\csc\left(\dfrac{1}{2}x - \dfrac{3\pi}{4}\right)$

B    Sketch a graph of each of the following.

**25.** $f(x) = e^{-x/2}\cos x$

**26.** $f(x) = e^{-0.4x}\sin x$

**27.** $f(x) = 0.6x^2\cos x$

**28.** $f(x) = e^{-x/4}\sin x$

**29.** $f(x) = x\sin x$

**30.** $f(x) = |x|\cos x$

**31.** $f(x) = 2^{-x}\sin x$

**32.** $f(x) = 2^{-x}\cos x$

● SYNTHESIS _____

Sketch a graph of each of the following functions.

**33.** $f(x) = |\tan x|$

**34.** $f(x) = |\csc x|$

**35.** $f(x) = (\ln x)(\sin x)$

**36.** $f(x) = \sin^2 x$

**37.** $f(x) = \sec^2 x$

**38.** $f(x) = (\tan x)(\csc x)$

**39.** *Damped oscillations.* Suppose that the motion of a spring is given by

$$d(t) = 6e^{-0.8t}\cos(6\pi t) + 4,$$

where $d$ is the distance, in inches, of a weight from the point at which the spring is attached to a ceiling, after $t$ seconds.

**a)** Sketch the graph over the interval $[0, 10]$.
**b)** How far do you think the spring is from the ceiling when the spring stops bobbing?

**40.** *Temperature during an illness.* A patient's temperature during an illness is given by

$$T(t) = 98.6° + 2°e^{-0.6t}\sin t,$$

where $T$ is the temperature after time $t$, in hours.

**a)** Sketch the graph over the interval $[0, 12]$.
**b)** What is the patient's normal temperature after the illness?

**41.** *Rotating beacon.* A police car is parked 10 ft from a wall. On top of the car is a beacon rotating in such a way that the light is at a distance $d(t)$ from point $Q$ after $t$ seconds, where

$$d(t) = 10\tan(2\pi t).$$

When $d$ is positive, the light is pointing north of $Q$, and when $d$ is negative, the light is pointing south of $Q$.

**a)** Sketch a graph of the function over the interval $[0, 2]$.
**b)** Explain the meaning of the values of $t$ for which the function is undefined.

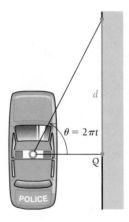

▨ TECHNOLOGY CONNECTION _____

Use a grapher to graph each of the following over the given interval and approximate the zeros.

**42.** $f(x) = \dfrac{\sin x}{x}$; $[-12, 12]$

**43.** $f(x) = \dfrac{\sin^2 x}{x}$; $[-4, 4]$

**44.** $f(x) = \ln|\sec x + \tan x|$; $[-12, 12]$

**45.** $f(x) = \dfrac{\cos x - 1}{x}$; $[-12, 12]$

**46.** $f(x) = x^3\sin x$; $[-5, 5]$

**47.** $f(x) = 3\cos^2 x\sin x + 1$; $[-2, 2]$

# SUMMARY AND REVIEW  6

### ● TERMS TO KNOW

Sine function, p. 354
Cosine function, p. 354
Tangent function, p. 354
Cotangent function, p. 354
Secant function, p. 354
Cosecant function, p. 354
Degrees, pp. 359, 374
Minutes, p. 359
Seconds, p. 359
Cofunctions, p. 362
Angle of elevation, p. 368
Angle of depression, p. 368
Bearing, p. 370
Angle, p. 374

Vertex, p. 374
Rotation, p. 374
Initial side, p. 374
Terminal side, p. 374
Standard position, p. 374
Coterminal angles, p. 374
Right angle, p. 376
Acute angle, p. 376
Obtuse angle, p. 376
Straight angle, p. 376
Complementary angles, p. 377
Supplementary angles, p. 377
Reference triangle, pp. 377, 384
Reference angle, p. 384

Unit circle, p. 391
Radians, p. 395
Arc length, p. 398
Central angle, p. 398
Linear speed, p. 399
Angular speed, p. 399
Circular functions, p. 406
Periodic function, p. 409
Period, p. 409
Identity, p. 410
Amplitude, p. 411
Basic identities, p. 422
Pythagorean identities, p. 423
Phase shift, p. 434

### ● REVIEW EXERCISES

**1.** Find the six trigonometric ratios for the specified angle.

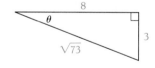

**2.** Convert 22.20° to degrees and minutes.

**3.** Convert 47°33′27″ to degrees and decimal parts of degrees.

**4.** Given that $\sin \theta = 0.6820$, $\cos \theta = 0.7314$, $\tan \theta = 0.9325$, $\cot \theta = 1.0724$, $\sec \theta = 1.3673$, and $\csc \theta = 1.4663$, find the six function values for $90° - \theta$.

Solve each of the following right triangles. Standard lettering has been used.

**5.** $a = 7.3$, $c = 8.6$

**6.** $a = 30.5$, $B = 51.17°$

**7.** One leg of a right triangle bears east. The hypotenuse is 734 cm long and bears N 57°23′ E. Find the perimeter of the triangle.

**8.** An observer's eye is 6 ft above the floor. A mural is being viewed. The bottom of the mural is at floor level. The observer looks downward 13° to see the bottom and upward 17° to see the top. How tall is the mural?

**9.** Find two positive angles and two negative angles coterminal with a 65° angle.

**10.** Find the six trigonometric function values for the angle $\theta$ as shown.

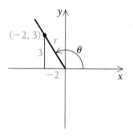

Find the following if they exist. Do not use a calculator.

**11.** $\sin 495°$

**12.** $\tan (-315°)$

**13.** $\cot 210°$

**14.** $\cos 150°$

**15.** $\sec (-270°)$

**16.** $\csc 600°$

Find each of the following function values.

**17.** $\tan 245°24′$

**18.** $\cos 125°42′$

**19.** $\sec 27.8°$

**20.** $\sin 556°25′$

**21.** $\cot (-33°05′)$

**22.** $\csc 2184°$

Find $\theta$ in the interval indicated.

**23.** $\cos \theta = -0.9044$, $(180°, 270°)$

**24.** $\tan \theta = 1.0791$, $(0°, 90°)$

**25.** Given that $\tan \theta = 2/\sqrt{5}$ and that the terminal side is in quadrant III, find the other five function values.

**26.** On a unit circle, mark and label the points determined by $7\pi/6$, $-3\pi/4$, $-\pi/3$, and $9\pi/4$.

**27.** The point $(\frac{3}{5}, -\frac{4}{5})$ is on a unit circle. Find the coordinates of its reflections across the $u$-axis, the $v$-axis, and the origin.

For angles of the following measures, state in which quadrant the terminal side lies, convert to radian measure in terms of $\pi$, and convert to radian measure not in terms of $\pi$.

**28.** $87°$          **29.** $145°$          **30.** $-30°$

Convert to degree measure.

**31.** $\dfrac{3\pi}{2}$          **32.** $3$

**33.** Find the length of an arc of a circle, given a central angle of $\pi/4$ and a radius of 7 cm.

**34.** An arc 18 m long on a circle of radius 8 m subtends an angle of how many radians? how many degrees, to the nearest degree?

**35.** A phonograph record revolves at 45 rpm. What is the linear velocity, in centimeters per minute, of a point 4 cm from the center?

**36.** An automobile wheel has a diameter of 26 in. If the car travels at a speed of 30 mph, what is the angular velocity, in radians per hour, of a point on the edge of the wheel?

Find each of the following function values, if they exist, using your knowledge of degrees.

**37.** $\cos \pi$          **38.** $\tan \dfrac{5\pi}{4}$          **39.** $\sin \dfrac{5\pi}{3}$

**40.** $\sin\left(-\dfrac{7\pi}{6}\right)$          **41.** $\tan \dfrac{\pi}{6}$          **42,** $\cos(-13\pi)$

Find each of the following function values, if they exist, using a calculator.

**43.** $\sin 24$          **44.** $\cos(-75)$          **45.** $\cot 16\pi$

**46.** $\tan \dfrac{3\pi}{7}$          **47.** $\sec 14.3$          **48.** $\cos\left(-\dfrac{\pi}{5}\right)$

Complete these Pythagorean identities.

**49.** $\sin^2 s + \cos^2 s = $ _____

**50.** $1 + \cot^2 s = $ _____

**51.** Sketch a graph of $y = \sin x$.

**52.** Sketch a graph of $y = \cot x$.

**53.** Sketch a graph of $y = \csc x$.

**54.** What are the period and the range of the sine function?

**55.** What are the amplitude and the domain of the sine function?

**56.** What are the period and the range of the cosecant function?

**57.** Sketch a graph of $y = \sin\left(x + \dfrac{\pi}{2}\right)$.

**58.** Sketch a graph of $y = 3 + \cos\left(x - \dfrac{\pi}{4}\right)$.

**59.** What is the phase shift of the function in Exercise 58?

**60.** What is the period of the function in Exercise 58?

**61.** Sketch a graph of $y = 3\cos x + \sin x$ for values of $x$ between 0 and $2\pi$.

Sketch the graph.

**62.** $y = 2\tan\left(x - \dfrac{\pi}{2}\right)$          **63.** $f(x) = e^{-0.7x}\cos x$

● **SYNTHESIS** _____

**64.** Does $5\sin x = 7$ have a solution for $x$? Why or why not?

**65.** For what values of $x$ in $(0, \pi/2]$ is $\sin x < x$ true?

**66.** Graph $y = 3\sin(x/2)$ and determine the domain, the range, and the period.

**67.** Find the domain of $y = \log(\cos x)$.

**68.** Given that $\sin x = 0.6144$ and that the terminal side is in quadrant II, find the other basic circular function values.

● **THINKING AND WRITING** _____

**1.** Compare the notations of radian and degree.

**2.** Describe the shape of the graph of the cosine function. How many maximum values are there of the cosine function? Where do they occur?

**3.** A graphing calculator has many advantages. Explain a disadvantage of such a calculator when graphing a function like

$$f(x) = \frac{\sin x}{x}.$$

## CHAPTER TEST  6

**1.** Convert $19.25°$ to degrees and minutes.

**2.** Convert $72°48'$ to degrees and decimal parts of degrees.

**3.** Given that $\sin \theta = 0.4540$, $\cos \theta = 0.8910$, $\tan \theta = 0.5095$, $\cot \theta = 1.963$, $\sec \theta = 1.122$, and $\csc \theta = 2.203$, find the six function values for $90° - \theta$.

Solve each of the following right triangles. Standard lettering has been used.

**4.** $A = 22.4°$, $b = 18.4$          **5.** $a = 11$, $c = 28$

**6.** A building 70 m high casts a shadow 100 m long. What is the angle of elevation of the sun?

7. Find the six trigonometric function values for the angle $\theta$ shown.

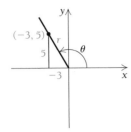

Find the following, if they exist. Do not use a calculator for Questions 8–11.

8. $\sin \dfrac{3\pi}{4}$

9. $\cos 390°$

10. $\tan 8\pi$

11. $\sec(-300°)$

12. $\cos(-213°48'13'')$

13. $\tan 1.228$

14. $\sin(-34\pi/19)$

15. $\sec 38°17'$

16. $\csc 3.4128$

17. $\cot(3472°)$

18. Given that $\sin \theta = -0.7450$, find $\theta$ between $180°$ and $270°$.

19. Given that $\sin \theta = \frac{1}{4}$ and that the terminal side is in quadrant II, find the other five function values.

For angles of the following measures, state in which quadrant the terminal side lies, convert to radian measure in terms of $\pi$, and convert to radian measure not in terms of $\pi$.

20. $-150°$

21. $117°$

22. $225°$

Convert to degree measure.

23. $\dfrac{7\pi}{4}$ radians

24. 2 radians

25. Find the length of an arc of a circle, given a central angle of $\pi/3$ and a radius of 10 cm.

26. An arc 8 in. long on a circle of radius 16 in. subtends an angle of how many radians? how many degrees, to the nearest degree?

27. A rock on a 4-ft string is rotated at 80 rpm. What is its linear speed, in feet per minute?

28. A pulley belt runs, uncrossed, around two pulleys of radii 10 cm and 4 cm, respectively. A point on the belt travels at a rate of 50 m/sec. Find the angular speed, in radians per second, of the smaller pulley.

Complete these Pythagorean identities.

29. $1 + \tan^2 s = $ _____

30. $\sin^2 s + \cos^2 s = $ _____

31. Sketch a graph of $y = \cos x$.

32. What is the range of the cosine function?

33. What is the period of the cosine function?

34. What is the domain of the sine function?

35. What is the amplitude of the sine function?

36. Sketch a graph of the secant function.

37. What is the period of the secant function?

38. What is the domain of the secant function?

39. In which quadrants are the signs of the sine and the secant different?

40. Sketch a graph of $y = 2 \cos(x - \pi/4)$.

41. What is the period of the function in Question 40?

42. What is the phase shift of the function in Question 40?

43. Sketch the graph of $y = 2 \cos x - \sin x$ for values between 0 and $2\pi$.

● **SYNTHESIS** _____

44. Given that $\cos(\pi/32) = 0.99518$, find each of the following.

a) $\sin(15\pi/32)$
b) $\sin(17\pi/32)$

45. What is the angle between the hands of a clock at 9:25?

**7**

There are a number of relationships among the trigonometric functions, given by identities, that are important in algebraic and trigonometric manipulation. A large part of this chapter is devoted to those identities. Among the manipulations in which the identities are used is the solving of *trigonometric equations.* ● We have not yet considered the inverses of the trigonometric functions beyond the use of calculators in reverse. We now look at those inverses. The chapter is an important one, both for problem solving and as a foundation for further work in mathematics. ●

# Trigonometric Identities, Inverse Functions, and Equations

## 7.1
## TRIGONOMETRIC MANIPULATIONS AND IDENTITIES

Trigonometric expressions such as $\sin 2x$ or $\tan(x - \pi)$ represent real numbers, in the same way that purely algebraic expressions do. Thus the laws of real numbers hold whether we are working with algebraic expressions or trigonometric expressions, or combinations thereof. We can factor, simplify, and manipulate trigonometric expressions in the same way that we manipulate strictly algebraic expressions.

The laws of real numbers can be combined with the fundamental identities considered in Chapter 6 to further enhance our ability to manipulate trigonometric expressions. The following is a list of these fundamental identities.

451

1. Multiply and simplify:

$$\sin x \,(\cot x + \csc x).$$

---

**Fundamental Identities**

$$\sin x = \frac{1}{\csc x} \qquad\qquad \csc x = \frac{1}{\sin x} \qquad\qquad \tan x = \frac{\sin x}{\cos x}$$

$$\cos x = \frac{1}{\sec x} \qquad\qquad \sec x = \frac{1}{\cos x} \qquad\qquad \cot x = \frac{\cos x}{\sin x}$$

$$\tan x = \frac{1}{\cot x} \qquad\qquad \cot x = \frac{1}{\tan x}$$

$$\sin^2 x + \cos^2 x = 1 \qquad 1 + \tan^2 x = \sec^2 x \qquad 1 + \cot^2 x = \csc^2 x$$

$$\sin(-x) = -\sin x \qquad \csc(-x) = -\csc x$$

$$\cos(-x) = \cos x \qquad\quad \sec(-x) = \sec x$$

$$\tan(-x) = -\tan x \qquad \cot(-x) = -\cot x$$

---

With these identities in mind, let us now consider an example.

## ◢ Multiplying and Factoring

**Example 1**    Multiply and simplify: $\cos y \,(\tan y - \sec y)$.

**Solution**

$$\cos y \,(\tan y - \sec y)$$

$$= \cos y \tan y - \cos y \sec y \qquad \text{Multiplying}$$

$$= \cos y \frac{\sin y}{\cos y} - \cos y \frac{1}{\cos y} \qquad \text{Recalling the identities } \tan x = \frac{\sin x}{\cos x} \text{ and}$$

$$\sec x = \frac{1}{\cos x} \text{ and substituting}$$

$$= \sin y - 1 \qquad \text{Simplifying} \qquad ◀$$

Recall that an **identity** is an equation that is true for all meaningful replacements of the variables. Thus we have proved in Example 1 that

$$\cos y \,(\tan y - \sec y) = \sin y - 1$$

is an identity. There is no general rule for doing a simplification that creates an identity as in Example 1 and no specific result, but it is often helpful to put everything in terms of sines and cosines. For the most part, the main goal of this section is to develop your skill by the practice you receive in manipulating trigonometric expressions. This kind of simplifying is very important for the rest of the trigonometry that you will study in this chapter and in other courses that you will take in mathematics and science.

2. Factor and simplify:

$$\sin^3 x + \sin x \cos^2 x.$$

**Example 2**    Factor and simplify: $\sin^2 x \cos^2 x + \cos^4 x$.

**Solution**

$$\sin^2 x \cos^2 x + \cos^4 x = \cos^2 x \,(\sin^2 x + \cos^2 x) \qquad \text{Factoring}$$

$$= \cos^2 x \qquad \text{Using } \sin^2 x + \cos^2 x = 1 \qquad ◀$$

**DO EXERCISES 1 AND 2.**

# Simplifying

**Example 3**    Simplify:

$$\frac{\cot(-\theta)}{\csc(-\theta)}.$$

**Solution**

$$\frac{\cot(-\theta)}{\csc(-\theta)} = \frac{\dfrac{\cos(-\theta)}{\sin(-\theta)}}{\dfrac{1}{\sin(-\theta)}}$$

$$= \frac{\cos(-\theta)}{\sin(-\theta)} \cdot \sin(-\theta)$$

$$= \cos(-\theta)$$

$$= \cos\theta$$

◀

**Example 4**    Simplify:

$$\frac{\sin x - \sin x \cos x}{\sin x + \sin x \tan x}.$$

**Solution**

$$\frac{\sin x - \sin x \cos x}{\sin x + \sin x \tan x} = \frac{\sin x (1 - \cos x)}{\sin x (1 + \tan x)} \qquad \text{Factoring}$$

$$= \frac{1 - \cos x}{1 + \tan x} \qquad \text{Simplifying}$$

◀

**Example 5**    Subtract and simplify:

$$\frac{2}{\sin x - \cos x} - \frac{3}{\sin x + \cos x}.$$

**Solution**

$$\frac{2}{\sin x - \cos x} - \frac{3}{\sin x + \cos x}$$

$$= \frac{2}{\sin x - \cos x} \cdot \frac{\sin x + \cos x}{\sin x + \cos x} - \frac{3}{\sin x + \cos x} \cdot \frac{\sin x - \cos x}{\sin x - \cos x}$$

$$= \frac{2(\sin x + \cos x) - 3(\sin x - \cos x)}{\sin^2 x - \cos^2 x}$$

$$= \frac{-\sin x + 5\cos x}{\sin^2 x - \cos^2 x}$$

◀

## TECHNOLOGY CONNECTION

A grapher can be used to provide a partial check of any trigonometric identity. First, graph the expression on the left side of the equals sign, using an appropriate domain. A viewing box of $[-2\pi, 2\pi] \times [-5, 5]$ will usually give satisfactory results, so long as the grapher is in RADIAN mode. Then graph the expression on the right side using the same axes. If the two graphs are indistinguishable, then you have a partial verification that the equation is an identity. Of course, you can never see the entire graph, so there can always be some doubt. If the two graphs are different, then the equation is not an identity.

It may be necessary to make some slight modifications in the equation, due to the limitations of some graphers. If the equation contains $\sec x$, $\csc x$, or $\cot x$, you may need to replace these terms with $1/\cos x$, $1/\sin x$, or $1/\tan x$, respectively.

For example, consider the identity in Example 1:

$$\cos y\,(\tan y - \sec y) = \sin y - 1.$$

Since this equation contains $\sec y$, it will usually be necessary to replace that term with $1/\cos y$ before graphing it. Thus the new equation is

$$\cos y\left(\tan y - \frac{1}{\cos y}\right) = \sin y - 1.$$

Now we graph each side. Using the viewing box of $[-2\pi, 2\pi] \times [-5, 5]$ that is suggested above, we first draw the graph of the expression on the left side. Of course, we first change the $y$'s to $x$'s, since most graphers require the form $y = f(x)$. After we enter

$$y = \cos x\left(\tan x - \frac{1}{\cos x}\right),$$

we graph it, with the result shown here:

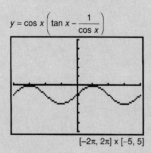

Then we graph the right side (after changing it to $y = \sin x - 1$). The result is a graph identical to the one shown above. When the two graphs are drawn on the same axes, it is impossible to distinguish between them.

To illustrate a case in which an equation is shown not to be an identity, we can use the equation

$$\sin^2 x \cos^2 x - \cos^4 x = \cos^2 x.$$

Since all the terms are either $\cos x$ or $\sin x$, no modifications are needed before graphing. The graph of the expression on the left side, $y = \sin^2 x \cos^2 x - \cos^4 x$, is shown here:

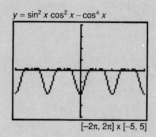

The expression on the right side, $y = \cos^2 x$, produces this graph:

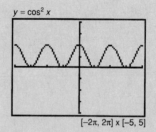

Clearly, the graphs are different. When graphed on the same set of axes, it is easy to tell them apart. Thus the expression $\sin^2 x \cos^2 x - \cos^4 x = \cos^2 x$ is *not* an identity.

The main limitation to this technique is that it can give only a partial check whether or not a given equation is an identity.

**TC 1.** Use a grapher to determine which of the following are identities.

A. $\sin x \cot x = \cos x$

B. $\dfrac{\sin^2 x}{\cos x} + \cos x = \sec x$

C. $\dfrac{1 - \sin x}{\cos x} = \dfrac{\cos x}{1 + \sin x}$

D. $(\sec x - \cos x)\csc^2 x = \cot x$

**TC 2.** Use a grapher to determine which of the following expressions can complete the identity

$$\cos x\,(1 + \sec x) = \ldots .$$

A. $1 + \cos x$

B. $\dfrac{1 + \cos x - \sin^2 x}{\cos x}$

C. $\cos x\,(\sec x + 1)$

D. $\sin x - \csc x$

E. $1 + \sin x \cot x$

**Example 6**   Add and simplify: $\dfrac{2}{\tan^3 x - 2\tan^2 x} + \dfrac{2}{\tan x - 2}$.

Solution

$$\dfrac{2}{\tan^3 x - 2\tan^2 x} + \dfrac{2}{\tan x - 2}$$

$$= \dfrac{2}{\tan^2 x\,(\tan x - 2)} + \dfrac{2}{\tan x - 2} \cdot \dfrac{\tan^2 x}{\tan^2 x}$$

$$= \dfrac{2 + 2\tan^2 x}{\tan^2 x\,(\tan x - 2)} = \dfrac{2(1 + \tan^2 x)}{\tan^2 x\,(\tan x - 2)}$$

$$= \dfrac{2\sec^2 x}{\tan^2 x\,(\tan x - 2)} \qquad \text{Recalling that } 1 + \tan^2 x = \sec^2 x$$

$$= \dfrac{2 \cdot \dfrac{1}{\cos^2 x}}{\dfrac{\sin^2 x}{\cos^2 x}\,(\tan x - 2)}$$

$$= \dfrac{2}{\sin^2 x\,(\tan x - 2)} \qquad \blacktriangleleft$$

**DO EXERCISES 3–6.**

▷ ## Radical Expressions

When radicals occur, the use of absolute value is sometimes necessary, but it can be difficult to determine when to use it. In the examples and exercises that follow, we will assume that all radicands are nonnegative.

**Example 7**   Multiply and simplify: $\sqrt{\sin^3 x \cos x} \cdot \sqrt{\cos x}$.

Solution

$$\sqrt{\sin^3 x \cos x} \cdot \sqrt{\cos x} = \sqrt{\sin^3 x \cos^2 x}$$

$$= \sqrt{\sin^2 x \cos^2 x \sin x}$$

$$= \sin x \cos x \sqrt{\sin x} \qquad \blacktriangleleft$$

**Example 8**   Rationalize the denominator: $\sqrt{\dfrac{2}{\tan x}}$.

Solution

$$\sqrt{\dfrac{2}{\tan x}} = \sqrt{\dfrac{2}{\tan x} \cdot \dfrac{\tan x}{\tan x}} = \sqrt{\dfrac{2\tan x}{\tan^2 x}} = \dfrac{\sqrt{2\tan x}}{\tan x} \qquad \blacktriangleleft$$

**DO EXERCISES 7 AND 8.**

Often in calculus, a substitution such as the following is a useful manipulation.

---

Simplify.

3. $\dfrac{\tan(-\theta)}{\sin(-\theta)}$

4. $\dfrac{\cos x + \sin x \cos x}{\cos x - \cos x \cot x}$

Add and simplify.

5. $\dfrac{2}{\sin x - \cos x} + \dfrac{3}{\sin x + \cos x}$

6. $\dfrac{2}{\cot^3 x - 2\cot^2 x} + \dfrac{2}{\cot x - 2}$

7. Multiply and simplify:

$$\sqrt{\tan x \sin^2 x} \cdot \sqrt{\tan x \sin x}.$$

8. Rationalize the numerator:

$$\sqrt{\dfrac{\cos x}{3}}.$$

**9.** Use the substitution $x = 2 \sin \theta$ to express $\sqrt{4 - x^2}$ as a trigonometric function without radicals. Assume that $-\pi/2 \le \theta \le \pi/2$. Then find $\tan \theta$ and $\cos \theta$.

**Example 9**    Use the substitution $x = a \tan \theta$ to express $\sqrt{a^2 + x^2}$ as a trigonometric function without radicals. Assume that $a > 0$ and that $-\pi/2 < \theta < \pi/2$. Then find $\sin \theta$ and $\cos \theta$.

**Solution**    Substituting $a \tan \theta$ for $x$, we get

$$\sqrt{a^2 + x^2} = \sqrt{a^2 + (a \tan \theta)^2} = \sqrt{a^2 + a^2 \tan^2 \theta}$$
$$= \sqrt{a^2(1 + \tan^2 \theta)} = \sqrt{a^2 \sec^2 \theta}$$
$$= \sqrt{a^2}\sqrt{\sec^2 \theta} = |a| \cdot |\sec \theta| = a \sec \theta.$$

The latter equality follows from the fact that $a > 0$ and that $\sec \theta > 0$ for values of $\theta$ for which $-\pi/2 < \theta < \pi/2$.

We can express the preceding result as

$$\sec \theta = \frac{\sqrt{a^2 + x^2}}{a}.$$

In terms of right triangles, we know that $\sec \theta$ is hypotenuse/adjacent, where $\theta$ is one of the acute angles. Now interpreting the hypotenuse to be $\sqrt{a^2 + x^2}$ and the adjacent side to be $a$, we can find the length of the side opposite $\theta$ to be $x$ using the Pythagorean theorem.

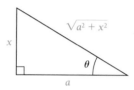

Then from the right triangle we see that

$$\sin \theta = \frac{x}{\sqrt{a^2 + x^2}} \quad \text{and} \quad \cos \theta = \frac{a}{\sqrt{a^2 + x^2}}.$$

**DO EXERCISE 9.**

 **EXERCISE SET 7.1**

**A**  Multiply and simplify.

**1.** $(\sin x - \cos x)(\sin x + \cos x)$

**2.** $(\tan y - \cot y)(\tan y + \cot y)$

**3.** $\tan x (\cos x - \csc x)$

**4.** $\cot x (\sin x + \sec x)$

**5.** $\cos y \sin y (\sec y + \csc y)$

**6.** $\tan y \sin y (\cot y - \csc y)$

**7.** $(\sin x + \cos x)(\csc x - \sec x)$

**8.** $(\sin x + \cos x)(\sec x + \csc x)$

**9.** $(\sin y - \cos y)^2$

**10.** $(\sin y + \cos y)^2$

**11.** $(1 + \tan x)^2$

**12.** $(1 + \cot x)^2$

**13.** $(\sin y - \csc y)^2$

**14.** $(\cos y + \sec y)^2$

**15.** $(\cos x - \sec x)(\cos^2 x + \sec^2 x + 1)$

**16.** $(\sin x + \csc x)(\sin^2 x + \csc^2 x - 1)$

**17.** $(\cot x - \tan x)(\cot^2 x + 1 + \tan^2 x)$

**18.** $(\cot y + \tan y)(\cot^2 y - 1 + \tan^2 y)$

**19.** $(1 - \sin x)(1 + \sin x)$

**20.** $(1 + \cos x)(1 - \cos x)$

Factor and simplify.

**21.** $\sin x \cos x + \cos^2 x$

**22.** $\sec x \csc x - \csc^2 x$

**23.** $\sin^2 \theta - \cos^2 \theta$

**24.** $\tan^2 \theta - \cot^2 \theta$

**25.** $\tan x + \sin (\pi - x)$

**26.** $\cot x - \cos (\pi - x)$

**27.** $\sin^4 x - \cos^4 x$

**28.** $\tan^4 x - \sec^4 x$

**29.** $3 \cot^2 y + 6 \cot y + 3$

**30.** $4 \sin^2 y + 8 \sin y + 4$

**31.** $\csc^4 x + 4 \csc^2 x - 5$

**32.** $-8 + \tan^4 x - 2 \tan^2 x$

**33.** $\sin^3 y + 27$

**34.** $1 - 125 \tan^3 y$

**35.** $\sin^3 v - \csc^3 v$

**36.** $\cos^3 u - \sec^3 u$

**B**   Simplify. If you have a grapher, verify your results.

**37.** $\dfrac{\sin^2 x \cos x}{\cos^2 x \sin x}$

**38.** $\dfrac{\cos^2 x \sin x}{\sin^2 x \cos x}$

**39.** $\dfrac{4 \sin x \cos^3 x}{18 \sin^2 x \cos x}$

**40.** $\dfrac{30 \sin^3 x \cos x}{6 \cos^2 x \sin x}$

**41.** $\dfrac{\cos^2 x - 2 \cos x + 1}{\cos x - 1}$

**42.** $\dfrac{\sin^2 x + 2 \sin x + 1}{\sin x + 1}$

**43.** $\dfrac{\cos^2 \alpha - 1}{\cos \alpha - 1}$

**44.** $\dfrac{\sin^2 \alpha - 1}{\sin \alpha + 1}$

**45.** $\dfrac{4 \tan x \sec x + 2 \sec x}{6 \sin x \sec x + 2 \sec x}$

**46.** $\dfrac{6 \tan x \sin x - 3 \sin x}{9 \sin^2 x + 3 \sin x}$

**47.** $\dfrac{\csc(-\theta)}{\cot(-\theta)}$

**48.** $\tan(-\beta) + \cot(-\beta)$

**49.** $\dfrac{\sin^4 x - \cos^4 x}{\sin^2 x - \cos^2 x}$

**50.** $\dfrac{\sec^4 x - \tan^4 x}{\sec^2 x + \tan^2 x}$

**51.** $\dfrac{2 \sin^2 x}{\cos^3 x} \cdot \left(\dfrac{\cos x}{2 \sin x}\right)^2$

**52.** $\dfrac{4 \cos^3 x}{\sin^2 x} \cdot \left(\dfrac{\sin x}{4 \cos x}\right)^2$

**53.** $\dfrac{3 \sin x}{\cos^2 x} \cdot \dfrac{\cos^2 x + \cos x \sin x}{\cos^2 x - \sin^2 x}$

**54.** $\dfrac{5 \cos x}{\sin^2 x} \cdot \dfrac{\sin^2 x - \sin x \cos x}{\sin^2 x - \cos^2 x}$

**55.** $\dfrac{\tan^2 \gamma}{\sec \gamma} \div \dfrac{3 \tan^3 \gamma}{\sec \gamma}$

**56.** $\dfrac{\cot^3 \phi}{\csc \phi} \div \dfrac{4 \cot^2 \phi}{\csc \phi}$

**57.** $\dfrac{1}{\sin^2 y - \cot^2 y} - \dfrac{2}{\cos y + \sin y}$

**58.** $\dfrac{3}{\cos y - \sin y} - \dfrac{2}{\sin^2 y - \cos^2 y}$

**59.** $\left(\dfrac{\sin x}{\cos x}\right)^2 - \dfrac{1}{\cos^2 x}$

**60.** $\left(\dfrac{\cot x}{\csc x}\right)^2 + \dfrac{1}{\csc^2 x}$

**61.** $\dfrac{\sin^2 x - 9}{2 \cos x + 1} \cdot \dfrac{10 \cos x + 5}{3 \sin x + 9}$

**62.** $\dfrac{9 \cos^2 x - 25}{2 \cos x - 2} \cdot \dfrac{\cos^2 x - 1}{6 \cos x - 10}$

**C**   Simplify. Assume that expressions in radicands are non-negative.

**63.** $\sqrt{\sin^2 x \cos x} \cdot \sqrt{\cos x}$

**64.** $\sqrt{\cos^2 x \sin x} \cdot \sqrt{\sin x}$

**65.** $\sqrt{\sin^3 y} + \sqrt{\sin y \cos^2 y}$

**66.** $\sqrt{\cos y \sin^2 y} - \sqrt{\cos^3 y}$

**67.** $\sqrt{\sin^2 x + 2 \cos x \sin x + \cos^2 x}$

**68.** $\sqrt{\tan^2 x - 2 \tan x \sin x + \sin^2 x}$

**69.** $(1 - \sqrt{\sin y})(\sqrt{\sin y} + 1)$

**70.** $(2 - \sqrt{\tan y})(\sqrt{\tan y} + 2)$

**71.** $\sqrt{\sin x}(\sqrt{2 \sin x} + \sqrt{\sin x \cos x})$

**72.** $\sqrt{\cos x}(\sqrt{2 \cos x} + \sqrt{\sin x \cos x})$

Rationalize the denominator.

**73.** $\sqrt{\dfrac{\sin x}{\cos x}}$

**74.** $\sqrt{\dfrac{\cos x}{\sin x}}$

**75.** $\sqrt{\dfrac{\sin x}{\cot x}}$

**76.** $\sqrt{\dfrac{\cos x}{\tan x}}$

**77.** $\sqrt{\dfrac{\cos^2 x}{2 \sin^2 x}}$

**78.** $\sqrt{\dfrac{\sin^2 x}{3 \cos^2 x}}$

**79.** $\sqrt{\dfrac{1 + \sin x}{1 - \sin x}}$

**80.** $\sqrt{\dfrac{1 - \cos x}{1 + \cos x}}$

Rationalize the numerator.

**81.** $\sqrt{\dfrac{\sin x}{\cos x}}$

**82.** $\sqrt{\dfrac{\cos x}{\sin x}}$

**83.** $\sqrt{\dfrac{\sin x}{\cot x}}$

**84.** $\sqrt{\dfrac{\cos x}{\tan x}}$

**85.** $\sqrt{\dfrac{\cos^2 x}{2 \sin^2 x}}$

**86.** $\sqrt{\dfrac{\sin^2 x}{3 \cos^2 x}}$

**87.** $\sqrt{\dfrac{1 + \sin x}{1 - \sin x}}$

**88.** $\sqrt{\dfrac{1 - \cos x}{1 + \cos x}}$

Use the given substitution to express the given radical expression as a trigonometric function without radicals. Assume that $a > 0$ and $0 < \theta < \pi/2$. Then find expressions for the indicated trigonometric functions.

**89.** Let $x = a \sin \theta$ in $\sqrt{a^2 - x^2}$. Then find $\cos \theta$ and $\tan \theta$.

**90.** Let $x = 2 \tan \theta$ in $\sqrt{4 + x^2}$. Then find $\sin \theta$ and $\cos \theta$.

**91.** Let $x = 3 \sec \theta$ in $\sqrt{x^2 - 9}$. Then find $\sin \theta$ and $\cos \theta$.

**92.** Let $x = a \sec \theta$ in $\sqrt{x^2 - a^2}$. Then find $\sin \theta$ and $\cos \theta$.

Use the given substitution to express the given radical expression as a trigonometric function without radicals. Assume that $0 < \theta < \pi/2$.

**93.** Let $x = \sin \theta$ in $\dfrac{x^2}{\sqrt{1 - x^2}}$.

**94.** Let $x = 4 \sec \theta$ in $\dfrac{\sqrt{x^2 - 16}}{x^2}$.

● **SYNTHESIS** _____

Show that each of the following is *not* an identity by finding a replacement or replacements for which each side of the equation does not name the same number.

**95.** $(\sin x + \cos x)^2 = \sin^2 x + \cos^2 x$

**96.** $\sqrt{\sin^2 \theta} = \sin \theta$

**97.** $\dfrac{\sin x}{\sin y} = \dfrac{x}{y}$

**98.** $\cos x = \sqrt{1 - \sin^2 x}$

**99.** $\cos(\alpha + \beta) = \cos\alpha + \cos\beta$

**100.** $\sin(-\theta) = \sin\theta$

**101.** $\cos(2\theta) = 2\cos\theta$

**102.** $\tan^2\theta + \cot^2\theta = 1$

**103.** $\ln(\sin\theta) = \sin(\ln\theta)$

# 7.2

# SUM AND DIFFERENCE IDENTITIES

## A Sum and Difference Identities

### Cosine Sum and Difference Identities

We now develop some important identities involving sums or differences of two numbers (or angles), beginning with an identity for the cosine of the difference of two numbers. We use the Greek letters $\alpha$ (alpha) and $\beta$ (beta) for these numbers.

Let us consider a real number $\alpha$ in the interval $[\pi/2, \pi]$ and a real number $\beta$ in the interval $[0, \pi/2]$. These determine points $A$ and $B$ on the unit circle as shown. The arc length $s$ is $\alpha - \beta$, and we know that $0 \leq s \leq \pi$. Recall that the coordinates of $A$ are $(\cos\alpha, \sin\alpha)$, and the coordinates of $B$ are $(\cos\beta, \sin\beta)$.

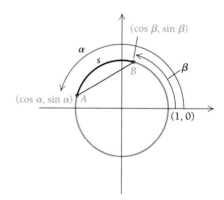

Using the distance formula, we can write an expression for the square of the distance $AB$:

$$AB^2 = (\cos\alpha - \cos\beta)^2 + (\sin\alpha - \sin\beta)^2.$$

This can be simplified as follows:

$$\begin{aligned}
AB^2 &= \cos^2\alpha - 2\cos\alpha\cos\beta + \cos^2\beta + \sin^2\alpha \\
&\quad - 2\sin\alpha\sin\beta + \sin^2\beta \\
&= (\sin^2\alpha + \cos^2\alpha) + (\sin^2\beta + \cos^2\beta) \\
&\quad - 2(\cos\alpha\cos\beta + \sin\alpha\sin\beta) \\
&= 2 - 2(\cos\alpha\cos\beta + \sin\alpha\sin\beta).
\end{aligned}$$

Now let us imagine rotating the circle above so that point $B$ is at $(1, 0)$.

Although the coordinates of point $A$ are now $(\cos s, \sin s)$, the distance $AB$ has not changed.

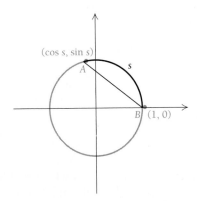

Again we use the distance formula to write an expression for the square of $AB$:

$$AB^2 = (\cos s - 1)^2 + (\sin s - 0)^2.$$

This simplifies as follows:

$$
\begin{aligned}
AB^2 &= \cos^2 s - 2 \cos s + 1 + \sin^2 s \\
&= (\sin^2 s + \cos^2 s) + 1 - 2 \cos s \\
&= 2 - 2 \cos s.
\end{aligned}
$$

Equating our two expressions for $AB^2$, we obtain

$$2 - 2(\cos \alpha \cos \beta + \sin \alpha \sin \beta) = 2 - 2 \cos s.$$

Solving this equation for $\cos s$ gives

$$\cos s = \cos \alpha \cos \beta + \sin \alpha \sin \beta. \tag{1}$$

But $s = \alpha - \beta$, so we have the equation

$$\cos(\alpha - \beta) = \cos \alpha \cos \beta + \sin \alpha \sin \beta. \tag{2}$$

This formula holds for any numbers $\alpha$ and $\beta$ for which $\alpha - \beta$ is the length of the shortest arc from $A$ to $B$—in other words, when $0 \leq \alpha - \beta \leq \pi$.

**Example 1**   Find $\cos\left(\dfrac{3\pi}{4} - \dfrac{\pi}{3}\right)$.

**Solution**

$$
\begin{aligned}
\cos\left(\frac{3\pi}{4} - \frac{\pi}{3}\right) &= \cos \frac{3\pi}{4} \cos \frac{\pi}{3} + \sin \frac{3\pi}{4} \sin \frac{\pi}{3} \\
&= -\frac{\sqrt{2}}{2} \cdot \frac{1}{2} + \frac{\sqrt{2}}{2} \cdot \frac{\sqrt{3}}{2} \\
&= \frac{\sqrt{2}}{4}(\sqrt{3} - 1), \quad \text{or} \quad \frac{\sqrt{6} - \sqrt{2}}{4}
\end{aligned}
$$

◀

**Example 2**   Find $\cos 15°$.

1. Find $\cos\left(\dfrac{\pi}{2} - \dfrac{\pi}{6}\right)$.

**Solution**

$$\cos 15° = \cos(45° - 30°)$$
$$= \cos 45° \cos 30° + \sin 45° \sin 30°$$
$$= \frac{\sqrt{2}}{2} \cdot \frac{\sqrt{3}}{2} + \frac{\sqrt{2}}{2} \cdot \frac{1}{2}$$
$$= \frac{\sqrt{2}}{4}(\sqrt{3} + 1), \quad \text{or} \quad \frac{\sqrt{6} + \sqrt{2}}{4} \qquad \blacktriangleleft$$

**DO EXERCISES 1 AND 2.**

Formula (1) above holds when $s$ is the length of the shortest arc from $A$ to $B$. Given any real numbers $\alpha$ and $\beta$, the length of the shortest arc from $A$ to $B$ is not always $\alpha - \beta$. In fact, it could be $\beta - \alpha$. However, since $\cos(-x) = \cos x$, we know that $\cos(\beta - \alpha) = \cos(\alpha - \beta)$. Thus, $\cos s$ is always equal to $\cos(\alpha - \beta)$.

Formula (2) holds for all real numbers $\alpha$ and $\beta$. That formula is thus the identity we sought.

2. Find $\cos 105°$ as $\cos(150° - 45°)$.

$$\cos(\alpha - \beta) = \cos\alpha\cos\beta + \sin\alpha\sin\beta$$

The cosine sum formula follows easily from the one we have just derived. Let us consider $\cos(\alpha + \beta)$. This is equal to $\cos[\alpha - (-\beta)]$, and by the identity above, we have

$$\cos(\alpha + \beta) = \cos[\alpha - (-\beta)] = \cos\alpha\cos(-\beta) + \sin\alpha\sin(-\beta).$$

But $\cos(-\beta) = \cos\beta$ and $\sin(-\beta) = -\sin\beta$, so the identity we seek is the following:

$$\cos(\alpha + \beta) = \cos\alpha\cos\beta - \sin\alpha\sin\beta.$$

**DO EXERCISE 3.**

3. Find $\cos\left(\dfrac{2\pi}{3} + \dfrac{\pi}{4}\right)$.

### Sine Sum and Difference Identities

To develop an identity for the sine of a sum, we first need two intermediary identities. Let us consider $\cos(\pi/2 - \theta)$. We can use the identity for the cosine of a difference to simplify as follows:

$$\cos\left(\frac{\pi}{2} - \theta\right) = \cos\frac{\pi}{2}\cos\theta + \sin\frac{\pi}{2}\sin\theta$$
$$= 0 \cdot \cos\theta + 1 \cdot \sin\theta = \sin\theta.$$

Thus we have developed the identity $\cos(\pi/2 - \theta) = \sin\theta$, which we can rewrite as

$$\sin\theta = \cos\left(\frac{\pi}{2} - \theta\right). \tag{1}$$

This identity holds for any real number $\theta$. From it we can obtain a similar identity for the sine function. Let $\alpha$ be any real number. Then replace $\theta$ in $\sin\theta = \cos(\pi/2 - \theta)$ by $\pi/2 - \alpha$. This gives us

$$\sin\left(\frac{\pi}{2} - \alpha\right) = \cos\left[\frac{\pi}{2} - \left(\frac{\pi}{2} - \alpha\right)\right] = \cos\alpha,$$

which yields the identity

$$\cos \alpha = \sin \left( \frac{\pi}{2} - \alpha \right). \tag{2}$$

This identity holds for any real number $\alpha$.

Using identities (1) and (2) and the identity for the cosine of a difference, we now obtain an identity for the sine of a sum. We start with identity (1) and substitute $\alpha + \beta$ for $\theta$, obtaining

$$\sin (\alpha + \beta)$$

$$= \cos \left[ \frac{\pi}{2} - (\alpha + \beta) \right]$$

$$= \cos \left[ \left( \frac{\pi}{2} - \alpha \right) - \beta \right] \qquad \text{Rewriting}$$

$$= \cos \left( \frac{\pi}{2} - \alpha \right) \cos \beta + \sin \left( \frac{\pi}{2} - \alpha \right) \sin \beta \qquad \begin{array}{l}\text{Using the identity for the} \\ \text{cosine of a difference}\end{array}$$

$$= \sin \alpha \cos \beta + \cos \alpha \sin \beta. \qquad \begin{array}{l}\text{Using identities (1)} \\ \text{and (2)}\end{array}$$

Thus the identity we seek is

$$\sin (\boldsymbol{\alpha} + \boldsymbol{\beta}) = \sin \boldsymbol{\alpha} \cos \boldsymbol{\beta} + \cos \boldsymbol{\alpha} \sin \boldsymbol{\beta}.$$

**DO EXERCISES 4 AND 5.** _____

To find a formula for the sine of a difference, we can use the identity just derived, substituting $-\beta$ for $\beta$. This gives us

$$\sin (\boldsymbol{\alpha} - \boldsymbol{\beta}) = \sin \boldsymbol{\alpha} \cos \boldsymbol{\beta} - \cos \boldsymbol{\alpha} \sin \boldsymbol{\beta}.$$

## Tangent Sum and Difference Identities

A formula for the tangent of a sum can be derived as follows, using identities already established:

$$\tan (\alpha + \beta) = \frac{\sin (\alpha + \beta)}{\cos (\alpha + \beta)}$$

$$= \frac{\sin \alpha \cos \beta + \cos \alpha \sin \beta}{\cos \alpha \cos \beta - \sin \alpha \sin \beta} \cdot \frac{\dfrac{1}{\cos \alpha \cos \beta}}{\dfrac{1}{\cos \alpha \cos \beta}}$$

$$= \frac{\dfrac{\sin \alpha \cos \beta}{\cos \alpha \cos \beta} + \dfrac{\cos \alpha \sin \beta}{\cos \alpha \cos \beta}}{\dfrac{\cos \alpha \cos \beta}{\cos \alpha \cos \beta} - \dfrac{\sin \alpha \sin \beta}{\cos \alpha \cos \beta}}$$

$$= \frac{\dfrac{\sin \alpha}{\cos \alpha} + \dfrac{\sin \beta}{\cos \beta}}{1 - \dfrac{\sin \alpha \sin \beta}{\cos \alpha \cos \beta}}$$

$$= \frac{\tan \alpha + \tan \beta}{1 - \tan \alpha \tan \beta}.$$

4. Find $\sin \left( \dfrac{\pi}{4} + \dfrac{\pi}{3} \right)$.

5. Simplify:

$$\sin \alpha \cos (-\beta) + \cos \alpha \sin (-\beta).$$

6. Derive the formula for $\tan(\alpha - \beta)$.

Similarly, a formula for the tangent of a difference can be established. A summary of the **sum and difference identities** follows.* These should be memorized.

---

**Sum and Difference Identities**

$$\sin(\alpha \pm \beta) = \sin\alpha\cos\beta \pm \cos\alpha\sin\beta$$
$$\cos(\alpha \pm \beta) = \cos\alpha\cos\beta \mp \sin\alpha\sin\beta$$
$$\tan(\alpha \pm \beta) = \frac{\tan\alpha \pm \tan\beta}{1 \mp \tan\alpha\tan\beta}$$

---

**DO EXERCISE 6.**

7. Find $\tan 75°$ as $\tan(45° + 30°)$.

**Example 3** Find $\tan 15°$.

**Solution**

$$\tan 15° = \tan(45° - 30°)$$
$$= \frac{\tan 45° - \tan 30°}{1 + \tan 45° \tan 30°}$$
$$= \frac{1 - \sqrt{3}/3}{1 + \sqrt{3}/3}$$
$$= \frac{3 - \sqrt{3}}{3 + \sqrt{3}}$$

**DO EXERCISE 7.**

Simplify.

8. $\sin\dfrac{\pi}{3}\sin\left(-\dfrac{\pi}{4}\right) + \cos\left(-\dfrac{\pi}{4}\right)\cos\dfrac{\pi}{3}$

## ▶ Simplification

**Example 4** Simplify: $\sin\left(-\dfrac{5\pi}{2}\right)\sin\dfrac{\pi}{2} + \cos\dfrac{\pi}{2}\cos\left(-\dfrac{5\pi}{2}\right)$.

**Solution** The expression is equal to

$$\cos\dfrac{\pi}{2}\cos\left(-\dfrac{5\pi}{2}\right) + \sin\dfrac{\pi}{2}\sin\left(-\dfrac{5\pi}{2}\right).$$

By the fourth identity in the list above, we can simplify to

9. $\cos 37° \cos 12° + \sin 12° \sin 37°$

$$\cos\left[\dfrac{\pi}{2} - \left(-\dfrac{5\pi}{2}\right)\right], \quad \text{or} \quad \cos\dfrac{6\pi}{2}, \quad \text{or} \quad \cos 3\pi,$$

which is $-1$.

**DO EXERCISES 8 AND 9.**

**Example 5** Simplify: $\sin\dfrac{\pi}{3}\cos\pi + \sin\pi\cos\dfrac{\pi}{3}$.

---
* There are six identities here, half of them obtained by using the signs shown in color.

**Solution**   The expression is equal to

$$\sin\frac{\pi}{3}\cos\pi + \cos\frac{\pi}{3}\sin\pi.$$

Thus by the first identity above, we can simplify to

$$\sin\left(\frac{\pi}{3} + \pi\right), \quad \text{or} \quad \sin\frac{4\pi}{3}, \quad \text{or} \quad -\frac{\sqrt{3}}{2}. \qquad \blacktriangleleft$$

**DO EXERCISE 10.**

**Example 6**   Assume that $\sin u = \frac{2}{3}$ and $\sin v = \frac{1}{3}$ and that $u$ and $v$ are between 0 and $\pi/2$. Then evaluate $\sin(u + v)$.

**Solution**   Using the identity for the sine of a sum, we have

$$\sin(u + v) = \sin u \cos v + \cos u \sin v$$
$$= \tfrac{2}{3}\cos v + \tfrac{1}{3}\cos u.$$

To finish, we need to know the values of $\cos v$ and $\cos u$. We first find $\cos v$. We know that $\sin^2 v + \cos^2 v = 1$. Solving for $\cos v$, we get

$$\cos v = \sqrt{1 - \sin^2 v}.$$

We know that the radical is positive because $v$ is between 0 and $\pi/2$. Then substituting $\frac{1}{3}$ for $\sin v$, we get

$$\cos v = \sqrt{1 - \left(\frac{1}{3}\right)^2} = \sqrt{1 - \frac{1}{9}} = \sqrt{\frac{8}{9}} = \frac{\sqrt{8}}{3} = \frac{2\sqrt{2}}{3}.$$

Similarly,

$$\cos u = \sqrt{1 - \sin^2 u} = \sqrt{1 - \left(\frac{2}{3}\right)^2} = \sqrt{1 - \frac{4}{9}} = \sqrt{\frac{5}{9}} = \frac{\sqrt{5}}{3}.$$

Substituting these values gives us

$$\sin(u + v) = \frac{2}{3}\cdot\frac{2\sqrt{2}}{3} + \frac{1}{3}\cdot\frac{\sqrt{5}}{3} = 4\frac{\sqrt{2}}{9} + \frac{\sqrt{5}}{9} = \frac{4\sqrt{2} + \sqrt{5}}{9}. \qquad \blacktriangleleft$$

**DO EXERCISE 11.**

10. Simplify:

$$\sin\frac{\pi}{2}\cos\frac{\pi}{3} - \sin\frac{\pi}{3}\cos\frac{\pi}{2}.$$

11. Assume that $\cos\alpha = \frac{3}{4}$ and $\cos\beta = \frac{1}{4}$ and that $\alpha$ and $\beta$ are between 0 and $\pi/2$. Then evaluate $\cos(\alpha - \beta)$.

● **EXERCISE SET 7.2**

**A**   Use the sum and difference identities to evaluate.

1. $\sin 75°$
   (*Hint:* $75° = 45° + 30°$.)

2. $\cos 75°$

3. $\sin 15°$
   (*Hint:* $15° = 45° - 30°$.)

4. $\cos 15°$

5. $\sin 105°$

6. $\cos 105°$

7. $\tan 75°$

8. $\cot 15°$

**B**   Simplify or evaluate.

9. $\sin 37° \cos 22° + \cos 37° \sin 22°$

10. $\cos 37° \cos 22° - \sin 22° \sin 37°$

11. $\dfrac{\tan 20° + \tan 32°}{1 - \tan 20° \tan 32°}$

**12.** $\dfrac{\tan 35° - \tan 12°}{1 + \tan 35° \tan 12°}$

Assume that $\sin u = \frac{3}{5}$ and $\sin v = \frac{4}{5}$ and that $u$ and $v$ are between 0 and $\pi/2$. Then evaluate.

**13.** $\sin(u + v)$          **14.** $\sin(u - v)$

**15.** $\cos(u + v)$          **16.** $\cos(u - v)$

**17.** $\tan(u + v)$          **18.** $\tan(u - v)$

▦ Assume that $\sin \theta = 0.6249$ and $\cos \phi = 0.1102$ and that $\theta$ and $\phi$ are both first-quadrant angles. Evaluate.

**19.** $\sin(\theta + \phi)$          **20.** $\cos(\theta + \phi)$

Simplify.

**21.** $\sin(\alpha + \beta) + \sin(\alpha - \beta)$

**22.** $\sin(\alpha + \beta) - \sin(\alpha - \beta)$

**23.** $\cos(\alpha + \beta) + \cos(\alpha - \beta)$

**24.** $\cos(\alpha + \beta) - \cos(\alpha - \beta)$

**25.** $\cos(u + v)\cos v + \sin(u + v)\sin v$

**26.** $\sin(u - v)\cos v + \cos(u - v)\sin v$

● **SYNTHESIS** _____

*Angles between lines.* One of the identities just developed gives us an easy way to find an angle formed by two lines. Consider two lines with equations: $l_1: y = m_1 x + b_1$ and $l_2: y = m_2 x + b_2$.

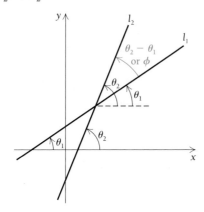

The slopes $m_1$ and $m_2$ are the tangents of the angles $\theta_1$ and $\theta_2$ that the lines form with the positive direction of the $x$-axis. Thus we have $m_1 = \tan \theta_1$ and $m_2 = \tan \theta_2$. To find the measure of the smallest angle—in this case, $\theta_2 - \theta_1$, or $\phi$ (phi)—through which $l_1$ can be rotated in the positive direction in order to get to $l_2$, we proceed as follows:

$$\tan \phi = \tan(\theta_2 - \theta_1) = \frac{\tan \theta_2 - \tan \theta_1}{1 + \tan \theta_2 \tan \theta_1}$$

$$= \frac{m_2 - m_1}{1 + m_2 m_1}.$$

This formula also holds when the lines are taken in the reverse order. When $\phi$ is an acute angle, $\tan \phi$ will be positive. When $\phi$ is obtuse, $\tan \phi$ will be negative.

Find the angle from $l_1$ to $l_2$.

**27.** $l_1: 2x = 3 - 2y,$
$l_2: x + y = 5$

**28.** $l_1: 3y = \sqrt{3}x + 3,$
$l_2: y = \sqrt{3}x + 2$

**29.** $l_1: y = 3,$
$l_2: x + y = 5$

**30.** $l_1: 2x + y - 4 = 0,$
$l_2: y - 2x + 5 = 0$

**31.** In a circus, a guy wire $A$ is attached to the top of a 30-ft pole. Wire $B$ is used for performers to walk up to the tight wire, 10 ft above the ground. Find the angle $\phi$ between the wires if they are attached to the ground 40 ft from the pole.

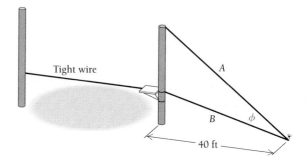

**32.** Given that $f(x) = \sin x$, show that
$$\frac{f(x + h) - f(x)}{h} = \sin x \left(\frac{\cos h - 1}{h}\right) + \cos x \left(\frac{\sin h}{h}\right).$$

**33.** Given that $f(x) = \cos x$, show that
$$\frac{f(x + h) - f(x)}{h} = \cos x \left(\frac{\cos h - 1}{h}\right) - \sin x \left(\frac{\sin h}{h}\right).$$

**34.** Given that $f(x) = \tan x$, show that
$$\frac{f(x + h) - f(x)}{h} = \frac{\sec^2 x}{1 - \tan x \tan h}\left(\frac{\tan h}{h}\right).$$

Find the slope of line $l_1$, where $m_2$ is the slope of line $l_2$ and $\phi$ is the smallest positive angle from $l_1$ to $l_2$.

**35.** $m_2 = \frac{4}{3}, \phi = 45°$

**36.** $m_2 = \frac{2}{3}, \phi = 30°$

**37.** Line $l_1$ contains the points $(-2, 4)$ and $(5, -1)$. Find the slope of line $l_2$ such that the angle from $l_1$ to $l_2$ is $45°$.

**38.** Line $l_1$ contains the points $(-3, 7)$ and $(-3, -2)$. Line $l_2$ contains $(0, -4)$ and $(2, 6)$. Find the smallest positive angle from $l_1$ to $l_2$.

**39.** Find an identity for $\sin 2\theta$. (*Hint:* $2\theta = \theta + \theta$.)

**40.** Find an identity for $\cos 2\theta$. (*Hint:* $2\theta = \theta + \theta$.)

**41.** Derive an identity for $\cot(\alpha + \beta)$ in terms of $\cot \alpha$ and $\cot \beta$.

**42.** Derive an identity for $\cot(\alpha - \beta)$ in terms of $\cot \alpha$ and $\cot \beta$.

Derive each of the following identities.

**43.** $\sin\left(x + \dfrac{3\pi}{2}\right) = -\cos x$

**44.** $\cos\left(x - \dfrac{3\pi}{2}\right) = -\sin x$

**45.** $\cos\left(x + \dfrac{3\pi}{2}\right) = \sin x$

**46.** $\sin\left(x - \dfrac{3\pi}{2}\right) = \cos x$

**47.** $\tan\left(x + \dfrac{\pi}{4}\right) = \dfrac{1 + \tan x}{1 - \tan x}$

**48.** $\tan\left(x - \dfrac{\pi}{4}\right) = \dfrac{\tan x - 1}{\tan x + 1}$

**49.** $\dfrac{\sin(\alpha + \beta)}{\cos(\alpha - \beta)} = \dfrac{\tan \alpha + \tan \beta}{1 + \tan \alpha \tan \beta}$

**50.** $\dfrac{\sin(\alpha + \beta)}{\sin(\alpha - \beta)} = \dfrac{\tan \alpha + \tan \beta}{\tan \alpha - \tan \beta}$

**51.** $\sin(\alpha + \beta) + \sin(\alpha - \beta) = 2 \sin \alpha \cos \beta$

**52.** $\cos(\alpha + \beta) \cdot \cos(\alpha - \beta) = \cos^2 \alpha - \sin^2 \beta$

● CHALLENGE

**53.** Find an identity for $\cos(\alpha + \beta)$ involving only cosines.

**54.** Find an identity for $\sin(\alpha + \beta + \gamma)$.

# 7.3
# COFUNCTION AND RELATED IDENTITIES

**OBJECTIVE**

You should be able to:

A State the cofunction identities for the sine and the cosine, and derive cofunction identities for the other functions. Use the cofunction identities and a given function value for an acute angle to find the function value for its complement.

## A Cofunction Identities

In Section 7.2, we developed the following two identities:

$$\cos\left(\frac{\pi}{2} - \theta\right) = \sin \theta \quad \text{and} \quad \sin\left(\frac{\pi}{2} - \theta\right) = \cos \theta.$$

We can now use these to prove another related identity:

$$\tan\left(\frac{\pi}{2} - \theta\right) = \frac{\sin\left(\frac{\pi}{2} - \theta\right)}{\cos\left(\frac{\pi}{2} - \theta\right)} = \frac{\cos \theta}{\sin \theta} = \cot \theta.$$

Prove each of the following identities.

**1.** $\cot\left(\dfrac{\pi}{2} - \theta\right) = \tan \theta$

**2.** $\sec\left(\dfrac{\pi}{2} - \theta\right) = \csc \theta$

**3.** $\csc\left(\dfrac{\pi}{2} - \theta\right) = \sec \theta$

**DO EXERCISES 1–3.**

Each of these identities yields a conversion to a cofunction. For this reason, we call them **cofunction identities.** They are summarized below.

**Cofunction Identities**

$$\sin\left(\frac{\pi}{2} - \theta\right) = \cos \theta \qquad \cos\left(\frac{\pi}{2} - \theta\right) = \sin \theta$$

$$\tan\left(\frac{\pi}{2} - \theta\right) = \cot \theta \qquad \cot\left(\frac{\pi}{2} - \theta\right) = \tan \theta$$

$$\sec\left(\frac{\pi}{2} - \theta\right) = \csc \theta \qquad \csc\left(\frac{\pi}{2} - \theta\right) = \sec \theta$$

These identities hold for all real numbers, and thus, for all degree measures, but if we restrict $\theta$ to values such that $0° < \theta < 90°$, then we have a special application to the acute angles of a right triangle.

4. Given that

$$\sin 75° = 0.9659,$$
$$\cos 75° = 0.2588,$$
$$\tan 75° = 3.732,$$
$$\cot 75° = 0.2679,$$
$$\sec 75° = 3.864,$$
$$\csc 75° = 1.035,$$

find the six function values for the complement of 75°.

In a right triangle, the acute angles are complementary, since the sum of all three angle measures is 180° and the right angle accounts for 90° of this total. Thus if one acute angle of a right triangle is $\theta$, the other is $90° - \theta$, or $\pi/2 - \theta$. Note that the sine of $\angle A$ is also the cosine of $\angle B$, its complement:

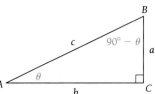

Similarly, the tangent of $\angle A$ is the cotangent of its complement, and the secant of $\angle A$ is the cosecant of its complement.

These pairs of functions are called **cofunctions**. The name **cosine** originally meant the sine of the complement, **cotangent** meant the tangent of the complement, and **cosecant** meant the secant of the complement.

**Example 1**    Given that

$$\sin 18° = 0.3090, \qquad \cos 18° = 0.9511,$$
$$\tan 18° = 0.3249, \qquad \cot 18° = 3.078,$$
$$\sec 18° = 1.051, \qquad \csc 18° = 3.236,$$

find the six function values for 72°.

**Solution**    Since 72° and 18° are complements, we have $\sin 72° = \cos 18°$, and so on, and the function values are

$$\sin 72° = 0.9511, \qquad \cos 72° = 0.3090,$$
$$\tan 72° = 3.078, \qquad \cot 72° = 0.3249,$$
$$\sec 72° = 3.236, \qquad \csc 72° = 1.051.$$

◀

**DO EXERCISE 4.**

## Other Cofunction Identities

Suppose we want to derive an identity for

$$\sin\left(\theta + \frac{\pi}{2}\right).$$

We have at least two ways to prove this identity. One way to do so is to use the identity for the sine of a sum developed in Section 7.2:

$$\sin\left(\theta + \frac{\pi}{2}\right) = \sin\theta\cos\frac{\pi}{2} + \cos\theta\sin\frac{\pi}{2}$$
$$= \sin\theta \cdot 0 + \cos\theta \cdot 1$$
$$= \cos\theta.$$

Another proof of this identity can be found by considering the left side in terms of $x$ in the form of a function, as follows:

$$g(x) = \sin\left(x + \frac{\pi}{2}\right) = \sin\left[x - \left(-\frac{\pi}{2}\right)\right].$$

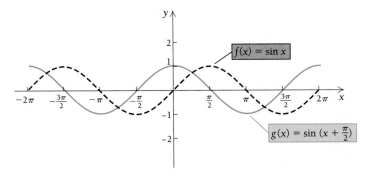

The graph of $g$ is found by translating the graph of $f(x) = \sin x$ to the left $\pi/2$ units. The translation is also a graph of the cosine function. Thus we obtain the identity

$$\sin\left(x + \frac{\pi}{2}\right) = \cos x.$$

**DO EXERCISE 5.**

In Margin Exercise 5, you gave a graphical proof of the following identity:

$$\cos\left(x - \frac{\pi}{2}\right) = \sin x.$$

We can also give a proof using the identity $\cos(-x) = \cos x$ and a cofunction identity:

$$\cos\left(x - \frac{\pi}{2}\right) = \cos\left[-\left(\frac{\pi}{2} - x\right)\right] = \cos\left(\frac{\pi}{2} - x\right) = \sin x.$$

Yet another proof can be given using the formula for the cosine of a difference.

**DO EXERCISE 6.**

Consider the following graph. The graph of $f(x) = \sin x$ has been translated $\pi/2$ units to the right, to obtain the graph of $g(x) = \sin(x - \pi/2)$. The latter is a reflection of the cosine function across the $x$-axis. In other words, it is a graph of $g(x) = -\cos x$. We thus obtain the following identity:

$$\sin\left(x - \frac{\pi}{2}\right) = -\cos x.$$

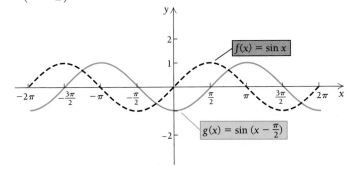

We have at least two other ways to prove this identity. One way to do so

5. a) Graph $y = \cos x$.
   b) Translate to obtain a graph of
   $$y = \cos\left(x - \frac{\pi}{2}\right).$$
   c) Graph $y = \sin x$.
   d) How do the graphs of parts (b) and (c) compare?
   e) Write the identity thus established.

6. Prove the identity
   $$\cos\left(x - \frac{\pi}{2}\right) = \sin x$$
   using the formula for the cosine of a difference.

7. Prove the identity

$$\sin\left(x - \frac{\pi}{2}\right) = -\cos x$$

using the formula for the sine of a difference.

8. Prove the identity

$$\cos\left(x + \frac{\pi}{2}\right) = -\sin x$$

a) by graphing. Explain your proof.

b) by using the formula for the cosine of a sum.

9. Find an identity for $\cot\left(x + \frac{\pi}{2}\right)$.

is to use the identity $\sin(-x) = -\sin x$ and a cofunction identity:

$$\sin\left(x - \frac{\pi}{2}\right) = \sin\left[-\left(\frac{\pi}{2} - x\right)\right] = -\sin\left(\frac{\pi}{2} - x\right) = -\cos x.$$

Yet another proof can be given using the formula for the sine of a difference.

**DO EXERCISE 7.**

By using the formula for the cosine of a sum or a graphical method, we can prove the following identity:

$$\cos\left(x + \frac{\pi}{2}\right) = -\sin x.$$

**DO EXERCISE 8.**

We now have established four more cofunction identities. We give those involving sine and cosine in the following list. Two of them are obtained by taking the top signs, the other two by taking the bottom signs.

---

**Cofunction Identities for Sine and Cosine**

$$\sin\left(x \pm \frac{\pi}{2}\right) = \pm\cos x, \qquad \cos\left(x \pm \frac{\pi}{2}\right) = \mp\sin x$$

---

A good memory device is to think of the "s" in "sine" as "same," meaning that the signs stay the same. For cosine, think of "c" for "change," meaning that the signs change. Cofunction identities for the other functions can be obtained from these easily, using the definitions of the other basic circular functions.

**Example 2**   Find an identity for $\tan\left(x + \frac{\pi}{2}\right)$.

**Solution**   By definition of the tangent function,

$$\tan\left(x + \frac{\pi}{2}\right) = \frac{\sin\left(x + \frac{\pi}{2}\right)}{\cos\left(x + \frac{\pi}{2}\right)}.$$

Using a cofunction identity, we obtain

$$\frac{\sin\left(x + \frac{\pi}{2}\right)}{\cos\left(x + \frac{\pi}{2}\right)} = \frac{\cos x}{-\sin x} = -\frac{\cos x}{\sin x} = -\cot x.$$

Thus the identity we seek is

$$\tan\left(x + \frac{\pi}{2}\right) = -\cot x.$$

**DO EXERCISE 9.**

**Example 3**    Find an identity for $\sec\left(x - \dfrac{\pi}{2}\right)$.

Solution    By definition of the secant function,

$$\sec\left(x - \frac{\pi}{2}\right) = \frac{1}{\cos\left(x - \dfrac{\pi}{2}\right)}.$$

Using a cofunction identity, we obtain

$$\frac{1}{\cos\left(x - \dfrac{\pi}{2}\right)} = \frac{1}{\sin x} = \csc x.$$

Thus the identity we seek is

$$\sec\left(x - \frac{\pi}{2}\right) = \csc x.$$

◀

**DO EXERCISE 10.**

Expressions such as

$$\sin\left(x \pm n \cdot \frac{\pi}{2}\right), \qquad \cos\left(x \pm n \cdot \frac{\pi}{2}\right), \quad \text{and} \quad \tan\left(x \pm n \cdot \frac{\pi}{2}\right),$$

where $n$ is an integer, can also be changed to expressions involving only $\sin x$, $\cos x$, $\tan x$, or $\cot x$ using sum and difference formulas for sine and cosine.

**Example 4**    Find an identity for $\cos\left(x + \dfrac{3\pi}{2}\right)$.

Solution    Using the cosine sum formula, we obtain

$$\begin{aligned}
\cos\left(x + \frac{3\pi}{2}\right) &= \cos x \cos\frac{3\pi}{2} - \sin x \sin\frac{3\pi}{2} \\
&= (\cos x)(0) - (\sin x)(-1) \\
&= \sin x.
\end{aligned}$$

◀

**Example 5**    Find an identity for $\tan(\theta - 270°)$.

Solution

$$\begin{aligned}
\tan(\theta - 270°) &= \frac{\sin(\theta - 270°)}{\cos(\theta - 270°)} \qquad \text{Using } \tan\theta = \frac{\sin\theta}{\cos\theta} \\
&= \frac{\sin\theta\cos 270° - \cos\theta\sin 270°}{\cos\theta\cos 270° + \sin\theta\sin 270°} \\
&= \frac{-\cos\theta \cdot (-1)}{\sin\theta \cdot (-1)} = -\frac{\cos\theta}{\sin\theta} \\
&= -\cot\theta
\end{aligned}$$

◀

**DO EXERCISES 11 AND 12.**

10. Find an identity for $\csc\left(x - \dfrac{\pi}{2}\right)$.

Find an identity for each of the following.

11. $\sin(x + \pi)$

12. $\tan(\theta - 450°)$

**TECHNOLOGY CONNECTION**

A grapher can be used to check identities involving cofunctions. The only drawback is that you must already know both sides of what you *think* is an identity. When using cofunctions, make sure the grapher is in the proper mode (RADIAN or DEGREE). You will be able to tell which mode is required by inspecting the angle that is added or subtracted in the cofunctions. If a term includes an angle involving $\pi$, then it is assumed that angles are being measured in RADIANS. If no $\pi$ appears, then the DEGREE system is being used (usually the degree symbol $°$ will also be used).

Use a grapher to determine whether each of the following equations is an identity.

**TC 1.** $\tan(x - \pi/2) = \cot x$

**TC 2.** $\cos(x - 450°) = \sin x$

**TC 3.** $\cos(900° - x) - \cos x$
$$= -2\cos x$$

**TC 4.** $\csc(3\pi/2 + x) + \sec(\pi - x)$
$$= 2\csc(3\pi/2 - x)$$

## •EXERCISE SET 7.3

**1.** Given that

$$\sin 65° = 0.9063, \qquad \cos 65° = 0.4226,$$
$$\tan 65° = 2.145, \qquad \cot 65° = 0.4663,$$
$$\sec 65° = 2.366, \qquad \csc 65° = 1.103,$$

find the six function values for 25°.

**2.** Given that

$$\sin 32° = 0.5299, \qquad \cos 32° = 0.8480,$$
$$\tan 32° = 0.6249, \qquad \cot 32° = 1.600,$$
$$\sec 32° = 1.179, \qquad \csc 32° = 1.887,$$

find the six function values for 58°.

**3.** Given that

$$\sin 22.5° = \frac{\sqrt{2 - \sqrt{2}}}{2}$$

and

$$\cos 22.5° = \frac{\sqrt{2 + \sqrt{2}}}{2},$$

find each of the following.

**a)** The other four function values for 22.5°
**b)** The six function values for 67.5°

**4.** Given that

$$\sin \frac{\pi}{12} = \frac{\sqrt{2 - \sqrt{3}}}{2}$$

and

$$\cos \frac{\pi}{12} = \frac{\sqrt{2 + \sqrt{3}}}{2},$$

find each of the following.

**a)** The other four function values for $\pi/12$
**b)** The six function values for $5\pi/12$

Find and prove an identity for each of the following.

**5.** $\tan\left(x - \dfrac{\pi}{2}\right)$          **6.** $\cot\left(x - \dfrac{\pi}{2}\right)$

**7.** $\csc\left(x + \dfrac{\pi}{2}\right)$          **8.** $\sec\left(x + \dfrac{\pi}{2}\right)$

**9.** $\sin(x - \pi)$          **10.** $\cos(\theta - 270°)$

Find an identity for each of the following.

**11.** $\tan(\theta + 270°)$          **12.** $\sin(\theta - 270°)$
**13.** $\sin(\theta + 450°)$          **14.** $\cos(\theta - 450°)$
**15.** $\sec(\pi - \theta)$          **16.** $\csc(\pi + \theta)$
**17.** $\tan(x - 4\pi)$          **18.** $\sec(x - 10\pi)$
**19.** $\cos\left(\dfrac{9\pi}{2} + x\right)$          **20.** $\cot\left(\dfrac{9\pi}{2} - x\right)$

**21.** $\csc(540° - \theta)$          **22.** $\tan(540° + \theta)$

**23.** $\cot\left(x - \dfrac{3\pi}{2}\right)$          **24.** $\sin\left(x + \dfrac{11\pi}{2}\right)$

● **SYNTHESIS**

**25.** Given that $\sin\theta = \frac{1}{3}$ and that the terminal side is in quadrant II.

**a)** Find the other function values for $\theta$.
**b)** Find the six function values for $\pi/2 - \theta$.
**c)** Find the six function values for $\pi + \theta$.
**d)** Find the six function values for $\pi - \theta$.
**e)** Find the six function values for $2\pi - \theta$.

**26.** Given that $\cos\theta = \frac{4}{5}$ and that the terminal side is in quadrant IV.

**a)** Find the other function values for $\theta$.
**b)** Find the six function values for $\pi/2 + \theta$.
**c)** Find the six function values for $\pi + \theta$.
**d)** Find the six function values for $\pi - \theta$.
**e)** Find the six function values for $2\pi - \theta$.

**27.** Given that $\sin 27° = 0.45399$.

**a)** Find the other function values for 27°.
**b)** Find the six function values for 63°.

**28.** Given that $\cot 54° = 0.72654$.

**a)** Find the other function values for 54°.
**b)** Find the six function values for 36°.

**29.** Given that $\cos 38° = 0.78801$, find the six function values for 128°.

**30.** Given that $\tan 73° = 3.27085$, find the six function values for 343°.

Simplify. If you have a grapher, verify your results.

**31.** $\sin\left(\dfrac{\pi}{2} - x\right)[\sec x - \cos x]$

**32.** $\cos\left(\dfrac{\pi}{2} - x\right)[\csc x - \sin x]$

**33.** $\sin x \cos y - \cos\left(x + \dfrac{\pi}{2}\right)\tan y$

**34.** $\cos\left(x - \dfrac{\pi}{2}\right)\tan y + \sin x \cot y$

**35.** $\cos(\pi - x) + \cot x \sin\left(x - \dfrac{\pi}{2}\right)$

**36.** $\sin(\pi - x) - \tan x \cos\left(\dfrac{\pi}{2} - x\right)$

**37.** $\dfrac{\cos^2 x - 1}{\sin\left(\dfrac{\pi}{2} - x\right) - 1}$

**38.** $\dfrac{\sin^2 x - 1}{\cos\left(\dfrac{\pi}{2} - x\right) + 1}$

**39.** $\dfrac{\sin x - \cos\left(x - \dfrac{\pi}{2}\right)\cos x}{-\sin x - \cos\left(x - \dfrac{\pi}{2}\right)\tan x}$

**40.** $\dfrac{\cos x - \sin\left(\dfrac{\pi}{2} - x\right)\sin x}{\cos x - \cos\left(\pi - x\right)\tan x}$

**41.** $\dfrac{\cos^2 x + 2\sin\left(x - \dfrac{\pi}{2}\right) + 1}{\sin\left(\dfrac{\pi}{2} - x\right) - 1}$

**42.** $\dfrac{\sin^2 x - 2\cos\left(x - \dfrac{\pi}{2}\right) + 1}{\cos\left(\dfrac{\pi}{2} - x\right) - 1}$

**43.** $\dfrac{\sin^2 y \cos\left(y + \dfrac{\pi}{2}\right)}{\cos^2 y \cos\left(\dfrac{\pi}{2} - y\right)}$

**44.** $\dfrac{\cos^2 y \sin\left(y + \dfrac{\pi}{2}\right)}{\sin^2 y \sin\left(\dfrac{\pi}{2} - y\right)}$

# 7.4
# SOME IMPORTANT IDENTITIES

Two important classes of trigonometric identities are known as the *double-angle identities* and the *half-angle identities*.

## A   Double-Angle Identities

To develop these identities, we will use the sum formulas from the preceding section. We first develop a formula for sin 2θ. Recall that

$$\sin(\alpha + \beta) = \sin\alpha\cos\beta + \cos\alpha\sin\beta.$$

We will consider a number $\theta$ and substitute it for both $\alpha$ and $\beta$ in this identity. We obtain

$$\sin(\theta + \theta) = \sin 2\theta$$
$$= \sin\theta\cos\theta + \cos\theta\sin\theta$$
$$= 2\sin\theta\cos\theta.$$

Our first double-angle identity is thus

$$\boldsymbol{\sin 2\theta = 2\sin\theta\cos\theta}.$$

**Example 1**   If $\sin\theta = \frac{3}{8}$ and $\theta$ is in the first quadrant, what is sin 2θ?

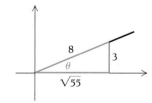

**Solution**   From the figure, we see that $\cos\theta = \sqrt{55}/8$. Thus,

$$\sin 2\theta = 2\sin\theta\cos\theta = 2\cdot\frac{3}{8}\cdot\frac{\sqrt{55}}{8} = \frac{3\sqrt{55}}{32}.$$

◀

**DO EXERCISE 1.**

You should be able to:

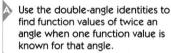 Use the double-angle identities to find function values of twice an angle when one function value is known for that angle.

Use the half-angle identities to find the function values of half an angle when one function value is known for that angle.

Simplify certain trigonometric expressions using the half-angle and the double-angle formulas. Also, derive some identities by simplifying.

1. Given that $\sin\theta = \frac{3}{5}$ and $\theta$ is in the first quadrant, what is sin 2θ?

2. Given that $\cos \theta = -\frac{5}{13}$ and $\theta$ is in the third quadrant, find $\sin 2\theta$, $\cos 2\theta$, and $\tan 2\theta$. Also, determine the quadrant in which $2\theta$ lies.

Double-angle identities for the cosine and tangent functions can be derived in much the same way as the identity above:

$$\cos (\alpha + \beta) = \cos \alpha \cos \beta - \sin \alpha \sin \beta$$
$$\cos 2\theta = \cos (\theta + \theta) = \cos \theta \cos \theta - \sin \theta \sin \theta$$
$$= \cos^2 \theta - \sin^2 \theta$$
$$\boldsymbol{\cos 2\theta = \cos^2 \theta - \sin^2 \theta.}$$

Now we derive an identity for $\tan 2\theta$:

$$\tan (\alpha + \beta) = \frac{\tan \alpha + \tan \beta}{1 - \tan \alpha \tan \beta}$$

$$\tan 2\theta = \tan (\theta + \theta) = \frac{\tan \theta + \tan \theta}{1 - \tan \theta \tan \theta} = \frac{2 \tan \theta}{1 - \tan^2 \theta}$$

$$\boldsymbol{\tan 2\theta = \frac{2 \tan \theta}{1 - \tan^2 \theta}.}$$

**Example 2**   Given that $\tan \theta = -\frac{3}{4}$ and $\theta$ is in the second quadrant, find $\sin 2\theta$, $\cos 2\theta$, $\tan 2\theta$, and the quadrant in which $2\theta$ lies.

Solution   By drawing a diagram as shown, we find that

$$\sin \theta = \frac{3}{5} \quad \text{and} \quad \cos \theta = -\frac{4}{5}.$$

Now,

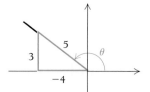

$$\sin 2\theta = 2 \sin \theta \cos \theta$$
$$= 2 \cdot \frac{3}{5} \cdot \left( -\frac{4}{5} \right) = -\frac{24}{25};$$
$$\cos 2\theta = \cos^2 \theta - \sin^2 \theta = \left( -\frac{4}{5} \right)^2 - \left( \frac{3}{5} \right)^2$$
$$= \frac{16}{25} - \frac{9}{25} = \frac{7}{25};$$
$$\tan 2\theta = \frac{2 \tan \theta}{1 - \tan^2 \theta} = \frac{2 \cdot \left( -\frac{3}{4} \right)}{1 - \left( -\frac{3}{4} \right)^2}$$
$$= \frac{-\frac{3}{2}}{1 - \frac{9}{16}} = -\frac{24}{7}.$$

Since $\sin 2\theta$ is negative and $\cos 2\theta$ is positive, we know that $2\theta$ is in quadrant IV. Note that $\tan 2\theta$ could have been found more easily in this case by simply dividing the values of $\sin 2\theta$ and $\cos 2\theta$.   ◄

**DO EXERCISE 2.**

Two other useful identities for $\cos 2\theta$ can be derived easily, as follows:

$$\cos 2\theta = \cos^2 \theta - \sin^2 \theta$$
$$= (1 - \sin^2 \theta) - \sin^2 \theta \qquad \text{Using } \sin^2 \theta + \cos^2 \theta = 1$$
$$= 1 - 2 \sin^2 \theta.$$

Similarly,

$$\cos 2\theta = \cos^2 \theta - \sin^2 \theta$$
$$= \cos^2 \theta - (1 - \cos^2 \theta) = 2 \cos^2 \theta - 1.$$

Solving these two identities for $\sin^2 \theta$ and $\cos^2 \theta$, respectively, we obtain two more identities that are often useful:

$$\sin^2 \theta = \frac{1 - \cos 2\theta}{2},$$

$$\cos^2 \theta = \frac{1 + \cos 2\theta}{2}.$$

Using division and the last two identities, we can easily deduce the following identity, which is also often useful:

$$\tan^2 \theta = \frac{1 - \cos 2\theta}{1 + \cos 2\theta}.$$

The following list of **double-angle identities** should be memorized.

---

**Double-Angle Identities**

$$\sin 2\theta = 2 \sin \theta \cos \theta, \qquad \sin^2 \theta = \frac{1 - \cos 2\theta}{2},$$

$$\cos 2\theta = \cos^2 \theta - \sin^2 \theta \qquad \cos^2 \theta = \frac{1 + \cos 2\theta}{2},$$
$$= 1 - 2 \sin^2 \theta$$
$$= 2 \cos^2 \theta - 1,$$

$$\tan 2\theta = \frac{2 \tan \theta}{1 - \tan^2 \theta}, \qquad \tan^2 \theta = \frac{1 - \cos 2\theta}{1 + \cos 2\theta}$$

---

From the basic identities (in the lists to be memorized), others can be obtained.

**Example 3**  Find a formula for $\sin 3\theta$ in terms of function values of $\theta$.

**Solution**

$$\sin 3\theta = \sin (2\theta + \theta)$$
$$= \sin 2\theta \cos \theta + \cos 2\theta \sin \theta$$
$$= (2 \sin \theta \cos \theta) \cos \theta + (2 \cos^2 \theta - 1) \sin \theta$$
$$= 2 \sin \theta \cos^2 \theta + 2 \sin \theta \cos^2 \theta - \sin \theta$$
$$= 4 \sin \theta \cos^2 \theta - \sin \theta \qquad \blacktriangleleft$$

**Example 4**  Find a formula for $\cos^3 x$ in terms of function values of $x$ or $2x$, raised only to the first power.

**Solution**

$$\cos^3 x = \cos^2 x \cos x$$
$$= \frac{1 + \cos 2x}{2} \cos x \qquad \blacktriangleleft$$

**DO EXERCISES 3 AND 4.**

3. Find a formula for $\cos 3\theta$ in terms of function values of $\theta$.

4. Find a formula for $\sin^3 x$ in terms of function values of $x$ or $2x$, raised only to the first power.

## B Half-Angle Identities

To develop these identities, we use three of the previously formed ones. As shown below, we take square roots and replace $\theta$ by $\phi/2$:

$$\sin^2 \theta = \frac{1 - \cos 2\theta}{2} \quad \longrightarrow \quad \left| \sin \frac{\phi}{2} \right| = \sqrt{\frac{1 - \cos \phi}{2}},$$

$$\cos^2 \theta = \frac{1 + \cos 2\theta}{2} \quad \longrightarrow \quad \left| \cos \frac{\phi}{2} \right| = \sqrt{\frac{1 + \cos \phi}{2}},$$

$$\tan^2 \theta = \frac{1 - \cos 2\theta}{1 + \cos 2\theta} \quad \longrightarrow \quad \left| \tan \frac{\phi}{2} \right| = \sqrt{\frac{1 - \cos \phi}{1 + \cos \phi}}.$$

The half-angle formulas are those on the right above. We can eliminate the absolute-value signs by introducing $\pm$ signs, with the understanding that our use of $+$ or $-$ depends on the quadrant in which the angle $\phi/2$ lies. We thus obtain these formulas in the following form.

---

**Half-Angle Identities**

$$\sin \frac{\phi}{2} = \pm \sqrt{\frac{1 - \cos \phi}{2}}$$

$$\cos \frac{\phi}{2} = \pm \sqrt{\frac{1 + \cos \phi}{2}}$$

$$\tan \frac{\phi}{2} = \pm \sqrt{\frac{1 - \cos \phi}{1 + \cos \phi}}$$

---

These formulas should be memorized.

There are two other formulas for $\tan(\phi/2)$ that are often useful. They can be obtained as follows:

$$\left| \tan \frac{\phi}{2} \right| = \sqrt{\frac{1 - \cos \phi}{1 + \cos \phi}} = \sqrt{\frac{1 - \cos \phi}{1 + \cos \phi} \cdot \frac{1 + \cos \phi}{1 + \cos \phi}}$$

$$= \sqrt{\frac{1 - \cos^2 \phi}{(1 + \cos \phi)^2}} = \sqrt{\frac{\sin^2 \phi}{(1 + \cos \phi)^2}}$$

$$= \frac{|\sin \phi|}{|1 + \cos \phi|}.$$

Now $1 + \cos \phi$ cannot be negative because $\cos \phi$ is never less than $-1$. Thus the absolute-value signs are not necessary in the denominator. As the following graph shows, $\tan(\phi/2)$ and $\sin \phi$ have the same sign for all $\phi$ for which $\tan(\phi/2)$ is defined.

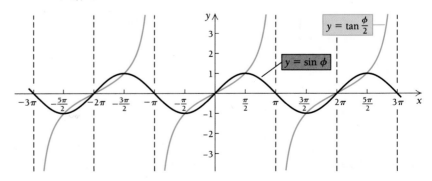

Thus we can dispense with the other absolute-value signs and obtain the formula we seek. A second formula can be obtained in a similar way.

$$\tan\frac{\phi}{2} = \frac{\sin\phi}{1+\cos\phi}, \qquad \tan\frac{\phi}{2} = \frac{1-\cos\phi}{\sin\phi}$$

These formulas have the advantage that they give the sign of $\tan(\phi/2)$ directly.

**Example 5**   Find $\sin 15°$.

Solution

$$\sin 15° = \sin\frac{30°}{2} = \pm\sqrt{\frac{1-\cos 30°}{2}}$$

$$= \pm\sqrt{\frac{1-(\sqrt{3}/2)}{2}} = \pm\sqrt{\frac{2-\sqrt{3}}{4}}$$

$$= \frac{\sqrt{2-\sqrt{3}}}{2}$$

The expression is positive, because $15°$ is in the first quadrant.   ◀

**Example 6**   Find $\tan\frac{\pi}{8}$.

Solution

$$\tan\frac{\pi}{8} = \tan\frac{\frac{\pi}{4}}{2} = \frac{\sin\frac{\pi}{4}}{1+\cos\frac{\pi}{4}} = \frac{\frac{\sqrt{2}}{2}}{1+\frac{\sqrt{2}}{2}}$$

$$= \frac{\sqrt{2}}{2+\sqrt{2}}$$

$$= \sqrt{2}-1 \qquad \text{Rationalizing the denominator}$$   ◀

**DO EXERCISES 5 AND 6.**

 ## Simplification

Many simplifications of trigonometric expressions are possible through the use of the identities that we have developed.

**Example 7**   Simplify:

$$\frac{1-\cos 2x}{4\sin x\cos x}.$$

Solution   We search the list of identities, looking for some substitution that will simplify the expression. In this case, one might note that the

---

5. Find $\cos 15°$.

6. Find $\tan\frac{\pi}{12}$.

7. Simplify:

$$\frac{2(\tan x + \tan^3 x)}{1-\tan^4 x}.$$

8. By simplifying, derive an identity:

$$\frac{\frac{1}{2}(\cos^2 x - \sin^2 x)}{\sin x\cos x}.$$

**TECHNOLOGY CONNECTION**

A grapher can also verify double-angle and half-angle identities.

**TC 1.** Use a grapher to determine which of the following expressions completes the identity

$$\frac{\cos 2x}{\cos x - \sin x} = \ldots.$$

A. $1 + \cos x$

B. $\cos x + \sin x$

C. $\cos x (\sec x + 1)$

D. $\sin x (\cot x + 1)$

**TC 2.** Use a grapher to determine which of the following expressions completes the identity

$$2 \cos^2 \frac{x}{2} = \ldots.$$

A. $\sin x (\csc x + \cot x)$

B. $\sin x - 2 \cos x$

C. $2 (\cos^2 x - \sin^2 x)$

D. $1 + \cos x$

denominator is $2(2 \sin x \cos x)$ and thus find a simplification, since $2 \sin x \cos x = \sin 2x$:

$$\frac{1 - \cos 2x}{4 \sin x \cos x} = \frac{1 - \cos 2x}{2(2 \sin x \cos x)} = \frac{1 - \cos 2x}{2 \sin 2x}.$$

Recalling that $(1 - \cos \phi)/\sin \phi = \tan (\phi/2)$ and letting $\phi = 2x$, we have

$$\frac{1 - \cos 2x}{\sin 2x} = \tan x.$$

Multiplying by $\frac{1}{2}$, we obtain

$$\frac{1 - \cos 2x}{2 \sin 2x} = \frac{1}{2} \tan x. \qquad \blacktriangleleft$$

We can derive an identity by simplifying an expression, as in the following example.

**Example 8**    By simplifying, derive an identity:

$$\frac{\sin x \cos x}{\frac{1}{2} \cos 2x}.$$

**Solution**    We can obtain $2 \sin x \cos x$ in the numerator by multiplying the expression by $\frac{2}{2}$:

$$\frac{2 \sin x \cos x}{2 \cdot \frac{1}{2} \cos 2x} = \frac{\sin 2x}{\cos 2x} \qquad \text{Using } \sin 2x = 2 \sin x \cos x$$

$$= \tan 2x.$$

We have thus derived the identity

$$\frac{\sin x \cos x}{\frac{1}{2} \cos 2x} = \tan 2x. \qquad \blacktriangleleft$$

**DO EXERCISES 7 AND 8 ON THE PRECEDING PAGE.**

---

**● EXERCISE SET**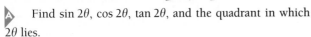

**A**    Find $\sin 2\theta$, $\cos 2\theta$, $\tan 2\theta$, and the quadrant in which $2\theta$ lies.

**1.** $\sin \theta = \frac{4}{5}$; $\theta$ in quadrant I

**2.** $\sin \theta = \frac{5}{13}$; $\theta$ in quadrant I

**3.** $\cos \theta = -\frac{4}{5}$; $\theta$ in quadrant III

**4.** $\cos \theta = -\frac{3}{5}$; $\theta$ in quadrant III

**5.** $\tan \theta = \frac{4}{3}$; $\theta$ in quadrant III

**6.** $\tan \theta = \frac{3}{4}$; $\theta$ in quadrant III

**7.** Find a formula for $\sin 4\theta$ in terms of function values of $\theta$.

**8.** Find a formula for $\cos 4\theta$ in terms of function values of $\theta$.

**9.** Find a formula for $\sin^4 \theta$ in terms of function values of $\theta$ or $2\theta$ or $4\theta$, raised only to the first power.

**10.** Find a formula for $\cos^4 \theta$ in terms of function values of $\theta$ or $2\theta$ or $4\theta$, raised only to the first power.

**B**    Find each of the following without using a calculator.

**11.** $\sin 75°$

(*Hint:* $75 = 150/2$.)

**12.** $\cos 75°$

**13.** $\tan 75°$

**14.** $\tan 67.5°$

(*Hint:* $67.5 = 135/2$.)

**15.** $\sin \dfrac{5\pi}{8}$

**16.** $\cos \dfrac{5\pi}{8}$

**17.** $\sin 112.5°$

**18.** $\cos 22.5°$

**19.** $\cos \dfrac{\pi}{8}$

**20.** $\tan 15°$

**21.** $\sin 22.5°$

**22.** $\tan 112.5°$

**A, B**   Given that $\sin \theta = 0.3416$ and $\theta$ is in the first quadrant, find each of the following.

**23.** $\sin 2\theta$

**24.** $\cos 2\theta$

**25.** $\sin 4\theta$

**26.** $\cos 4\theta$

**27.** $\sin \dfrac{\theta}{2}$

**28.** $\cos \dfrac{\theta}{2}$

**C**   Simplify.

**29.** $\dfrac{\sin 2x}{2 \sin x}$

**30.** $\dfrac{\sin 2x}{2 \cos x}$

**31.** $1 - 2 \sin^2 \dfrac{x}{2}$

**32.** $2 \cos^2 \dfrac{x}{2} - 1$

**33.** $2 \sin \dfrac{x}{2} \cos \dfrac{x}{2}$

**34.** $2 \sin 2x \cos 2x$

**35.** $\cos^2 \dfrac{x}{2} - \sin^2 \dfrac{x}{2}$

**36.** $\cos^4 x - \sin^4 x$

**37.** $(\sin x + \cos x)^2 - \sin 2x$

**38.** $(\sin x - \cos x)^2 + \sin 2x$

**39.** $2 \sin^2 \dfrac{x}{2} + \cos x$

**40.** $2 \cos^2 \dfrac{x}{2} - \cos x$

**41.** $(-4 \cos x \sin x + 2 \cos 2x)^2$
$$+ (2 \cos 2x + 4 \sin x \cos x)^2$$

**42.** $(-4 \cos 2x + 8 \cos x \sin x)^2$
$$+ (8 \sin x \cos x + 4 \cos 2x)^2$$

**43.** $2 \sin x \cos^3 x + 2 \sin^3 x \cos x$

**44.** $2 \sin x \cos^3 x - 2 \sin^3 x \cos x$

By simplifying, derive an identity. If you have a grapher, verify your results.

**45.** $(\sin x + \cos x)^2$

**46.** $\cos^4 x - \sin^4 x$

**47.** $\dfrac{2 \cot x}{\cot^2 x - 1}$

**48.** $\dfrac{2 - \sec^2 x}{\sec^2 x}$

**49.** $2 \sin^2 2x + \cos 4x$

**50.** $\dfrac{1 + \sin 2x + \cos 2x}{1 + \sin 2x - \cos 2x}$

● **SYNTHESIS** _____

Find $\sin \theta$, $\cos \theta$, and $\tan \theta$ under the given conditions.

**51.** $\sin 2\theta = \dfrac{1}{5}, \dfrac{\pi}{2} \le 2\theta \le \pi$

**52.** $\cos 2\theta = \dfrac{7}{12}, \dfrac{3\pi}{2} \le 2\theta \le 2\pi$

**53.** $\tan \dfrac{\theta}{2} = -\dfrac{1}{4}, \dfrac{3\pi}{2} \le \theta \le 2\pi$

**54.** $\tan \dfrac{\theta}{2} = -\dfrac{5}{3}, \pi < \theta \le \dfrac{3\pi}{2}$

**55. Nautical mile.**   *Latitude* is used to measure north–south location on the earth between the equator and the poles. For example, the north end of Chicago has latitude 42°N. (See the figure.) In Great Britain, the **nautical mile** is defined as the length of a minute of arc of the earth's radius. Since the earth is flattened at the poles, a British nautical mile varies with latitude. In fact, it is given, in feet, by the function

$$N(\phi) = 6066 - 31 \cos 2\phi,$$

where $\phi$ is the latitude in degrees.

**a)** What is the length of a British nautical mile at the north end of Chicago?

**b)** What is the length of a British nautical mile at the North Pole?

**c)** Express $N(\phi)$ in terms of $\cos \phi$ only. That is, eliminate the double angle.

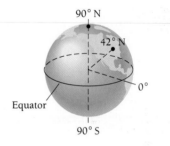

**56. Acceleration due to gravity.**   The acceleration due to gravity is often denoted by $g$ in a formula such as $S = \frac{1}{2}gt^2$, where $S$ is the distance that an object falls in time $t$. It has to do with the physics of motion near the earth's surface and is usually considered constant. In fact, however, $g$ is not constant, but varies slightly with latitude. If $\phi$ stands for latitude, in degrees, $g$ is given with good approximation by the formula

$$g = 9.78049(1 + 0.005288 \sin^2 \phi - 0.000006 \sin^2 2\phi),$$

where $g$ is measured in meters per second per second at sea level.

**a)** Chicago has latitude 42°N. Find $g$.

**b)** Philadelphia has latitude 40°N. Find $g$.

**c)** Express $g$ in terms of $\sin \phi$ only. That is, eliminate the double angle.

**d)** Where on earth is $g$ greatest? least?

● **CHALLENGE** _____

**57.** Graph: $f(x) = \cos^2 x - \sin^2 x$.

**58.** Graph: $f(x) = |\sin x \cos x|$.

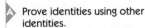

# 7.5

# PROVING TRIGONOMETRIC IDENTITIES

## Basic Identities

We have proved a great many identities in the last chapter and in this chapter. Here we consider the proving of identities more intensely. You will be amazed how many trigonometric identities we can prove, and how the proofs of some identities enable us to prove others. The proofs we considered in Sections 7.1–7.4 are good examples.

A first step in learning to prove identities is to have at hand as many identities as possible that you have already learned. The following list contains many such identities. Although it is not necessary to learn them all, the more you learn, the better your identity-proving skill will be.

---

**Basic Identities**

$$\sin x = \frac{1}{\csc x} \qquad \csc x = \frac{1}{\sin x} \qquad \tan x = \frac{\sin x}{\cos x}$$

$$\cos x = \frac{1}{\sec x} \qquad \sec x = \frac{1}{\cos x} \qquad \cot x = \frac{\cos x}{\sin x}$$

$$\tan x = \frac{1}{\cot x} \qquad \cot x = \frac{1}{\tan x}$$

$$\sin(-x) = -\sin x \qquad \cos(-x) = \cos x \qquad \tan(-x) = -\tan x$$

$$\csc(-x) = -\csc x \qquad \sec(-x) = \sec x \qquad \cot(-x) = -\cot x$$

---

**Pythagorean Identities**

$$\sin^2 x + \cos^2 x = 1$$

$$1 + \tan^2 x = \sec^2 x$$

$$1 + \cot^2 x = \csc^2 x$$

---

**Cofunction Identities**

$$\sin\left(x \pm \frac{\pi}{2}\right) = \pm\cos x \qquad \sin\left(\frac{\pi}{2} \pm x\right) = \cos x$$

$$\cos\left(x \pm \frac{\pi}{2}\right) = \mp\sin x \qquad \cos\left(\frac{\pi}{2} \pm x\right) = \mp\sin x$$

---

**Sum and Difference Identities**

$$\sin(\alpha \pm \beta) = \sin\alpha\cos\beta \pm \cos\alpha\sin\beta$$

$$\cos(\alpha \pm \beta) = \cos\alpha\cos\beta \mp \sin\alpha\sin\beta$$

$$\tan(\alpha \pm \beta) = \frac{\tan\alpha \pm \tan\beta}{1 \mp \tan\alpha\tan\beta}$$

**Double-Angle Identities**

$$\sin 2x = 2 \sin x \cos x$$

$$\cos 2x = \cos^2 x - \sin^2 x$$

$$= 1 - 2 \sin^2 x$$

$$= 2 \cos^2 x - 1$$

$$\tan 2x = \frac{2 \tan x}{1 - \tan^2 x}$$

$$\sin^2 x = \frac{1 - \cos 2x}{2}$$

$$\cos^2 x = \frac{1 + \cos 2x}{2}$$

$$\tan^2 x = \frac{1 - \cos 2x}{1 + \cos 2x}$$

**Half-Angle Identities**

$$\sin \frac{x}{2} = \pm \sqrt{\frac{1 - \cos x}{2}}$$

$$\cos \frac{x}{2} = \pm \sqrt{\frac{1 + \cos x}{2}}$$

$$\tan \frac{x}{2} = \pm \sqrt{\frac{1 - \cos x}{1 + \cos x}}$$

$$= \frac{\sin x}{1 + \cos x}$$

$$= \frac{1 - \cos x}{\sin x}$$

## Proving Identities

### The Logic of Proving Identities

We now outline two methods for proving identities.

*Method 1.   Start with either the left or the right side of an identity and deduce the other side.* For example, suppose you are trying to prove that the equation $P = Q$ is an identity. You might try to produce a string of statements like the following, which start at $P$ and end with $Q$:

$$P = S_1$$
$$= S_2$$
$$\vdots$$
$$= Q.$$

*Method 2.   Work with each side separately until you deduce the same expression.* For example, suppose you are trying to prove that $P = Q$ is an identity. You might be able to produce two strings of statements like the following, each ending with the same statement $S$.

$$P = S_1 \qquad Q = P_1$$
$$= S_2 \qquad\quad = P_2$$
$$\vdots \qquad\qquad \vdots$$
$$= S. \qquad\quad = S.$$

The number of steps in each string might be different, but in each case the result is $S$.

Prove each of the following identities.

**1.** $\dfrac{\csc t - 1}{t \csc t} = \dfrac{1 - \sin t}{t}$

**2.** $\sec \theta \csc \theta = \tan \theta + \cot \theta$

Prove each of the following identities.

**3.** $(\sin \theta - \cos \theta)^2 = 1 - \sin 2\theta$

**4.** $(\sec u - \tan u)(1 + \sin u) = \cos u$

Examples 1–3 illustrate Method 1, and Examples 4 and 5 illustrate Method 2.

**Example 1**  Prove the identity

$$\frac{\sec t - 1}{t \sec t} = \frac{1 - \cos t}{t}.$$

**Proof.**  We use Method 1, starting with the left side and deducing the right side. We note that the left side involves $\sec t$, whereas the right side involves $\cos t$, so it might be wise to make use of an identity that involves these two expressions. That basic identity is $\sec t = 1/\cos t$.

$$\frac{\sec t - 1}{t \sec t} = \frac{\dfrac{1}{\cos t} - 1}{t \dfrac{1}{\cos t}} \qquad \text{Substituting } \frac{1}{\cos t} \text{ for } \sec t$$

$$= \frac{\left(\dfrac{1}{\cos t} - 1\right)}{\dfrac{t}{\cos t}} \cdot \frac{\cos t}{\cos t} \qquad \text{Multiplying by 1 in order to simplify the complex rational expression}$$

$$= \frac{1 - \cos t}{t} \qquad \text{Multiplying and simplifying}$$

We started with the left side and deduced the right side, so the proof is complete. ◀

**DO EXERCISES 1 AND 2.**

**Example 2**  Prove the identity $1 + \sin 2\theta = (\sin \theta + \cos \theta)^2$.

**Proof.**  We again use Method 1. This time we start with the right side and deduce the left side.

$(\sin \theta + \cos \theta)^2$

$= \sin^2 \theta + 2 \sin \theta \cos \theta + \cos^2 \theta \qquad \text{Squaring}$

$= 1 + 2 \sin \theta \cos \theta \qquad \text{Recalling the identity } \sin^2 \theta + \cos^2 \theta = 1 \text{ and substituting}$

$= 1 + \sin 2\theta \qquad \text{Recalling the identity } \sin 2\theta = 2 \sin \theta \cos \theta \text{ and substituting}$

We started with the right side and deduced the left side, so the proof is complete. ◀

**DO EXERCISES 3 AND 4.**

**Example 3**  Prove the identity

$$\frac{\sin \theta - \cos \theta}{\sin \theta + \cos \theta} = -\frac{\cos 2\theta}{1 + \sin 2\theta}.$$

**Proof.**  We start with the left side and deduce the right side. In the first step, we multiply by 1, where the symbol for 1 is formed from the conjugate of the numerator of the original expression.

$$\frac{\sin\theta - \cos\theta}{\sin\theta + \cos\theta} = \frac{\sin\theta - \cos\theta}{\sin\theta + \cos\theta} \cdot \frac{\sin\theta + \cos\theta}{\sin\theta + \cos\theta}$$    Multiplying by 1

$$= \frac{\sin^2\theta - \cos^2\theta}{(\sin\theta + \cos\theta)^2}$$

$$= \frac{\sin^2\theta - \cos^2\theta}{1 + \sin 2\theta}$$    Substituting, using the identity of Example 2

$$= \frac{-(\cos^2\theta - \sin^2\theta)}{1 + \sin 2\theta}$$

$$= -\frac{\cos 2\theta}{1 + \sin 2\theta}$$    Recalling the identity $\cos^2\theta - \sin^2\theta = \cos 2\theta$ and substituting

We started with the left side and deduced the right side, so the proof is complete.

**DO EXERCISE 5.**

**5.** Prove the identity

$$\frac{\sin t}{1 + \cos t} = \frac{1 - \cos t}{\sin t}.$$

**Example 4**    Prove the identity $\tan^2 x - \sin^2 x = \sin^2 x \tan^2 x$.

*Proof.*    For this proof, we are going to work with each side separately using Method 2. We try to deduce the same expression. In practice, when carrying out this method of proof, you might work on one side for awhile, then work on the other side separately, and then go back to the other side. That is, you bounce back and forth until you arrive at the same expression. Let us start with the left side:

$$\tan^2 x - \sin^2 x = \frac{\sin^2 x}{\cos^2 x} - \sin^2 x$$    Recalling the identity $\tan x = \frac{\sin x}{\cos x}$ and substituting

$$= \frac{\sin^2 x}{\cos^2 x} - \sin^2 x \cdot \frac{\cos^2 x}{\cos^2 x}$$    Multiplying by 1 in order to subtract

$$= \frac{\sin^2 x - \sin^2 x \cos^2 x}{\cos^2 x}$$    Carrying out the subtraction

$$= \frac{\sin^2 x (1 - \cos^2 x)}{\cos^2 x}$$    Factoring

$$= \frac{\sin^2 x \sin^2 x}{\cos^2 x}$$    Recalling the identity $\sin^2 x + \cos^2 x = 1$ or $1 - \cos^2 x = \sin^2 x$ and substituting

$$= \frac{\sin^4 x}{\cos^2 x}.$$

At this point, we stop and work with the right side of the original identity and try to end with the same expression that we ended with on the left side:

$$\sin^2 x \tan^2 x = \sin^2 x \frac{\sin^2 x}{\cos^2 x}$$    Recalling the identity $\tan x = \frac{\sin x}{\cos x}$ and substituting

$$= \frac{\sin^4 x}{\cos^2 x}.$$

6. Prove the identity
$$\cot^2 x - \cos^2 x = \cos^2 x \cot^2 x.$$

From each side we have deduced the same expression, so the proof is complete. ◀

**DO EXERCISE 6.** _____

**Example 5** Prove the identity
$$\frac{\sin 2\theta}{\sin \theta} - \frac{\cos 2\theta}{\cos \theta} = \sec \theta.$$

*Proof.* We are again using Method 2, beginning with the left side:

$$\frac{\sin 2\theta}{\sin \theta} - \frac{\cos 2\theta}{\cos \theta}$$

$$= \frac{2 \sin \theta \cos \theta}{\sin \theta} - \frac{\cos^2 \theta - \sin^2 \theta}{\cos \theta} \qquad \begin{array}{l}\text{Using the identities} \\ \sin 2\theta = 2 \sin \theta \cos \theta \text{ and} \\ \cos 2\theta = \cos^2 \theta - \sin^2 \theta \\ \text{and substituting}\end{array}$$

$$= 2 \cos \theta - \frac{\cos^2 \theta - \sin^2 \theta}{\cos \theta} \qquad \text{Simplifying}$$

$$= \frac{2 \cos^2 \theta}{\cos \theta} - \frac{\cos^2 \theta - \sin^2 \theta}{\cos \theta} \qquad \begin{array}{l}\text{Multiplying } 2 \cos \theta \text{ by 1,} \\ \text{or } \cos \theta / \cos \theta\end{array}$$

$$= \frac{2 \cos^2 \theta - \cos^2 \theta + \sin^2 \theta}{\cos \theta} \qquad \text{Carrying out the subtraction}$$

$$= \frac{\cos^2 \theta + \sin^2 \theta}{\cos \theta}$$

$$= \frac{1}{\cos \theta}. \qquad \begin{array}{l}\text{Recalling the identity} \\ \sin^2 \theta + \cos^2 \theta = 1\end{array}$$

7. Prove the identity
$$\frac{\sin 2\theta + \sin \theta}{\cos 2\theta + \cos \theta + 1} = \tan \theta.$$

At this point, we stop and work with the right side of the original identity and try to end with the same expression:

$$\sec \theta = \frac{1}{\cos \theta}. \qquad \text{Recalling a basic identity}$$

From each side we have deduced the same expression, so the proof is complete. ◀

**DO EXERCISE 7.** _____

---

**Hints for Providing Identities**

1. Use Methods 1 or 2 previously outlined.
2. Work with the more complex side first.
3. Carry out the algebraic manipulations, such as adding, subtracting, multiplying, or factoring.
4. Multiplying by 1 can often be helpful when rational expressions are involved.
5. Converting all expressions to sines and cosines is often helpful.
6. Try something! Put your pencil to work and get involved. You will be amazed at how often this leads to success.

## Sum–Product Identities

On occasion, it is convenient to convert a product of trigonometric expressions to a sum, or the reverse. The following identities are useful in this connection. Proofs are left as exercises.

---

**Sum–Product Identities**

$$\sin u \cdot \cos v = \frac{1}{2}[\sin(u+v) + \sin(u-v)]$$

$$\cos u \cdot \sin v = \frac{1}{2}[\sin(u+v) - \sin(u-v)]$$

$$\cos u \cdot \cos v = \frac{1}{2}[\cos(u-v) + \cos(u+v)]$$

$$\sin u \cdot \sin v = \frac{1}{2}[\cos(u-v) - \cos(u+v)]$$

$$\sin x + \sin y = 2 \sin\frac{x+y}{2} \cos\frac{x-y}{2}$$

$$\sin x - \sin y = 2 \cos\frac{x+y}{2} \sin\frac{x-y}{2}$$

$$\cos y + \cos x = 2 \cos\frac{x+y}{2} \cos\frac{x-y}{2}$$

$$\cos y - \cos x = 2 \sin\frac{x+y}{2} \sin\frac{x-y}{2}$$

---

**DO EXERCISE 8.**

**Example 6**  Find an identity for $\cos\theta + \cos 5\theta$.

**Solution**  We will use the identity

$$\cos y + \cos x = 2 \cos\frac{x+y}{2} \cos\frac{x-y}{2}.$$

Here $x = 5\theta$ and $y = \theta$. Thus,

$$\cos\theta + \cos 5\theta = 2 \cos\frac{5\theta + \theta}{2} \cos\frac{5\theta - \theta}{2}$$

$$= 2 \cos 3\theta \cos 2\theta. \qquad \triangleleft$$

**DO EXERCISE 9.**

---

8. Prove the first identity in the list of sum–product identities. (*Hint:* Start with the right side and use the sine of a sum and the sine of a difference.)

9. Find an identity for

$$2 \sin 7\theta \cos 5\theta.$$

---

 **TECHNOLOGY CONNECTION**

Almost any identity can be checked on a grapher. The exception is an identity that involves two different angles, such as $u$ and $v$, or $x$ and $y$, in the sum–product identities. Because we must enter each side of the identity as a function of the single variable $x$, there is no way to graph any function that has two variables.

Match each expression in column A with the expression from column B that would result in an identity.

| COLUMN A | COLUMN B |
|---|---|
| **TC 1.** $1 - \dfrac{\sin^2 x}{1 + \cos x}$ | A. $\sin x$ |
| **TC 2.** $\dfrac{\sin x \sec x}{1 + \tan^2 x}$ | B. $\sin x \cos x + 1$ |
| **TC 3.** $\frac{1}{2}\sec^2 x \cos x \sin 2x$ | C. $\cos x$ |
| **TC 4.** $\dfrac{(\sin x \cos x + 1)(\cos^2 x \tan x - 1)}{\cos^2 x \tan x - 1}$ | D. $\dfrac{\cos x}{\csc x}$ |

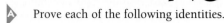

**•EXERCISE SET 7.5**

Prove each of the following identities.

**1.** $\csc x - \cos x \cot x = \sin x$

**2.** $\sec x - \sin x \tan x = \cos x$

**3.** $\dfrac{1 + \cos \theta}{\sin \theta} + \dfrac{\sin \theta}{\cos \theta} = \dfrac{\cos \theta + 1}{\sin \theta \cos \theta}$

**4.** $\dfrac{1}{\sin \theta \cos \theta} - \dfrac{\cos \theta}{\sin \theta} = \dfrac{\sin \theta \cos \theta}{1 - \sin^2 \theta}$

**5.** $\dfrac{1 - \sin x}{\cos x} = \dfrac{\cos x}{1 + \sin x}$

**6.** $\dfrac{1 - \cos x}{\sin x} = \dfrac{\sin x}{1 + \cos x}$

**7.** $\dfrac{1 + \tan \theta}{1 + \cot \theta} = \dfrac{\sec \theta}{\csc \theta}$

**8.** $\dfrac{\cot \theta - 1}{1 - \tan \theta} = \dfrac{\csc \theta}{\sec \theta}$

**9.** $\dfrac{\sin x + \cos x}{\sec x + \csc x} = \dfrac{\sin x}{\sec x}$

**10.** $\dfrac{\sin x - \cos x}{\sec x - \csc x} = \dfrac{\cos x}{\csc x}$

**11.** $\dfrac{1 + \tan \theta}{1 - \tan \theta} + \dfrac{1 + \cot \theta}{1 - \cot \theta} = 0$

**12.** $\dfrac{\cos^2 \theta + \cot \theta}{\cos^2 \theta - \cot \theta} = \dfrac{\cos^2 \theta \tan \theta + 1}{\cos^2 \theta \tan \theta - 1}$

**13.** $\dfrac{1 + \cos 2\theta}{\sin 2\theta} = \cot \theta$

**14.** $\dfrac{2 \tan \theta}{1 + \tan^2 \theta} = \sin 2\theta$

**15.** $\sec 2\theta = \dfrac{\sec^2 \theta}{2 - \sec^2 \theta}$

**16.** $\cot 2\theta = \dfrac{\cot^2 \theta - 1}{2 \cot \theta}$

**17.** $\dfrac{\sin (\alpha + \beta)}{\cos \alpha \cos \beta} = \tan \alpha + \tan \beta$

**18.** $\dfrac{\cos (\alpha - \beta)}{\cos \alpha \sin \beta} = \tan \alpha + \cot \beta$

**19.** $1 - \cos 5\theta \cos 3\theta - \sin 5\theta \sin 3\theta = 2 \sin^2 \theta$

**20.** $2 \sin \theta \cos^3 \theta + 2 \sin^3 \theta \cos \theta = \sin 2\theta$

**21.** $\dfrac{\tan \theta + \sin \theta}{2 \tan \theta} = \cos^2 \dfrac{\theta}{2}$

**22.** $\dfrac{\tan \theta - \sin \theta}{2 \tan \theta} = \sin^2 \dfrac{\theta}{2}$

**23.** $\cos^4 x - \sin^4 x = \cos 2x$

**24.** $\dfrac{\cos^4 x - \sin^4 x}{1 - \tan^4 x} = \cos^4 x$

**25.** $\dfrac{\tan 3\theta - \tan \theta}{1 + \tan 3\theta \tan \theta} = \dfrac{2 \tan \theta}{1 - \tan^2 \theta}$

**26.** $\left(\dfrac{1 + \tan \theta}{1 - \tan \theta}\right)^2 = \dfrac{1 + \sin 2\theta}{1 - \sin 2\theta}$

**27.** $\dfrac{\cos^3 x - \sin^3 x}{\cos x - \sin x} = \dfrac{2 + \sin 2x}{2}$

**28.** $\dfrac{\sin^3 t + \cos^3 t}{\sin t + \cos t} = \dfrac{2 - \sin 2t}{2}$

**29.** $\sin (\alpha + \beta) \sin (\alpha - \beta) = \sin^2 \alpha - \sin^2 \beta$

**30.** $\cos (\alpha + \beta) \cos (\alpha - \beta) = \cos^2 \alpha - \sin^2 \beta$

**31.** $\cos (\alpha + \beta) + \cos (\alpha - \beta) = 2 \cos \alpha \cos \beta$

**32.** $\sin (\alpha + \beta) + \sin (\alpha - \beta) = 2 \sin \alpha \cos \beta$

**33.** $\sin^2 x - \cos^2 x = 1 - 2 \cos^2 x$

**34.** $\cos^2 x (1 - \sec^2 x) = -\sin^2 x$

**35.** $\sin^2 \theta = \cos^2 \theta (\sec^2 \theta - 1)$

**36.** $\tan \theta (\tan \theta + \cot \theta) = \sec^2 \theta$

**37.** $\dfrac{\tan x}{\sec x - \cos x} = \dfrac{\sec x}{\tan x}$

**38.** $\dfrac{\tan x + \sin x}{1 + \sec x} = \sin x$

**39.** $\dfrac{\cos \theta + \sin \theta}{\cos \theta} = 1 + \tan \theta$

**40.** $\dfrac{1 + \tan^2 \theta}{\csc^2 \theta} = \tan^2 \theta$

**41.** $\dfrac{\tan^2 x}{1 + \tan^2 x} = \sin^2 x$

**42.** $\dfrac{1 + \cos^2 x}{\sin^2 x} = 2 \csc^2 x - 1$

**43.** $\dfrac{\cot x + \tan x}{\csc x} = \sec x$

**44.** $\dfrac{\csc x - \sin x}{\cot x} = \cos x$

**45.** $\dfrac{\tan \theta + \cot \theta}{\csc \theta} = \sec \theta$

**46.** $\dfrac{\csc \theta - \sin \theta}{\cos^2 \theta} = \csc \theta$

**47.** $\dfrac{\sin x}{1 + \cos x} + \dfrac{1 + \cos x}{\sin x} = 2 \csc x$

**48.** $\dfrac{1 + \sin x}{1 - \sin x} + \dfrac{\sin x - 1}{1 + \sin x} = 4 \sec x \tan x$

**49.** $\cos \theta (1 + \csc \theta) - \cot \theta = \cos \theta$

**50.** $\cos^2 \theta \cot^2 \theta = \cot^2 \theta - \cos^2 \theta$

**51.** $\dfrac{\tan x + \cot x}{\sec x + \csc x} = \dfrac{1}{\cos x + \sin x}$

**52.** $\dfrac{\cos^2 x - 1}{1 - \sec^2 x} = \dfrac{1}{\tan^2 x + 1}$

**53.** $(\sec \theta - \tan \theta)(1 + \csc \theta) = \cot \theta$

**54.** $\tan \theta - \cot \theta = (\sec \theta - \csc \theta)(\sin \theta + \cos \theta)$

**55.** $(\sec x + \tan x)(1 - \sin x) = \cos x$

**56.** $\csc x - \cot x = \dfrac{1}{\csc x + \cot x}$

**57.** $\cos^2 \theta - \sin^2 \theta = \cos^4 \theta - \sin^4 \theta$

**58.** $2 \sin^2 \theta \cos^2 \theta + \cos^4 \theta = 1 - \sin^4 \theta$

**59.** $\dfrac{\cos \theta}{1 - \cos \theta} = \dfrac{1 + \sec \theta}{\tan^2 \theta}$

**60.** $\dfrac{\cot \theta}{\csc \theta - 1} = \dfrac{\csc \theta + 1}{\cot \theta}$

**61.** $\dfrac{\sin x - \cos x}{\cos^2 x} = \dfrac{\tan^2 x - 1}{\sin x + \cos x}$

**62.** $\dfrac{1 + \sin x}{1 - \sin x} = (\sec x + \tan x)^2$

**63.** $1 + \sec^4 \theta = \tan^4 \theta + 2 \sec^2 \theta$

**64.** $\sec^4 \theta - \tan^2 \theta = \tan^4 \theta + \sec^2 \theta$

**65.** $\dfrac{1 + \tan x}{1 - \tan x} = \dfrac{\cot x + 1}{\cot x - 1}$

**66.** $1 + \sec x = \csc x (\sin x + \tan x)$

**67.** $\dfrac{\sin \theta \tan \theta}{\sin \theta + \tan \theta} = \dfrac{\tan \theta - \sin \theta}{\tan \theta \sin \theta}$

**68.** $\dfrac{\sin^3 \theta - \cos^3 \theta}{\sin \theta - \cos \theta} = \sin \theta \cos \theta + 1$

**69.** $\dfrac{\cos x + 1}{\sin x} + \dfrac{\sin x}{\cos x + 1} = \dfrac{2}{\sin x}$

**70.** $2 \sec^2 x + \dfrac{\csc x}{1 - \csc x} = \dfrac{\csc x}{\csc x + 1}$

**71.** $\sec \theta \csc \theta + \dfrac{\cot \theta}{\tan \theta - 1} = \dfrac{\tan \theta}{1 - \cot \theta} - 1$

**72.** $\cos \theta + \dfrac{\sin \theta}{\cot \theta - 1} = \dfrac{\cos \theta}{1 - \tan \theta} - \sin \theta$

**73.** $\cot x + \csc x = \dfrac{\sin x}{1 - \cos x}$

**74.** $\dfrac{\cos x + \cot x}{1 + \csc x} = \cos x$

**75.** $\tan \theta + \cot \theta = \dfrac{1}{\cot \theta \sin^2 \theta}$

**76.** $\sec^2 \theta - \csc^2 \theta = \dfrac{\tan \theta - \cot \theta}{\sin \theta \cos \theta}$

**77.** $2 \cos^2 x - 1 = \cos^4 x - \sin^4 x$

**78.** $\sec^4 x - 4 \tan^2 x = (1 - \tan^2 x)^2$

**79.** $(\cos x + \sin x)(1 - \sin x \cos x) = \cos^3 x + \sin^3 x$

**80.** $\dfrac{\tan^2 x + \sec^2 x}{\sec^4 x} = 1 - \sin^4 x$

**81.** Prove the second, third, and fourth of the sum–product identities using the sum and difference formulas for sine and cosine.

**82.** Prove the last four of the sum–product identities. (*Hint:* Use the first four sum–product identities.)

Use the sum–product identities to find identities for each of the following.

**83.** $\sin 3\theta - \sin 5\theta$

**84.** $\sin 7x - \sin 4x$

**85.** $\sin 8\theta + \sin 5\theta$

**86.** $\cos \theta - \cos 7\theta$

**87.** $\sin 7u \sin 5u$

**88.** $2 \sin 7\theta \cos 3\theta$

**89.** $7 \cos \theta \sin 7\theta$

**90.** $\cos 2t \sin t$

**91.** $\cos 55° \sin 25°$

**92.** $7 \cos 5\theta \cos 7\theta$

Use the sum–product identities to prove each of the following.

**93.** $\sin 4\theta + \sin 6\theta = \cot \theta (\cos 4\theta - \cos 6\theta)$

**94.** $\tan 2x (\cos x + \cos 3x) = \sin x + \sin 3x$

**95.** $\cot 4x(\sin x + \sin 4x + \sin 7x)$
$= \cos x + \cos 4x + \cos 7x$

**96.** $\tan \dfrac{x + y}{2} = \dfrac{\sin x + \sin y}{\cos x + \cos y}$

**97.** $\cot \dfrac{x + y}{2} = \dfrac{\sin y - \sin x}{\cos x - \cos y}$

**98.** $\tan \dfrac{\theta + \phi}{2} \tan \dfrac{\phi - \theta}{2} = \dfrac{\cos \theta - \cos \phi}{\cos \theta + \cos \phi}$

**99.** $\tan \dfrac{\theta + \phi}{2} (\sin \theta - \sin \phi)$

$= \tan \dfrac{\theta - \phi}{2} (\sin \theta + \sin \phi)$

**100.** $\sin 2\theta + \sin 4\theta + \sin 6\theta = 4 \cos \theta \cos 2\theta \sin 3\theta$

● **SYNTHESIS** _____

Prove each of the following identities. If you have a grapher, verify your results.

**101.** $\ln |\sec x| = -\ln |\cos x|$

**102.** $\ln |\tan x| = -\ln |\cot x|$

**103.** $\ln |\tan x| = \ln |\sin x| - \ln |\cos x|$

**104.** $\ln |\csc x| = -\ln |\sin x|$

**105.** $\ln e^{\sin t} = \sin t$

**106.** $e^{\ln |\cos t|} = |\cos t|$

**107.** $\ln |\csc \theta - \cot \theta| = -\ln |\csc \theta + \cot \theta|$

**108.** $\ln |\sec \theta + \tan \theta| = -\ln |\sec \theta - \tan \theta|$

**109.** Show that $\log (\cos x - \sin x) + \log (\cos x + \sin x) = \log \cos 2x$.

**110.** The following equation occurs in the study of mechanics:

$$\sin\theta = \frac{I_1\cos\phi}{\sqrt{(I_1\cos\phi)^2 + (I_2\sin\phi)^2}}.$$

It can happen that $I_1 = I_2$. Assuming that this happens, simplify the equation.

**111.** In the theory of alternating current, the following equation occurs:

$$R = \frac{1}{\omega C(\tan\theta + \tan\phi)}.$$

Show that this equation is equivalent to

$$R = \frac{\cos\theta\cos\phi}{\omega C\sin(\theta + \phi)}.$$

● CHALLENGE

**112.** In electrical theory, the following equations occur:

$$E_1 = \sqrt{2}E_t\cos\left(\theta + \frac{\pi}{P}\right)$$

and

$$E_2 = \sqrt{2}E_t\cos\left(\theta - \frac{\pi}{P}\right).$$

Assuming that these equations hold, show that

$$\frac{E_1 + E_2}{2} = \sqrt{2}E_t\cos\theta\cos\frac{\pi}{P}$$

and

$$\frac{E_1 - E_2}{2} = -\sqrt{2}E_t\sin\theta\sin\frac{\pi}{P}.$$

---

# 7.6

# INVERSES OF THE TRIGONOMETRIC FUNCTIONS

In this section, we develop inverse trigonometric functions. It may be helpful for you to review the material on inverse functions in Section 5.1.

The graphs of the sine, cosine, tangent, and cotangent functions follow. Do these functions have inverses that are functions? They do if these functions are one-to-one, which means that they pass the horizontal-line test.

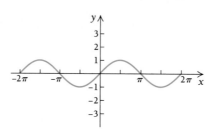

The sine function

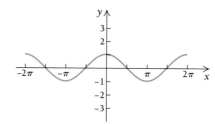

The cosine function

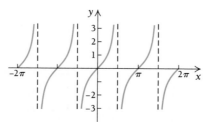

The tangent function

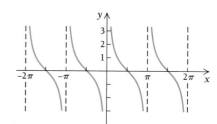

The cotangent function

Note that each function has a horizontal line that crosses the graph more than once. Therefore, none of them has an inverse that is a function.

Recall that to obtain the inverse of any relation, we interchange the first and second members of each ordered pair in the relation. If a relation is defined in terms of, say, $x$ and $y$, interchanging $x$ and $y$ produces an equation of the inverse relation. The graphs of an equation and its inverse are reflections of each other across the line $y = x$. Let us examine the inverses of each of the four trigonometric functions graphed above. The graphs are as follows.

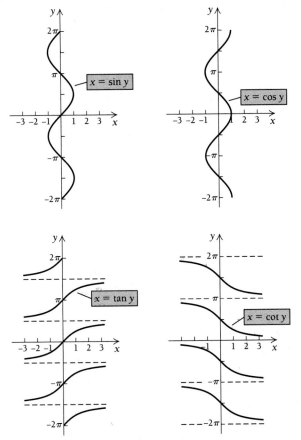

We can check again to see whether these are graphs of functions by using the vertical-line test. In each case, there is a vertical line that crosses the graph more than once, so each fails to be a function.

Let us consider specifically the graph of the inverse of $y = \sin x$, which is $x = \sin y$. Consider the input $x = 1/2$. On the graph of the inverse of the sine function, we draw a vertical line at $x = 1/2$, as shown.

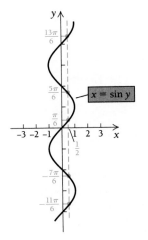

It intersects the graph at points whose y-value is such that $1/2 = \sin y$. Some numbers whose sine is $1/2$ are $\pi/6, 5\pi/6, -7\pi/6$, and so on. From the graph, we see that $\pi/6$ plus any multiple of $2\pi$ is such a number. Also, $5\pi/6$ plus any multiple of $2\pi$ is such a number. The complete set of values is given by $\pi/6 + 2k\pi$, $k$ an integer, and $5\pi/6 + 2k\pi$, $k$ an integer. Indeed, the vertical line crosses the curve at many points, verifying that the inverse of the sine function is not a function.

## Restricting Ranges to Define Inverse Functions

Recall that a function like $f(x) = x^2$ does not have an inverse that is a function, but by restricting the domain of $f$ to nonnegative numbers, we have a new squaring function, $f(x) = x^2$, $x \geq 0$, that has an inverse, $f^{-1}(x) = \sqrt{x}$, $x \geq 0$. This is equivalent to restricting the range of the inverse relation to exclude ordered pairs that contain negative numbers.

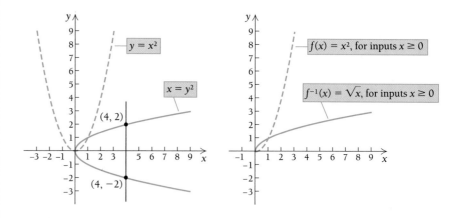

In a similar manner, we can define new trigonometric functions whose inverses are functions. We do this by restricting the ranges of the inverses. This can be done in many ways, but we choose restrictions that are fairly standard in mathematics. For the inverse sine function, we choose a range close to the origin that allows all inputs in the interval $[-1, 1]$ to have function values in the interval $[-\pi/2, \pi/2]$. For the inverse cosine function, we choose a range close to the origin that allows all inputs in the interval $[-1, 1]$ to have function values in the interval $[0, \pi]$. For the inverse tangent function, the domain is the set of all real numbers and the range is the interval $(-\pi/2, \pi/2)$. For the inverse cotangent function, the domain is the set of all real numbers and the range is the interval $(0, \pi)$.

We denote the inverse trigonometric functions as follows:

---

**DEFINITION**                    **The Inverse Sine Function**

$y = \sin^{-1} x$,   or   $y = \arcsin x$,   where $x = \sin y$.

The domain $= [-1, 1]$ and the range $= [-\pi/2, \pi/2]$.

---

The notation $\arcsin x$ arises because it is the length of an arc on the unit circle for which the sine is $x$. The notation $\sin^{-1} x$ is *not* exponential notation. It does not mean $1/\sin x$! Either of the two kinds of notation above can be read "the inverse sine of $x$" or "the arc sine of $x$" or "the number (or angle) whose sine is $x$." Notation is chosen similarly for the inverse of the other trigonometric functions: $\cos^{-1} x$, or arccos $x$, $\tan^{-1} x$, or arctan $x$, and so on.

The following diagrams show the restricted ranges for the inverse trigonometric functions on a unit circle. These ranges should be memorized. The missing endpoints indicate inputs that are not in the domain of the original function.

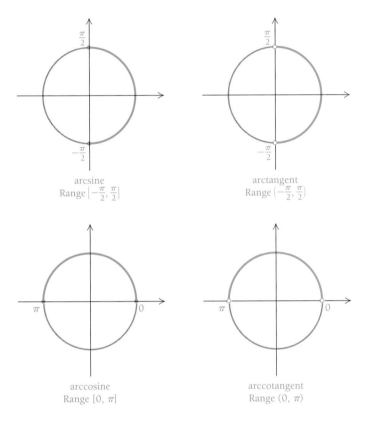

arcsine
Range $\left[-\frac{\pi}{2}, \frac{\pi}{2}\right]$

arctangent
Range $\left(-\frac{\pi}{2}, \frac{\pi}{2}\right)$

arccosine
Range $[0, \pi]$

arccotangent
Range $(0, \pi)$

The graphs of the inverse trigonometric functions are as follows.

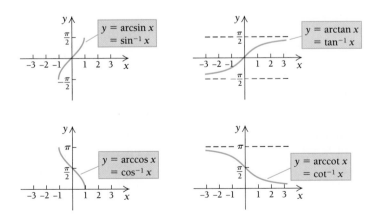

$y = \arcsin x$
$= \sin^{-1} x$

$y = \arctan x$
$= \tan^{-1} x$

$y = \arccos x$
$= \cos^{-1} x$

$y = \text{arccot } x$
$= \cot^{-1} x$

**DO EXERCISES 1–4. (EXERCISE 4 IS ON THE FOLLOWING PAGE.)**

In Margin Exercises 1–4, the graph shown is an inverse of a trigonometric function. In each case, shade the restricted range so the inverse is a function.

1.

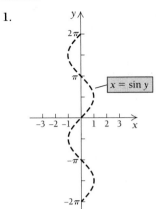

$x = \sin y$

2.

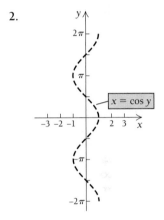

$x = \cos y$

3.

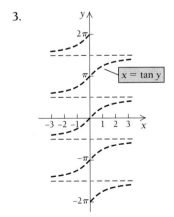

$x = \tan y$

**4.**

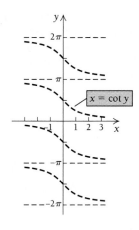

Find each of the following.

5. $\arcsin \dfrac{\sqrt{3}}{2}$

6. $\cos^{-1}\left(-\dfrac{\sqrt{2}}{2}\right)$

7. $\operatorname{arccot}(-1)$

8. $\tan^{-1}(-1)$

Now let us find some function values.

**Example 1**   Find $\sin^{-1}\frac{1}{2}$.

**Solution**   In the restricted range, as shown in the figure, the only number whose sine is $\frac{1}{2}$ is $\pi/6$. Therefore, $\sin^{-1}\frac{1}{2} = \pi/6$. In degrees, we have $\sin^{-1}\frac{1}{2} = 30°$.

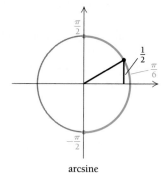

arcsine

**Example 2**   Find $\arcsin \dfrac{\sqrt{2}}{2}$.

**Solution**   In the restricted range, as shown in the figure, the only number whose sine is $\sqrt{2}/2$ is $\pi/4$. Thus, $\arcsin(\sqrt{2}/2) = \pi/4$, or $45°$.

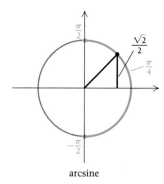

arcsine

**Example 3**   Find $\cos^{-1}(-\frac{1}{2})$.

**Solution**   The only number whose cosine is $-\frac{1}{2}$ in the restricted range $[0, \pi]$ is $2\pi/3$. Thus, $\cos^{-1}(-\frac{1}{2}) = 2\pi/3$, or $120°$.

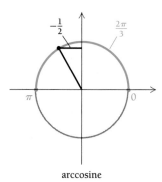

arccosine

**Example 4**   Find $\tan^{-1} 1$.

**Solution**   The only number in the restricted range $(-\pi/2, \pi/2)$ whose tangent is 1 is $\pi/4$, or $45°$.

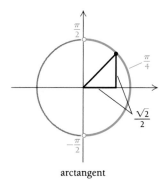

arctangent

**DO EXERCISES 5–8.**

We can also use a calculator to find inverse trigonometric function values. Some calculators give inverse function values in both radians and degrees. Some give values only in degrees. The key strokes involved in finding inverse function values vary with the calculator. Be sure to read the instructions for the calculator you are using, or try a simple value such as $\sin^{-1} 1$. If you get 90, you know that the values are in degrees. If you get 1.570796327, which is about $\pi/2$, you know the values are in radians.

**Example 5**   Find $\cos^{-1}(-0.925678)$ in degrees, using a calculator.

**Solution**   Using the inverse trigonometric keys on a calculator, we find that
$$\cos^{-1}(-0.925678) \approx 157.77°.$$

**DO EXERCISES 9–12.**

Find each of the following using a calculator.

9.  $\cos^{-1} 0.63254$

10.  $\arcsin(-0.10203)$

11.  $\arctan 1.34568$

12.  $\tan^{-1}(-22.467)$

The following is a summary of the domains and the ranges of the trigonometric functions together with a summary of the domains and the ranges of the inverse trigonometric functions. For completeness, we have included the arcsecant and the arccosecant, though there is a lack of uniformity on their definitions in the mathematical literature.

| Function | Domain | Range | Inverse Function | Domain | Range |
|---|---|---|---|---|---|
| sin | All reals, $(-\infty, \infty)$ | $[-1, 1]$ | $\sin^{-1}$ | $[-1, 1]$ | $\left[-\dfrac{\pi}{2}, \dfrac{\pi}{2}\right]$ |
| cos | All reals, $(-\infty, \infty)$ | $[-1, 1]$ | $\cos^{-1}$ | $[-1, 1]$ | $[0, \pi]$ |
| tan | All reals except $k\pi/2$, $k$ odd | All reals, $(-\infty, \infty)$ | $\tan^{-1}$ | All reals, $(-\infty, \infty)$ | $\left(-\dfrac{\pi}{2}, \dfrac{\pi}{2}\right)$ |
| cot | All reals except $k\pi$ | All reals, $(-\infty, \infty)$ | $\cot^{-1}$ | All reals, $(-\infty, \infty)$ | $(0, \pi)$ |
| sec | All reals except $k\pi/2$, $k$ odd | $(-\infty, -1] \cup [1, \infty)$ | $\sec^{-1}$ | $(-\infty, -1] \cup [1, \infty)$ | $\left[0, \dfrac{\pi}{2}\right) \cup \left[\pi, \dfrac{3\pi}{2}\right)$ |
| csc | All reals except $k\pi$ | $(-\infty, -1] \cup [1, \infty)$ | $\csc^{-1}$ | $(-\infty, -1] \cup [1, \infty)$ | $\left(0, \dfrac{\pi}{2}\right] \cup \left(-\pi, -\dfrac{\pi}{2}\right]$ |

●**EXERCISE SET** 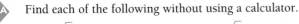 **7.6**

Find each of the following without using a calculator.

1.  $\arcsin \dfrac{\sqrt{2}}{2}$

2.  $\arcsin \dfrac{\sqrt{3}}{2}$

3.  $\cos^{-1} \dfrac{\sqrt{2}}{2}$

4.  $\cos^{-1} \dfrac{\sqrt{3}}{2}$

5.  $\sin^{-1}\left(-\dfrac{\sqrt{2}}{2}\right)$

6.  $\sin^{-1}\left(-\dfrac{\sqrt{3}}{2}\right)$

7.  $\arccos\left(-\dfrac{\sqrt{2}}{2}\right)$

8.  $\arccos\left(-\dfrac{\sqrt{3}}{2}\right)$

**9.** $\arctan \sqrt{3}$

**10.** $\arctan \dfrac{\sqrt{3}}{3}$

**11.** $\cot^{-1} 1$

**12.** $\cot^{-1} \sqrt{3}$

**13.** $\arctan\left(-\dfrac{\sqrt{3}}{3}\right)$

**14.** $\arctan(-\sqrt{3})$

**15.** $\text{arccot}(-1)$

**16.** $\text{arccot}(-\sqrt{3})$

**17.** $\text{arcsec } 1$

**18.** $\text{arcsec } 2$

**19.** $\csc^{-1} 1$

**20.** $\csc^{-1} 2$

**21.** $\arcsin\left(-\dfrac{\sqrt{2}}{2}\right)$

**22.** $\arcsin \dfrac{1}{2}$

**23.** $\cos^{-1} \dfrac{1}{2}$

**24.** $\arccos \dfrac{\sqrt{2}}{2}$

**25.** $\arcsin\left(-\dfrac{\sqrt{3}}{2}\right)$

**26.** $\sin^{-1}\left(-\dfrac{1}{2}\right)$

**27.** $\cos^{-1}\left(-\dfrac{1}{2}\right)$

**28.** $\cos^{-1} 0$

**29.** $\tan^{-1} 1$

**30.** $\tan^{-1} 0$

**31.** $\text{arccot}\left(-\dfrac{\sqrt{3}}{3}\right)$

**32.** $\text{arccot}(-\sqrt{3})$

Use a calculator to find each of the following in degrees.

**33.** $\arcsin 0.3907$

**34.** $\arcsin 0.9613$

**35.** $\sin^{-1}(-0.619867)$

**36.** $\sin^{-1}(-0.867314)$

**37.** $\arccos 0.7990$

**38.** $\arccos 0.9265$

**39.** $\cos^{-1}(-0.981028)$

**40.** $\cos^{-1}(-0.271568)$

**41.** $\tan^{-1} 0.3673$

**42.** $\tan^{-1} 1.091$

**43.** $\cot^{-1} 1.265$

**44.** $\cot^{-1} 0.4770$

**45.** $\sec^{-1} 1.1677$

**46.** $\sec^{-1}(-1.4402)$

**47.** $\text{arccsc}(-6.2774)$

**48.** $\text{arccsc } 1.11123$

**49.** $\arcsin 0.2334$

**50.** $\arcsin 0.4514$

**51.** $\sin^{-1}(-0.6361)$

**52.** $\sin^{-1}(-0.8192)$

**53.** $\arcsin(-0.8886)$

**54.** $\arccos(-0.2935)$

**55.** $\tan^{-1}(-0.4087)$

**56.** $\tan^{-1}(-0.2410)$

**57.** $\cot^{-1}(-5.936)$

**58.** $\cot^{-1}(-1.319)$

**59.** $\cot^{-1}(-23)$

**60.** $\tan^{-1}(158)$

●  **SYNTHESIS** _____

**61.** Graph the function $y = \sec^{-1} x$.

**62.** Graph the function $y = \csc^{-1} x$.

Show that each of the following is *not* an identity.

**63.** $\sin^{-1} x = (\sin x)^{-1}$

**64.** $\cos^{-1} x = (\cos x)^{-1}$

**65.** $\tan^{-1} x = (\tan x)^{-1}$

**66.** $\cot^{-1} x = (\cot x)^{-1}$

**67.** $\tan^{-1} x = \dfrac{\sin^{-1} x}{\cos^{-1} x}$

**68.** $(\cos^{-1} x)^2 + (\sin^{-1} x)^2 = 1$

**69.** A guy wire is attached to the top of a 50-ft pole and stretched to a point that is $b$ ft from the bottom of the pole. Show that the angle of inclination $\theta$ of the wire to the top is given by

$$\theta = \tan^{-1}\left(\dfrac{50}{b}\right).$$

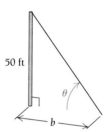

**70.** An airplane at an altitude of 2000 ft is flying toward an island. The straight-line distance from the airplane to the island is $h$ ft. Show that $\theta$, which is called the *angle of depression*, is given by

$$\theta = \sin^{-1}\left(\dfrac{2000}{h}\right).$$

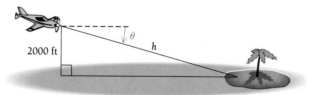

**71. a)** Use your calculator to approximate the following expression:

$$16 \tan^{-1} \dfrac{1}{5} - 4 \tan^{-1} \dfrac{1}{239}.$$

**b)** What number does this expression seem to approximate?

# 7.7
# COMPOSITION OF TRIGONOMETRIC FUNCTIONS AND THEIR INVERSES

## A  Immediate Simplification

Various compositions of trigonometric functions and their inverses often arise in practice. For example, we might want to try to simplify expressions such as

$$\sin(\sin^{-1} x) \quad \text{or} \quad \sin\left(\operatorname{arccot} \frac{x}{2}\right).$$

In the expression on the left, we are finding "the sine of a number whose sine is $x$." Recall from Section 5.1 that if a function $f$ has an inverse that is also a function, then

$$f(f^{-1}(x)) = x, \quad \text{for all } x \text{ in the domain of } f^{-1},$$

and

$$f^{-1}(f(x)) = x, \quad \text{for all } x \text{ in the domain of } f.$$

Thus, if $f(x) = \sin x$ and $f^{-1}(x) = \sin^{-1} x$, then $\sin(\sin^{-1} x) = x$, if $x$ is in the domain of $\sin^{-1}$, which is any number in the interval $[-1, 1]$. Similar results hold for the other trigonometric functions. These yield the following theorem.

---

### THEOREM 1

a) $\sin(\sin^{-1} x) = x$, for any $x$ in the domain of $\sin^{-1}$.
b) $\cos(\cos^{-1} x) = x$, for any $x$ in the domain of $\cos^{-1}$.
c) $\tan(\tan^{-1} x) = x$, for any $x$ in the domain of $\tan^{-1}$.
d) $\cot(\cot^{-1} x) = x$, for any $x$ in the domain of $\cot^{-1}$.
e) $\sec(\sec^{-1} x) = x$, for any $x$ in the domain of $\sec^{-1}$.
f) $\csc(\csc^{-1} x) = x$, for any $x$ in the domain of $\csc^{-1}$.

---

**Example 1**  Find $\cos\left(\cos^{-1} \dfrac{\sqrt{3}}{2}\right)$.

**Solution**  Since $\sqrt{3}/2$ is in the interval $[-1, 1]$, it follows that

$$\cos\left(\cos^{-1} \frac{\sqrt{3}}{2}\right) = \frac{\sqrt{3}}{2}. \qquad \blacktriangleleft$$

**Example 2**  Find $\sin(\sin^{-1} 12.3)$.

**Solution**  Since 12.3 is not in the interval $[-1, 1]$, we cannot evaluate this expression. Thinking one step at a time, we know that there is no number whose sine is 12.3. Since we cannot find $\sin^{-1} 12.3$, we cannot evaluate the expression. $\blacktriangleleft$

**DO EXERCISES 1–4.**

---

**OBJECTIVES**

You should be able to:

**A** Simplify expressions such as $\sin(\sin^{-1} x)$ and $\sin^{-1}(\sin x)$.

**B** Simplify expressions involving compositions such as $\sin(\cos^{-1} \frac{1}{2})$, without using a calculator or a table, and simplify expressions such as $\sin \arctan(a/b)$ by drawing a triangle or triangles and reading off appropriate ratios.

Find each of the following.

1. $\sin\left(\sin^{-1} \dfrac{1}{2}\right)$

2. $\tan(\tan^{-1} 1)$

3. $\cos\left(\arccos \dfrac{\sqrt{2}}{2}\right)$

4. $\sin[\sin^{-1}(-5.7)]$

Find each of the following.

5. $\cos^{-1}\left(\cos\dfrac{2\pi}{3}\right)$

6. $\arctan\left(\tan\dfrac{3\pi}{4}\right)$

7. $\sin^{-1}\left(\sin\dfrac{\pi}{6}\right)$

8. $\arccos\left[\cos\left(-\dfrac{2\pi}{3}\right)\right]$

Find each of the following.

9. $\sin(\arctan 1)$

10. $\cos\left(\arcsin\dfrac{1}{2}\right)$

11. $\cos^{-1}\left(\sin\dfrac{\pi}{6}\right)$

12. $\sin^{-1}\left(\tan\dfrac{\pi}{4}\right)$

13. $\cos(\sin^{-1} 0)$

Now let us consider an expression like $\sin^{-1}(\sin x)$. We might also suspect that this is equal to $x$ for any $x$, but this is not true unless $x$ is in the range of the $\sin^{-1}$ function. Note that in order to define $\sin^{-1}$, we had to restrict the domain of the sine function. In so doing, we restricted the range of the inverse sine function.

**Example 3** Find $\sin^{-1}\left(\sin\dfrac{3\pi}{4}\right)$.

Solution We first find $\sin(3\pi/4)$. It is $\sqrt{2}/2$. Next we find $\sin^{-1}(\sqrt{2}/2)$. It is $\pi/4$. So

$$\sin^{-1}\left(\sin\dfrac{3\pi}{4}\right) = \dfrac{\pi}{4}.$$ ◀

**DO EXERCISES 5–8.**

For any $x$ in the range of the arcsine function, $[-\pi/2, \pi/2]$, we do have $\arcsin(\sin x) = x$, or $\sin^{-1}(\sin x) = x$. Similar conditions hold for the other functions.

---

**THEOREM 2**

The following are true for any $x$ in the range of the inverse function.

a) $\sin^{-1}(\sin x) = x$, for any $x$ in the range of $\sin^{-1}$.
b) $\cos^{-1}(\cos x) = x$, for any $x$ in the range of $\cos^{-1}$.
c) $\tan^{-1}(\tan x) = x$, for any $x$ in the range of $\tan^{-1}$.
d) $\cot^{-1}(\cot x) = x$, for any $x$ in the range of $\cot^{-1}$.
e) $\sec^{-1}(\sec x) = x$, for any $x$ in the range of $\sec^{-1}$.
f) $\csc^{-1}(\csc x) = x$, for any $x$ in the range of $\csc^{-1}$.

---

**B** **Simplifying Other Compositions**

Now we find some other function compositions.

**Example 4** Find $\sin[\arctan(-1)]$.

Solution We first find $\arctan(-1)$. It is $-\pi/4$. Now we find the sine of this, which is $\sin(-\pi/4) = -\sqrt{2}/2$, so $\sin[\arctan(-1)] = -\sqrt{2}/2$. ◀

**Example 5** Find $\cos^{-1}\left(\sin\dfrac{\pi}{2}\right)$.

Solution We first find $\sin(\pi/2)$. It is 1. Now we find $\cos^{-1} 1$. It is 0, so $\cos^{-1}[\sin(\pi/2)] = 0$. ◀

**DO EXERCISES 9–13.**

Now let us consider

$\cos(\arcsin \frac{3}{5})$.

Without using a calculator, we cannot find arcsin $\frac{3}{5}$. However, we can still evaluate the entire expression without using a calculator. We are looking for an angle $\theta$ such that arcsin $\frac{3}{5} = \theta$, or sin $\theta = \frac{3}{5}$.

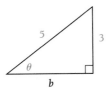

We sketch a triangle, as shown above. The angle $\theta$ in this triangle is an angle whose sine is $\frac{3}{5}$ (it is arcsin $\frac{3}{5}$). We wish to find the cosine of this angle. Since the triangle is a right triangle, we can find the length of the base, $b$. It is 4. Thus we know that cos $\theta = b/5$, or $\frac{4}{5}$. Therefore,

$$\cos\left(\arcsin \tfrac{3}{5}\right) = \tfrac{4}{5}.$$

**Example 6**  Find $\sin\left(\operatorname{arccot}\dfrac{x}{2}\right)$.

**Solution**  We draw a right triangle whose legs have lengths $x$ and 2, so that cot $\theta = x/2$. If $x$ is negative, we get the triangle in standard position shown on the left below. If $x$ is positive, we get the triangle shown on the right.

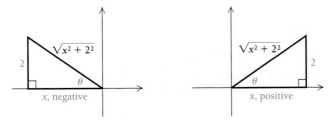

We find the length of the hypotenuse and then read off the sine ratio. In either case, we get

$$\sin\left(\operatorname{arccot}\dfrac{x}{2}\right) = \dfrac{2}{\sqrt{x^2 + 2^2}}, \quad \text{or} \quad \dfrac{2}{\sqrt{x^2 + 4}}. \quad \blacktriangleleft$$

**Example 7**  Find cos (arctan $p$).

**Solution**  We draw two right triangles, whose legs have lengths $p$ and 1, so that tan $\theta = p/1$.

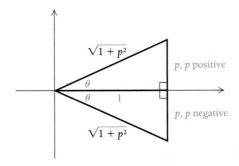

Find each of the following.

**14.** $\cos\left(\arctan\dfrac{b}{3}\right)$

**15.** $\sin(\arctan 2)$

$\left(\textit{Hint: } 2 = \dfrac{2}{1}.\right)$

**16.** $\tan(\arcsin t)$

Evaluate.

**17.** $\cos\left(\sin^{-1}\dfrac{\sqrt{3}}{2} - \cos^{-1}\dfrac{1}{2}\right)$

**18.** $\tan\left(\dfrac{1}{2}\arcsin\dfrac{3}{5}\right)$

(*Hint:* Let $\arcsin\frac{3}{5} = u$ and use a half-angle formula.)

We find the length of the other side and then read off the cosine ratio. In either case, we get

$$\cos(\arctan p) = \frac{1}{\sqrt{1+p^2}}.$$

The idea in Examples 6 and 7 is to first sketch a triangle, or triangles, two of whose sides have the appropriate ratio. Then use the Pythagorean theorem to find the length of the third side and read off the desired ratio.

**DO EXERCISES 14–16.**

In some cases, we use certain identities to evaluate expressions.

**Example 8**    Evaluate $\sin(\sin^{-1}\frac{1}{2} + \cos^{-1}\frac{4}{5})$.

**Solution**    We simplify this expression by letting

$$u = \sin^{-1}\tfrac{1}{2} \quad \text{and} \quad v = \cos^{-1}\tfrac{4}{5}.$$

Then we have $\sin(u + v)$. By a sum formula, this is equivalent to

$$\sin u \cos v + \cos u \sin v.$$

Now we make the reverse substitutions and obtain

$$\sin(\sin^{-1}\tfrac{1}{2}) \cdot \cos(\cos^{-1}\tfrac{4}{5}) + \cos(\sin^{-1}\tfrac{1}{2}) \cdot \sin(\cos^{-1}\tfrac{4}{5}).$$

This immediately simplifies to

$$\tfrac{1}{2} \cdot \tfrac{4}{5} + \cos(\sin^{-1}\tfrac{1}{2}) \cdot \sin(\cos^{-1}\tfrac{4}{5}).$$

Now $\cos(\sin^{-1}\frac{1}{2})$ simplifies to $\sqrt{3}/2$. To find $\sin(\cos^{-1}\frac{4}{5})$, we need a triangle. We set up the triangle so that $\cos^{-1}\frac{4}{5} = \theta$, or $\cos\theta = \frac{4}{5}$. Then we find $\sin\theta$.

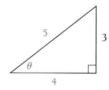

We see that $\sin(\cos^{-1}\frac{4}{5}) = \frac{3}{5}$. Our expression is now simplified to

$$\frac{1}{2}\cdot\frac{4}{5} + \frac{\sqrt{3}}{2}\cdot\frac{3}{5},$$

or

$$\frac{4 + 3\sqrt{3}}{10}.$$

Thus,

$$\sin\left(\sin^{-1}\frac{1}{2} + \cos^{-1}\frac{4}{5}\right) = \frac{4 + 3\sqrt{3}}{10}.$$

**DO EXERCISES 17 AND 18.**

**● EXERCISE SET 7.7**

**A** Evaluate or simplify.

**1.** $\sin(\arcsin 0.3)$

**2.** $\cos(\arccos 0.2)$

**3.** $\tan[\tan^{-1}(-4.2)]$

**4.** $\cot[\cot^{-1}(-1.5)]$

**5.** $\arcsin\left(\sin\dfrac{2\pi}{3}\right)$

**6.** $\arccos\left(\cos\dfrac{3\pi}{2}\right)$

**7.** $\sin^{-1}\left[\sin\left(-\dfrac{3\pi}{4}\right)\right]$

**8.** $\cos^{-1}\left[\cos\left(-\dfrac{\pi}{4}\right)\right]$

**9.** $\sin^{-1}\left(\sin\dfrac{\pi}{5}\right)$

**10.** $\cos^{-1}\left(\cos\dfrac{\pi}{7}\right)$

**11.** $\tan^{-1}\left(\tan\dfrac{2\pi}{3}\right)$

**12.** $\cot^{-1}\left(\cot\dfrac{2\pi}{3}\right)$

**B** Evaluate or simplify.

**13.** $\sin(\arctan\sqrt{3})$

**14.** $\sin\left(\arctan\dfrac{\sqrt{3}}{3}\right)$

**15.** $\cos\left(\arcsin\dfrac{\sqrt{3}}{2}\right)$

**16.** $\cos\left(\arcsin\dfrac{\sqrt{2}}{2}\right)$

**17.** $\tan\left(\cos^{-1}\dfrac{\sqrt{2}}{2}\right)$

**18.** $\tan\left(\cos^{-1}\dfrac{\sqrt{3}}{2}\right)$

**19.** $\cos^{-1}\left(\sin\dfrac{\pi}{3}\right)$

**20.** $\cos^{-1}(\sin\pi)$

**21.** $\arcsin\left(\cos\dfrac{\pi}{6}\right)$

**22.** $\arcsin\left(\cos\dfrac{\pi}{4}\right)$

**23.** $\sin^{-1}\left(\tan\dfrac{\pi}{4}\right)$

**24.** $\sin^{-1}\left[\tan\left(-\dfrac{\pi}{4}\right)\right]$

**25.** $\sin\left(\arctan\dfrac{x}{2}\right)$

**26.** $\sin\left(\arctan\dfrac{a}{3}\right)$

**27.** $\tan\left(\cos^{-1}\dfrac{3}{x}\right)$

**28.** $\cot\left(\sin^{-1}\dfrac{5}{y}\right)$

**29.** $\cot\left(\sin^{-1}\dfrac{a}{b}\right)$

**30.** $\tan\left(\cos^{-1}\dfrac{p}{q}\right)$

**31.** $\cos\left(\tan^{-1}\dfrac{\sqrt{2}}{3}\right)$

**32.** $\cos\left(\tan^{-1}\dfrac{\sqrt{3}}{4}\right)$

**33.** $\tan(\arcsin 0.1)$

**34.** $\tan(\arcsin 0.2)$

**35.** $\cot[\cos^{-1}(-0.2)]$

**36.** $\cot[\cos^{-1}(-0.3)]$

**37.** $\sin(\text{arccot } y)$

**38.** $\cos(\text{arccot } x)$

**39.** $\cos(\arctan t)$

**40.** $\sin(\arctan t)$

**41.** $\cot(\sin^{-1} y)$

**42.** $\tan(\cos^{-1} y)$

**43.** $\sin(\cos^{-1} x)$

**44.** $\cos(\sin^{-1} x)$

**45.** $\tan\left(\arcsin\dfrac{p}{\sqrt{p^2+9}}\right)$

**46.** $\csc\left(\tan^{-1}\dfrac{\sqrt{25-p^2}}{p}\right)$

**47.** $\tan\left(\dfrac{1}{2}\arcsin\dfrac{4}{5}\right)$

**48.** $\tan\left(\dfrac{1}{2}\arcsin\dfrac{1}{2}\right)$

**49.** $\cos\left(\dfrac{1}{2}\arcsin\dfrac{1}{2}\right)$

**50.** $\cos\left(\dfrac{1}{2}\arcsin\dfrac{\sqrt{3}}{2}\right)$

**51.** $\sin\left(2\cos^{-1}\dfrac{3}{5}\right)$

**52.** $\sin\left(2\cos^{-1}\dfrac{1}{2}\right)$

**53.** $\cos\left(2\sin^{-1}\dfrac{5}{13}\right)$

**54.** $\cos\left(2\cos^{-1}\dfrac{4}{5}\right)$

**55.** $\sin\left(\sin^{-1}\dfrac{1}{2}+\cos^{-1}\dfrac{3}{5}\right)$

**56.** $\sin\left(\sin^{-1}\dfrac{1}{2}-\cos^{-1}\dfrac{4}{5}\right)$

**57.** $\cos\left(\sin^{-1}\dfrac{\sqrt{2}}{2}+\cos^{-1}\dfrac{3}{5}\right)$

**58.** $\cos\left(\sin^{-1}\dfrac{4}{5}-\cos^{-1}\dfrac{1}{2}\right)$

**59.** $\sin(\sin^{-1} x+\cos^{-1} y)$

**60.** $\sin(\sin^{-1} x-\cos^{-1} y)$

**61.** $\cos(\sin^{-1} x+\cos^{-1} y)$

**62.** $\cos(\sin^{-1} x-\cos^{-1} y)$

**63.** ▦ $\sin(\sin^{-1} 0.6032+\cos^{-1} 0.4621)$

**64.** ▦ $\cos(\sin^{-1} 0.7325-\cos^{-1} 0.4838)$

**● SYNTHESIS** _____

**65.** Suppose that $\theta=\sin^{-1} x$. Find expressions in terms of $x$ for $\sin\theta$, $\cos\theta$, $\tan\theta$, $\cot\theta$, $\sec\theta$, and $\csc\theta$.

**66.** Suppose that $\theta=\arccos x$. Find expressions in terms of $x$ for $\sin\theta$, $\cos\theta$, $\tan\theta$, $\cot\theta$, $\sec\theta$, and $\csc\theta$.

**67.** Suppose that $\theta=\tan^{-1} x$. Find expressions in terms of $x$ for $\sin\theta$, $\cos\theta$, $\tan\theta$, $\cot\theta$, $\sec\theta$, and $\csc\theta$.

**68.** Suppose that $\theta=\text{arccot } x$. Find expressions in terms of $x$ for $\sin\theta$, $\cos\theta$, $\tan\theta$, $\cot\theta$, $\sec\theta$, and $\csc\theta$.

Prove each of the following identities.

**69.** $\sin^{-1} x+\cos^{-1} x=\dfrac{\pi}{2}$

**70.** $\tan^{-1} x+\cot^{-1} x=\dfrac{\pi}{2}$

**71.** $\sin^{-1} x=\tan^{-1}\dfrac{x}{\sqrt{1-x^2}}$

**72.** $\tan^{-1} x=\sin^{-1}\dfrac{x}{\sqrt{x^2+1}}$

**73.** For $x\geq 0$,
$$\arcsin x=\arccos\sqrt{1-x^2}.$$

**74.** For $x > 0$,

$$\arccos x = \arctan \frac{\sqrt{1 - x^2}}{x}.$$

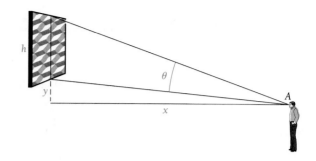

**75.** An observer's eye is at a point $A$, looking at a mural of height $h$, with the bottom of the mural $y$ feet above the eye. The eye is $x$ feet from the wall. Write an expression for $\theta$ in terms of $x$, $y$, and $h$.

**76.** ▣ Evaluate the expression found in Exercise 75 when $x = 20$ ft, $y = 7$ ft, and $h = 25$ ft.

**1.** Solve. Give answers in both degrees and radians.

$$2\cos x = 1$$

# 7.8

# TRIGONOMETRIC EQUATIONS

## Ⓐ  Simple Equations

When an equation contains a trigonometric expression with a variable, such as $\sin x$, it is called a **trigonometric equation.** We have worked with many trigonometric equations that are identities. Now we consider equations that may not be identities. To solve such an equation, we find all replacements for the variable that make the equation true.

**Example 1**    Solve: $2\sin x = 1$.

**Solution**    We first solve for $\sin x$: $\sin x = \frac{1}{2}$. Now we note that the solutions are those numbers having a sine of $\frac{1}{2}$. We look for them. The unit circle is helpful.

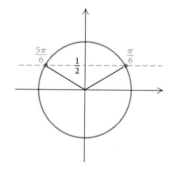

There are just two points on it for which the sine is $\frac{1}{2}$, as shown. They are the points for $\pi/6$ and $5\pi/6$. These numbers, plus any multiple of $2\pi$, are the solutions

$$\frac{\pi}{6} + 2k\pi \quad \text{or} \quad \frac{5\pi}{6} + 2k\pi,$$

where $k$ is any integer. In degrees, the solutions are

$$30° + k \cdot 360° \quad \text{or} \quad 150° + k \cdot 360°,$$

where $k$ is any integer.                                                                         ◀

Note that what we did in Example 1 was comparable to solving $x = \sin^{-1} \frac{1}{2}$, only now we get *all* the numbers whose sines are $\frac{1}{2}$, not just $\pi/6$.

**DO EXERCISE 1.**

**Example 2**    Solve: $4\cos^2 x = 1$.

Solution   We first solve for $\cos x$:

$$\cos^2 x = \tfrac{1}{4}$$
$$\cos x = \pm\tfrac{1}{2}.$$

Now we use the unit circle to find those numbers having a cosine of $\tfrac{1}{2}$ or $-\tfrac{1}{2}$. The solutions are $\pi/3$, $2\pi/3$, $4\pi/3$, $5\pi/3$, plus any multiple of $2\pi$. In degrees, the solutions are $60°$, $120°$, $240°$, $300°$, plus any multiple of $360°$.

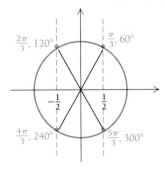

**DO EXERCISE 2.**

In most practical applications, it is usually sufficient to find just the solutions from $0$ to $2\pi$ or from $0°$ to $360°$. We then remember that any multiple of $2\pi$, or $360°$, can be added to obtain the rest of the solutions.

The following example illustrates that when we look for solutions in the interval $[0, 2\pi)$, or $[0°, 360°)$, we must be cautious.

Example 3   Solve $2\sin 2x = -1$ in the interval $[0, 2\pi)$.

Solution   We first solve for $\sin 2x$:

$$2\sin 2x = -1$$
$$\sin 2x = -\tfrac{1}{2}.$$

We are looking for solutions $x$ to the equation for which

$$0 \le x < 2\pi.$$

Multiplying by 2, we get

$$0 \le 2x < 4\pi,$$

which is the interval we consider to solve $\sin 2x = -\tfrac{1}{2}$.

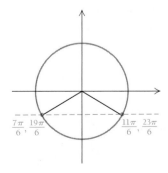

Using the unit circle, we find points $2x$ for which $\sin 2x = -\tfrac{1}{2}$ and $0 \le 2x < 4\pi$. These values of $2x$ are $7\pi/6$, $11\pi/6$, $19\pi/6$, and $23\pi/6$. Thus

2. Find all solutions in $[0, 2\pi)$. Give answers in both degrees and radians.

$$4\sin^2 x = 1$$

3. Find all solutions in $[0, 2\pi)$. Leave answers in terms of $\pi$.

$$2\cos 2x = 1$$

the desired values of $x$ in $[0, 2\pi)$ are half of these. Therefore,

$$x = \frac{7\pi}{12}, \frac{11\pi}{12}, \frac{19\pi}{12}, \frac{23\pi}{12}.$$

◀

**DO EXERCISE 3.**

## ▶ **Using a Calculator**

In solving some trigonometric equations, it is necessary to use a calculator. Answers can then be found in radians or degrees, depending on how the calculator is set. We will usually find answers in degrees.

**Example 4**   Solve $2 + \sin x = 2.5263$ in $[0, 360°)$.

**Solution**   We first solve for $\sin x$:

$$2 + \sin x = 2.5263$$
$$\sin x = 0.5263.$$

From a calculator, we find the reference angle, $x \approx 31.76°$. Since $\sin x$ is positive, the solutions are to be found in the first and second quadrants. The solutions are $31.76°$ and $148.24°$.

Solve in $[0°, 360°)$.

4. $2 + \cos x = 2.7660$

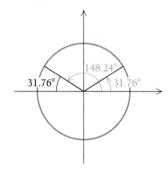

◀

**Example 5**   Solve $\sin x - 1 = -1.6381$ in $[0°, 360°)$.

**Solution**   We first solve for $\sin x$:

$$\sin x - 1 = -1.6381$$
$$\sin x = -0.6381.$$

Using a calculator, we find that the reference angle is $39.65°$. Since $\sin x$ is negative, the solutions are to be found in the third and fourth quadrants. The solutions are $219.65°$ and $320.35°$.

5. $\cos x - 1 = -1.7660$

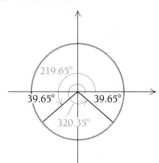

◀

**DO EXERCISES 4 AND 5.**

## Using Algebraic Techniques

In solving trigonometric equations, we can expect to apply some algebra before concerning ourselves with the trigonometric part. In the next examples, we begin by factoring.

**Example 6** Solve $8 \cos^2 \theta - 2 \cos \theta = 1$ in $[0°, 360°)$.

**Solution** Since we will be using the principle of zero products, we first obtain a 0 on one side of the equation:

$$8 \cos^2 \theta - 2 \cos \theta - 1 = 0$$
$$(4 \cos \theta + 1)(2 \cos \theta - 1) = 0 \qquad \text{Factoring}$$

$4 \cos \theta + 1 = 0 \qquad\qquad \text{or} \quad 2 \cos \theta - 1 = 0 \qquad \begin{array}{l}\text{Principle of}\\ \text{zero products}\end{array}$

$\cos \theta = -\frac{1}{4} = -0.25 \qquad \text{or} \qquad \cos \theta = \frac{1}{2}$

$\theta = 104.48°, 255.52° \quad \text{or} \qquad \theta = 60°, 300°.$

The solutions in $[0°, 360°)$ are $104.48°$, $255.52°$, $60°$, and $300°$. ◀

It may be helpful in accomplishing the algebraic part of solving trigonometric equations to make substitutions as in Chapter 2. If we use such an approach, the algebraic part of Example 6 would look like this:

$$8 \cos^2 \theta - 2 \cos \theta = 1.$$

We would then let $u = \cos \theta$:

$$8u^2 - 2u = 1$$
$$(4u + 1)(2u - 1) = 0$$
$$4u + 1 = 0 \quad \text{or} \quad 2u - 1 = 0$$
$$u = -\frac{1}{4} \quad \text{or} \qquad u = \frac{1}{2}.$$

We then substitute $\cos \theta$ for $u$ and solve for $\theta$.

**Example 7** Solve $2 \sin^2 \phi + \sin \phi = 0$ in $[0, 2\pi)$.

**Solution**

$$2 \sin^2 \phi + \sin \phi = 0$$
$$\sin \phi (2 \sin \phi + 1) = 0 \qquad \text{Factoring}$$
$$\sin \phi = 0 \quad \text{or} \quad 2 \sin \phi + 1 = 0 \qquad \text{Principle of zero products}$$

$$\sin \phi = 0 \quad \text{or} \qquad \sin \phi = -\frac{1}{2}$$

$$\phi = 0, \pi \quad \text{or} \qquad \phi = \frac{7\pi}{6}, \frac{11\pi}{6}$$

The solutions in $[0, 2\pi)$ are $0$, $\pi$, $7\pi/6$, and $11\pi/6$. ◀

**DO EXERCISES 6 AND 7.**

If a trigonometric equation is quadratic but difficult or impossible to factor, we use the quadratic formula.

Solve in $[0°, 360°)$.

6. $8 \cos^2 \theta + 2 \cos \theta = 1$

7. $2 \cos^2 \phi + \cos \phi = 0$

8. Solve in $[0°, 360°)$ using the quadratic formula:

$$10 \cos^2 x - 10 \cos x - 7 = 0.$$

**Example 8**    Solve $10 \sin^2 x - 12 \sin x - 7 = 0$ in $[0°, 360°)$.

**Solution**    It may help to make the substitution $u = \sin x$, in order to obtain the equation $10u^2 - 12u - 7 = 0$. We then use the quadratic formula to find $u$, or $\sin x$:

$$\sin x = \frac{12 \pm \sqrt{144 + 280}}{20} \qquad \text{Using the quadratic formula}$$

$$= \frac{12 \pm \sqrt{424}}{20}$$

$$= \frac{12 \pm 2\sqrt{106}}{20} = \frac{6 \pm \sqrt{106}}{10}$$

$$\approx \frac{6 \pm 10.296}{10}$$

$$\sin x \approx 1.6296 \quad \text{or} \quad \sin x \approx -0.4296.$$

Since sines are never greater than 1, the first of the equations has no solution. Using the other equation, we find the reference angle to be $25.44°$. Since $\sin x$ is negative, the solutions are to be found in the third and fourth quadrants. Thus the solutions are $205.44°$ and $334.56°$. ◀

**DO EXERCISE 8.**

---

 **TECHNOLOGY CONNECTION**

We have used a grapher to find the solutions of equations by zooming in on their points of intersection. We now solve equations that involve trigonometric functions. For example, if we wish to find the solutions for the equation $5 \cos^2 x = 2$, we graph the left side as one function and the right side as another function and look for point(s) of intersection. The first function is $y = 5 \cos^2 x$, and the second function is $y = 2$. These functions are then graphed together on the same set of axes. If we want to find $x$ in radian measure, the viewing box in RADIAN mode might be $[-\pi, \pi] \times [-5, 5]$. If we want to find $x$ in degrees, the same viewing box in DEGREE mode would be $[-180, 180] \times [-5, 5]$. Either way, the result would be the following:

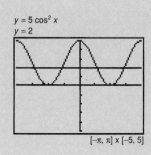

$y = 5 \cos^2 x$
$y = 2$

$[-\pi, \pi] \times [-5, 5]$

By zooming in on the points of intersection, we find that their coordinates (rounded to two decimal places) are $(-2.26, 2.00)$, $(-0.89, 2.00)$, $(0.89, 2.00)$, and $(2.26, 2.00)$, in RADIAN mode. If we were using DEGREE mode, the same four points would have the coordinates $(-129.23°, 2.00)$, $(-50.77°, 2.00)$, $(50.77°, 2.00)$, and $(129.23°, 2.00)$.

Use a grapher to find the solutions of each of the following trigonometric equations on the specified domain. Use the appropriate mode (RADIAN or DEGREE) to match each domain.

**TC 1.** $3 \sin^2 x - 4 \sin x = 5 \cos x$; $[-180°, 180°]$

**TC 2.** $5 \sin x \cos x = 2 + \cos^2 x$; $[0, \pi/2]$

**TC 3.** $10 \cos^2 x \sin^2 x = 4 + \sin x$; $[-90°, 90°]$

**TC 4.** $(\cos x + 2)(\sin x - 3) = 6 \cos x$; $[0, 2\pi]$

## ● EXERCISE SET 7.8

**A, B** Solve, finding all solutions.

**1.** $\sin x = \dfrac{\sqrt{3}}{2}$

**2.** $\cos x = \dfrac{\sqrt{3}}{2}$

**3.** $\cos x = \dfrac{1}{\sqrt{2}}$

**4.** $\tan x = \sqrt{3}$

**5.** $\sin x = 0.3448$

**6.** $\cos x = 0.6406$

Solve, finding all solutions in $[0, 2\pi)$ or $[0°, 360°)$.

**7.** $\cos x = -0.5495$

**8.** $\sin x = -0.4279$

**9.** $2 \sin x + \sqrt{3} = 0$

**10.** $\sqrt{3} \tan x + 1 = 0$

**11.** $2 \tan x + 3 = 0$

**12.** $4 \sin x - 1 = 0$

**C** Solve, finding all solutions in $[0, 2\pi)$ or $[0°, 360°)$. If you have a grapher, verify your results.

**13.** $4 \sin^2 x - 1 = 0$

**14.** $2 \cos^2 x = 1$

**15.** $\cot^2 x - 3 = 0$

**16.** $\csc^2 x - 4 = 0$

**17.** $2 \sin^2 x + \sin x = 1$

**18.** $2 \cos^2 x + 3 \cos x = -1$

**19.** $\cos^2 x + 2 \cos x = 3$

**20.** $2 \sin^2 x - \sin x = 3$

**21.** $4 \sin^3 x - \sin x = 0$

**22.** $2 \cos^2 x - \sqrt{3} \cos x = 0$

**23.** $2 \sin^2 \theta + 7 \sin \theta = 4$

**24.** $2 \sin^2 \theta - 5 \sin \theta + 2 = 0$

**25.** $6 \cos^2 \phi + 5 \cos \phi + 1 = 0$

**26.** $2 \sin^2 \phi + \sin \phi - 1 = 0$

**27.** $2 \sin t \cos t + 2 \sin t - \cos t - 1 = 0$

**28.** $2 \sin t \tan t + \tan t - 2 \sin t - 1 = 0$

**29.** $\cos 2x \sin x + \sin x = 0$

**30.** $\sin 2x \cos x - \sin x = 0$

**31.** $5 \sin^2 x - 8 \sin x = 3$

**32.** $\cos^2 x + 6 \cos x + 4 = 0$

**33.** $2 \tan^2 x = 3 \tan x + 7$

**34.** $3 \sin^2 x = 3 \sin x + 2$

**35.** $7 = \cot^2 x + 4 \cot x$

**36.** $3 \tan^2 x + 2 \tan x = 7$

● SYNTHESIS

Solve, restricting solutions to $[0, 2\pi)$ or $[0°, 360°)$. If you have a grapher, verify your results.

**37.** $|\sin x| = \dfrac{\sqrt{3}}{2}$

**38.** $|\cos x| = \dfrac{1}{2}$

**39.** $\sqrt{\tan x} = \sqrt[4]{3}$

**40.** $12 \sin x - 7\sqrt{\sin x} + 1 = 0$

**41.** $16 \cos^4 x - 16 \cos^2 x + 3 = 0$

**42.** $\ln (\cos x) = 0$

**43.** $e^{\sin x} = 1$

**44.** $\sin (\ln x) = -1$

**45.** $e^{\ln(\sin x)} = 1$

**46.** A guy wire is attached to the top of a 50-ft pole and stretched to a point that is $b$ ft from the bottom of the pole.

   **a)** Suppose that the angle of inclination of the wire to the top is $\theta$. Show that

$$\tan \theta = \frac{50}{b}.$$

   **b)** For what angle of inclination $\theta$ does $b = 40$ ft?

**47.** An airplane is flying at an altitude of 2000 ft toward an island. The straight-line distance from the airplane to the island is $h$ ft.

   **a)** Suppose that the angle of depression is $\theta$. Show that

$$\sin \theta = \frac{2000}{h}.$$

   **b)** For what angle of depression $\theta$ does $h = 3000$ ft?

**48.** Solve graphically: $\sin x = \tan (x/2)$.

# 7.9

# IDENTITIES IN SOLVING TRIGONOMETRIC EQUATIONS

**OBJECTIVE**

You should be able to:

**A** Solve trigonometric equations requiring the use of identities.

**A** When a trigonometric equation involves more than one function, we can use identities to put it in terms of a single function. This can usually be done in several ways. In Example 1, we illustrate by solving the same equation using two different methods.

1. Solve $\sin x - \cos x = 1$, as in Example 1 (method 1).

**Example 1**

*Method 1.* Solve the equation $\sin x + \cos x = 1$.

Solution To express $\cos x$ in terms of $\sin x$, we use the identity

$$\sin^2 x + \cos^2 x = 1.$$

From this, we obtain $\cos x = \pm\sqrt{1 - \sin^2 x}$. Now we substitute in the original equation:

$$\sin x \pm \sqrt{1 - \sin^2 x} = 1.$$

This is a radical equation. We get the radical alone on one side and then square both sides and simplify:

$$\pm\sqrt{1 - \sin^2 x} = 1 - \sin x$$
$$(\pm\sqrt{1 - \sin^2 x})^2 = (1 - \sin x)^2$$
$$1 - \sin^2 x = 1 - 2\sin x + \sin^2 x$$
$$-2\sin^2 x + 2\sin x = 0.$$

Factoring and using the principle of zero products gives us

$$-2\sin x (\sin x - 1) = 0$$
$$-2\sin x = 0 \quad \text{or} \quad \sin x - 1 = 0$$
$$\sin x = 0 \quad \text{or} \quad \sin x = 1.$$

The values of $x$ in $[0, 2\pi)$ satisfying these are

$$x = 0, \qquad x = \pi, \quad \text{or} \quad x = \frac{\pi}{2}.$$

Now we check these in the original equation. We find that $\pi$ does not check but the other values do. Thus the solutions are 0 and $\pi/2$. ◀

---

> CAUTION! It is important to check when solving trigonometric equations. Values are often obtained that are not solutions of the original equation.

---

**DO EXERCISE 1.**

**Example 1**

*Method 2.* Solve $\sin x + \cos x = 1$.

Solution This time we will square both sides, obtaining

$$\sin^2 x + 2\sin x \cos x + \cos^2 x = 1.$$

Since $\sin^2 x + \cos^2 x = 1$, we can simplify to

$$2\sin x \cos x = 0.$$

Now we use the identity $2\sin x \cos x = \sin 2x$, and obtain

$$\sin 2x = 0.$$

We are looking for solutions $x$ to the equation for which

$$0 \le x < 2\pi.$$

Multiplying by 2, we get

$$0 \le 2x < 4\pi,$$

the interval we consider to solve $\sin 2x = 0$. These values of $2x$ are $0$, $\pi$, $2\pi$, and $3\pi$. Thus the desired values of $x$ in $[0, 2\pi)$ satisfying this equation are

$$x = 0, \qquad x = \frac{\pi}{2}, \qquad x = \pi, \quad \text{or} \quad x = \frac{3\pi}{2}.$$

Now we check these in the original equation. We find that $\pi$ and $3\pi/2$ do not check but the other values do. Thus the solutions are $0$ and $\pi/2$. ◀

**DO EXERCISE 2.**

2. Solve $\sin x - \cos x = 1$, as in Example 1 (method 2).

**Example 2**   Solve:  $2 \cos^2 x \tan x - \tan x = 0$.

**Solution**   First, we factor:

$$2 \cos^2 x \tan x - \tan x = 0$$
$$\tan x (2 \cos^2 x - 1) = 0.$$

Now we use the identity $2 \cos^2 x - 1 = \cos 2x$:

$$\tan x \cos 2x = 0$$

$$\tan x = 0 \qquad \text{or} \quad \cos 2x = 0 \qquad \text{\small Principle of zero products}$$

$$x = 0, \pi \quad \text{or} \qquad 2x = \frac{\pi}{2}, \frac{3\pi}{2}, \frac{5\pi}{2}, \frac{7\pi}{2}$$

$$x = 0, \pi \quad \text{or} \qquad x = \frac{\pi}{4}, \frac{3\pi}{4}, \frac{5\pi}{4}, \frac{7\pi}{4}.$$

All values check. The solutions in $[0, 2\pi)$ are $0$, $\pi$, $\pi/4$, $3\pi/4$, $5\pi/4$, and $7\pi/4$. ◀

3. Solve:  $\tan^2 x \cos x - \cos x = 0$.

**DO EXERCISE 3.**

**Example 3**   Solve:  $\cos 2x + \sin x = 1$.

**Solution**   We first use the identity $\cos 2x = 1 - 2 \sin^2 x$, to get

$$1 - 2 \sin^2 x + \sin x = 1$$
$$-2 \sin^2 x + \sin x = 0$$
$$\sin x (1 - 2 \sin x) = 0 \qquad \text{\small Factoring}$$

$$\sin x = 0 \qquad \text{or} \quad 1 - 2 \sin x = 0 \qquad \text{\small Principle of zero products}$$

$$\sin x = 0 \qquad \text{or} \qquad \sin x = \frac{1}{2}$$

$$x = 0, \pi \quad \text{or} \qquad x = \frac{\pi}{6}, \frac{5\pi}{6}.$$

All values check. The solutions in $[0, 2\pi)$ are $0$, $\pi$, $\pi/6$, and $5\pi/6$. ◀

4. Solve:  $\sin 2x + \cos x = 0$.

**DO EXERCISE 4.**

**Example 4**   Solve:  $\sin 5\theta \cos 2\theta - \cos 5\theta \sin 2\theta = \sqrt{2}/2$.

5. Solve:

$\sin 3\theta \cos \theta - \cos 3\theta \sin \theta = \frac{1}{2}$.

**Solution** We use the identity

$\sin(\alpha - \beta) = \sin \alpha \cos \beta - \cos \alpha \sin \beta$

to get $\sin 3\theta = \sqrt{2}/2$. We are looking for solutions to the equation for which

$0 \le \theta < 2\pi$.

Multiplying by 3, we get

$0 \le 3\theta < 6\pi$,

which is the interval we consider to solve $\sin 3\theta = \sqrt{2}/2$. These values of $3\theta$ are

$$3\theta = \frac{\pi}{4}, \frac{3\pi}{4}, \frac{9\pi}{4}, \frac{11\pi}{4}, \frac{17\pi}{4}, \frac{19\pi}{4}.$$

Then

$$\theta = \frac{\pi}{12}, \frac{\pi}{4}, \frac{3\pi}{4}, \frac{11\pi}{12}, \frac{17\pi}{12}, \frac{19\pi}{12}.$$

Let us check $\pi/12$:

$$\sin \frac{5\pi}{12} \cos \frac{\pi}{6} - \cos \frac{5\pi}{12} \sin \frac{\pi}{6}.$$

We again use the difference identity to obtain $\sin(5\pi/12 - \pi/6)$. This is equal to $\sin(3\pi/12)$, or $\sin(\pi/4)$, which is equal to $\sqrt{2}/2$. Thus $\pi/12$ checks. All of the above answers check. ◀

**DO EXERCISE 5.** _____

**Example 5** Solve: $\tan^2 x + \sec x - 1 = 0$.

**Solution** We use the identity $1 + \tan^2 x = \sec^2 x$. Substituting, we then get

$\sec^2 x - 1 + \sec x - 1 = 0$,

or

$\sec^2 x + \sec x - 2 = 0$

$(\sec x + 2)(\sec x - 1) = 0$     Factoring

$\sec x = -2$    or    $\sec x = 1$     Principle of zero products

$x = \frac{2\pi}{3}, \frac{4\pi}{3}$    or     $x = 0$.

All of these values check. The solutions in $[0, 2\pi)$ are $0$, $2\pi/3$, and $4\pi/3$. ◀

**Example 6** Solve: $\sin x = \sin(2x - \pi)$.

**Solution** We can use either the identity $\sin(y - \pi) = -\sin y$ or the identity $\sin(\alpha - \beta) = \sin \alpha \cos \beta - \cos \alpha \sin \beta$ with the right-hand side. We get

$\sin x = -\sin 2x$.

Next, we use the identity $\sin 2x = 2 \sin x \cos x$, and obtain

$$\sin x = -2 \sin x \cos x,$$

or

$$2 \sin x \cos x + \sin x = 0$$

$$\sin x (2 \cos x + 1) = 0 \qquad \text{Factoring}$$

$$\sin x = 0 \quad \text{or} \quad \cos x = -\frac{1}{2} \qquad \text{Principle of zero products}$$

$$x = 0, \pi \quad \text{or} \quad x = \frac{2\pi}{3}, \frac{4\pi}{3}.$$

All of these values check, so the solutions in $[0, 2\pi)$ are $0, \pi, 2\pi/3$, and $4\pi/3$.

◢

**DO EXERCISE 6.**

**6.** Solve: $\cos x = \cos (\pi - 2x)$.

---

**EXERCISE SET** **7.9**

A   Find all solutions of the following equations in $[0, 2\pi)$.

**1.** $\tan x \sin x - \tan x = 0$

**2.** $2 \sin x \cos x + \sin x = 0$

**3.** $2 \sec x \tan x + 2 \sec x + \tan x + 1 = 0$

**4.** $2 \csc x \cos x - 4 \cos x - \csc x + 2 = 0$

**5.** $\sin 2x - \cos x = 0$

**6.** $\cos 2x - \sin x = 1$

**7.** $\sin 2x \sin x - \cos x = 0$

**8.** $\sin 2x \cos x - \sin x = 0$

**9.** $\sin 2x + 2 \sin x \cos x = 0$

**10.** $\cos 2x \sin x + \sin x = 0$

**11.** $\cos 2x \cos x + \sin 2x \sin x = 1$

**12.** $\sin 2x \sin x - \cos 2x \cos x = -\cos x$

**13.** $\sin 4x - 2 \sin 2x = 0$

**14.** $\sin 4x + 2 \sin 2x = 0$

**15.** $\sin 2x + 2 \sin x - \cos x - 1 = 0$

**16.** $\sin 2x + \sin x + 2 \cos x + 1 = 0$

**17.** $\sec^2 x = 4 \tan^2 x$

**18.** $\sec^2 x - 2 \tan^2 x = 0$

**19.** $\sec^2 x + 3 \tan x - 11 = 0$

**20.** $\tan^2 x + 4 = 2 \sec^2 x + \tan x$

**21.** $\cot x = \tan (2x - 3\pi)$

**22.** $\tan x = \cot (2x + \pi)$

**23.** $\cos (\pi - x) + \sin \left( x - \frac{\pi}{2} \right) = 1$

**24.** $\sin (\pi - x) + \cos \left( \frac{\pi}{2} - x \right) = 1$

**25.** $\dfrac{\cos^2 x - 1}{\sin \left( \dfrac{\pi}{2} - x \right) - 1} = \dfrac{\sqrt{2}}{2} + 1$

**26.** $\dfrac{\sin^2 x - 1}{\cos \left( \dfrac{\pi}{2} - x \right) + 1} = \dfrac{\sqrt{2}}{2} - 1$

**27.** $2 \cos x + 2 \sin x = \sqrt{6}$

**28.** $2 \cos x + 2 \sin x = \sqrt{2}$

**29.** $\sqrt{3} \cos x - \sin x = 1$

**30.** $\sqrt{2} \cos x - \sqrt{2} \sin x = 2$

**31.** $\sec^2 x + 2 \tan x = 6$

**32.** $6 \tan^2 x = 5 \tan x + \sec^2 x$

**33.** $3 \cos 2x + \sin x = 1$

**34.** $5 \cos 2x + \sin x = 4$

●   SYNTHESIS

**35.** ***Temperature during an illness.*** The temperature $T$ of a patient during a 12-day illness is given by

$$T(t) = 101.6° + 3° \sin \left( \frac{\pi}{8} t \right).$$

Find the times $t$ during the illness at which the patient's temperature was $103°$.

**36.** ***Satellite location.*** A satellite circles the earth in such a manner that it is $y$ miles from the equator (north or south, height from the surface not considered) $t$ minutes after its launch, where

$$y = 5000 \left[ \cos \frac{\pi}{45} (t - 10) \right].$$

At what times $t$ in the interval $[0, 240]$, the first 4 hr, is the satellite 3000 mi north of the equator?

**37. Nautical mile.** (See Exercise 55 in Exercise Set 7.4.) In Great Britain, the **nautical mile** is defined as the length of a minute of arc of the earth's radius. Since the earth is flattened at the poles, a British nautical mile varies with latitude. In fact, it is given, in feet, by the function

$$N(\phi) = 6066 - 31 \cos 2\phi,$$

where $\phi$ is the latitude in degrees. At what latitude north is the length of a British nautical mile found to be 6040 ft?

**38. Acceleration due to gravity.** (See Exercise 56 in Exercise Set 7.4.) The acceleration due to gravity is often denoted by $g$ in a formula such as $S = \frac{1}{2}gt^2$, where $S$ is the distance that an object falls in time $t$. It has to do with the physics of motion near the earth's surface and is usually considered constant. In fact, however, $g$ is not constant, but varies slightly with latitude. If $\phi$ stands for latitude, in degrees, $g$ is given with good approximation by the formula

$$g = 9.78049(1 + 0.005288 \sin^2 \phi - 0.000006 \sin^2 2\phi),$$

where $g$ is measured in meters per second per second at sea level. At what latitude north does $g = 9.8$?

Solve.

**39.** $\arccos x = \arccos \frac{3}{5} - \arcsin \frac{4}{5}$

**40.** $\sin^{-1} x = \tan^{-1} \frac{1}{3} + \tan^{-1} \frac{1}{2}$

Solve graphically.

**41.** $\sin x - \cos x = \cot x$

**42.** $x \sin x = 1$

**43.** Suppose that $\sin x = 5 \cos x$. Find $\sin x \cos x$.

# 7

## SUMMARY AND REVIEW

- **TERMS TO KNOW**

Identity, p. 452
Sum and difference identities, p. 462
Cofunction identities, p. 465
Double-angle identities, p. 473
Half-angle identities, p. 474

Basic identities, p. 478
Pythagorean identities, p. 478
Sum–product identities, p. 483
Inverse trigonometric functions,
   p. 486

Arcsine function, p. 488
Arccosine function, p. 489
Arctangent function, p. 489
Trigonometric equations, p. 498

- **REVIEW EXERCISES**

**1.** Find an identity for $\cot (x - \pi)$.

Complete these Pythagorean identities.

**2.** $\sin^2 x + \cos^2 x = $ _____

**3.** $1 + \cot^2 x = $ _____

Complete these cofunction identities.

**4.** $\cos \left( x + \dfrac{\pi}{2} \right) = $ _____

**5.** $\cos \left( \dfrac{\pi}{2} - x \right) = $ _____

**6.** $\sin \left( x - \dfrac{\pi}{2} \right) = $ _____

**7.** Express $\tan x$ in terms of $\sec x$.

Simplify.

**8.** $\cos x (\tan x + \cot x)$

**9.** $\dfrac{\csc x (\sin^2 x + \cos^2 x \tan x)}{\sin x + \cos x}$

**10.** Rationalize the denominator: $\sqrt{\dfrac{\tan x}{\sec x}}$.

**11.** Rationalize the numerator: $\sqrt{\dfrac{\tan x}{\sec x}}$.

**12.** Given that $\sin \theta = 0.6820$, $\cos \theta = 0.7314$, $\tan \theta = 0.9325$, $\cot \theta = 1.0724$, $\sec \theta = 1.3673$, and $\csc \theta = 1.4663$, find the six function values for $90° - \theta$.

Use the sum and difference formulas to write equivalent expressions. You need not simplify.

**13.** $\cos \left( x + \dfrac{3\pi}{2} \right)$        **14.** $\tan (45° - 30°)$

**15.** Simplify: $\cos 27° \cos 16° + \sin 27° \sin 16°$.

**16.** Find $\cos 165°$ exactly.

**17.** Given that $\tan \alpha = \sqrt{3}$ and $\sin \beta = \sqrt{2}/2$ and that $\alpha$ and $\beta$ are between 0 and $\pi/2$, evaluate $\tan (\alpha - \beta)$ exactly.

**18.** Find $\tan 2\theta$, $\cos 2\theta$, and $\sin 2\theta$ and the quadrant in which $2\theta$ lies, where $\cos \theta = -\frac{3}{5}$ and $\theta$ is in quadrant III.

**19.** Find $\sin (\pi/8)$ without using a table or a calculator.

**20.** Simplify: $\dfrac{\sin 2\theta}{\sin^2 \theta}$.

Prove each of the following identities.

**21.** $\tan 2\theta = \dfrac{2 \tan \theta}{1 - \tan^2 \theta}$

**22.** $\dfrac{\sec x - \cos x}{\tan x} = \sin x$

**23.** Find, in radians, $\sin^{-1} \frac{1}{2}$.

**24.** Use a calculator to find, in degrees, arccot 0.1584.

Find each of the following without the use of a table or a calculator.

**25.** $\sin^{-1} \left( -\dfrac{\sqrt{2}}{2} \right)$

**26.** $\text{arccot} \left( -\sqrt{3} \right)$

Evaluate or simplify.

**27.** $\tan \left( \arctan \dfrac{7}{8} \right)$

**28.** $\cos^{-1} \left[ \cos \left( -\dfrac{\pi}{3} \right) \right]$

**29.** $\cos^{-1} \left( \sin \dfrac{2\pi}{3} \right)$

**30.** $\sin \left( \arctan \dfrac{b}{5} \right)$

Solve, finding all solutions in $[0, 2\pi)$.

**31.** $\sin^2 x - 7 \sin x = 0$

**32.** $2 \cos^2 x - 5 \cos x + 2 = 0$

**33.** $\csc^2 x - 2 \cot^2 x = 0$

**34.** $2 \cot^2 x = 3 \cot x + 1$

Solve, finding all solutions in $[0, 2\pi)$.

**35.** $|\sin x| = \frac{1}{2}$

**36.** $\cos (\ln 2x) = -1$

Graph.

**37.** $y + 1 = 2 \cos^2 x$

**38.** $f(x) = 2 \sin^{-1} \left( x + \dfrac{\pi}{2} \right)$

● THINKING AND WRITING _____

**1.** Prove the identity $2 \cos^2 x - 1 = \cos^4 x - \sin^4 x$ in three ways:

  **a)** Start with the left side and deduce the right (Method 1).

  **b)** Start with the right side and deduce the left (Method 1).

  **c)** Work with each side separately until you deduce the same expression (Method 2).

Then determine the most efficient method and explain why.

**2.** Why are the ranges of the inverse trigonometric functions restricted?

**3.** Explain why $\tan (x + 450°)$ cannot be simplified using the tangent sum formula, but can be simplified using the sine and cosine sum formulas.

---

# CHAPTER TEST 7

Use the sum and difference formulas to write equivalent expressions. You need not simplify.

**1.** $\cos (\pi - x)$

**2.** $\tan (83° + 15°)$

**3.** Simplify: $\sin 40° \cos 5° - \cos 40° \sin 5°$.

**4.** Find $\cos 105°$ exactly.

**5.** Given that $\cos \alpha = \sqrt{2}/2$ and $\sin \beta = 1/2$ and that $\alpha$ and $\beta$ are between 0 and $\pi/2$, evaluate $\tan (\alpha - \beta)$ exactly.

**6.** Find an identity for $\sec (x - 450°)$.

Complete these identities.

**7.** $\sin^2 x + \cos^2 x =$ _____

**8.** $1 + \cot^2 x =$ _____

**9.** $\sin \left( x + \dfrac{\pi}{2} \right) =$ _____

**10.** $\cos \left( \dfrac{\pi}{2} - x \right) =$ _____

**11.** $\cos \left( x - \dfrac{\pi}{2} \right) =$ _____

**12.** Express $\csc x$ in terms of $\cot x$.

Simplify.

**13.** $\dfrac{\sqrt{\sec^2 x - 1}}{\sin x}$

**14.** $\dfrac{\tan^2 x \csc^2 x - 1}{\csc x \tan^2 x \sin x}$

**15.** Rationalize the denominator: $\sqrt{\dfrac{\sec x}{\csc x}}$.

**16.** Find $\sin 2\theta$, $\cos 2\theta$, and $\tan 2\theta$ and the quadrant in which $2\theta$ lies, where $\sin \theta = \frac{12}{13}$ and $\theta$ is in quadrant II.

**17.** Find $\tan (5\pi/12)$ exactly.

**18.** Simplify: $4 \sin 2x \cos 2x$.

**19.** Prove the identity

$$\frac{1 - \cos 2\theta}{\sin 2\theta} = \tan \theta.$$

**20.** Find, in radians, $\arcsin \left(-\sqrt{2}/2\right)$.

**21.** Use a calculator to find, in degrees, $\tan^{-1} 0.931304$.

**22.** Find $\cos^{-1}\left(\sqrt{3}/2\right)$.

**23.** Find $\operatorname{arccot}(-1)$.

**24.** Find $\sin\left(\tan^{-1}\sqrt{3}\right)$.

**25.** Find $\arccos\left(\cos \dfrac{\pi}{4}\right)$.

Find all solutions of the following equations in $[0, 2\pi)$.

**26.** $2 \sin^2 x - 5 \sin x = 3$

**27.** $2 \cos^3 x - \cos x = 0$

**28.** $\tan^2 x + 3 \tan x = 5$

**29.** Given that $\sin \theta = 0.4540$, $\cos \theta = 0.8910$, $\tan \theta = 0.5095$, $\cot \theta = 1.963$, $\sec \theta = 1.122$, and $\csc \theta = 2.203$, find the six function values for $90° - \theta$.

● **SYNTHESIS** _____

**30.** Solve $\ln (\sin x) = 0$. Restrict the solutions to $[0, 2\pi)$.

Triangle trigonometry is important in applications such as surveying and navigation. In this chapter, we continue the study of triangle trigonometry that we began in Chapter 7. The trigonometric functions can also be used to solve triangles that are not right (*oblique triangles*). • The idea of a *vector* is related to the study of triangles. A vector is a quantity that has a direction. Vectors have many practical applications in the physical sciences. • We conclude the chapter with a continuation of the study of complex numbers begun in Chapter 2. Complex numbers have applications in both electricity and engineering. •

# 8
# Triangles, Vectors, and Applications

# 8.1
## THE LAW OF SINES

The trigonometric functions can also be used to solve triangles that are not right triangles. Such triangles are called **oblique.** Any triangle, right or oblique, can be solved if at least one side and any other two measures are known. The four possible situations are as follows:

1. Two angles and any side of a triangle are known (AAS and ASA).
2. Two sides of a triangle and an angle opposite one of them are known (SSA). In this case, there may be no solution, one solution, or two solutions. The latter is known as the *ambiguous case.*
3. Two sides of a triangle and the included angle are known (SAS).
4. All three sides of the triangle are known (SSS).

In order to solve oblique triangles, we need to derive some properties. One is the *law of sines* and the other is the *law of cosines.* The law of sines applies to the first two situations listed above. The law of cosines, which we will develop in Section 8.2, applies to the last two situations.

## The Law of Sines

We consider any oblique triangle. It may or may not have an obtuse angle. Although we look at both cases, the derivations are essentially the same. In the following figures, $h$ is the same length.

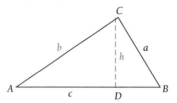 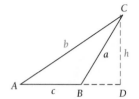

The triangles are lettered in the standard way, with angles $A$, $B$, and $C$ and the sides opposite them $a$, $b$, and $c$, respectively. We have drawn an altitude from vertex $C$. It has length $h$. In either triangle we now have, from triangle $ADC$,

$$\frac{h}{b} = \sin A, \quad \text{or}$$

$$h = b \sin A.$$

From triangle $DBC$ on the left above, we have

$$\frac{h}{a} = \sin B, \quad \text{or}$$

$$h = a \sin B.$$

On the right above, we have

$$\frac{h}{a} = \sin (180° - B) = \sin B, \quad \text{or}$$

$$h = a \sin B.$$

So in either triangle we now have

$$h = a \sin B \quad \text{and} \quad h = b \sin A.$$

It follows then that

$$a \sin B = b \sin A.$$

We divide by $\sin A \sin B$ to obtain

$$\frac{a}{\sin A} = \frac{b}{\sin B}.$$

There is no danger of dividing by 0 here because we are dealing with triangles, whose angles are never 0° or 180°.

If we were to consider an altitude from vertex $A$ in the triangles shown above, the same argument would give us

$$\frac{b}{\sin B} = \frac{c}{\sin C}.$$

We combine these results to obtain the law of sines, which holds for right triangles as well as oblique triangles.

> **THEOREM 1**          **The Law of Sines**
>
> In any triangle $ABC$,
>
> $$\frac{a}{\sin A} = \frac{b}{\sin B} = \frac{c}{\sin C}.$$
>
> (The sides are proportional to the sines of the opposite angles.)

## ▶ Solving Triangles (AAS and ASA)

When two angles and a side of any triangle are known, the law of sines can be used to solve the triangle.

**Example 1**   In triangle $ABC$, $a = 4.56$, $A = 43°$, and $C = 57°$. Solve the triangle.

Solution   We first draw a sketch.

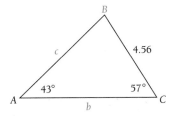

We then find $B$, as follows:

$$B = 180° - (43° + 57°) = 80°.$$

We can now find the other two sides, using the law of sines:

$$\frac{c}{\sin C} = \frac{a}{\sin A}$$

$$\frac{c}{\sin 57°} = \frac{4.56}{\sin 43°}$$

$$c = \frac{4.56 \sin 57°}{\sin 43°} \qquad \text{Solving for } c$$

$$c = 5.61;$$

$$\frac{b}{\sin B} = \frac{a}{\sin A}$$

$$\frac{b}{\sin 80°} = \frac{4.56}{\sin 43°}$$

$$b = \frac{4.56 \sin 80°}{\sin 43°} \qquad \text{Solving for } b$$

$$b = 6.58.$$

We have now found the unknown parts of the triangle: $B = 80°$, $c = 5.61$, and $b = 6.58$.   ◀

1. Solve this triangle.

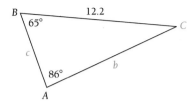

 **Solving Triangles (SSA)**

When two sides of a triangle and an angle opposite one of them are known, the law of sines can be used to solve the triangle. However, there may be no solution, one solution, or two solutions.

Suppose that $a$, $b$, and $A$ are given. Then the various possibilities are as shown in the five cases below. Note that $a < b$ in cases I, II, and III, $a = b$ in case IV, and $a > b$ in case V. Since there are two solutions in case III, this possibility is referred to as the **ambiguous case.**

Case I:  No solution

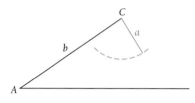

$a < b$, side $a$ is too short to reach the base.

Case II:  One solution

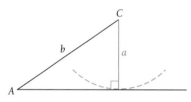

$a < b$, side $a$ just reaches the base and is perpendicular to it.

Case III:  Two solutions (Ambiguous case)

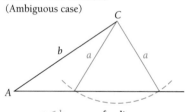

$a < b$, an arc of radius $a$ meets the base at two points.

Case IV:  One solution

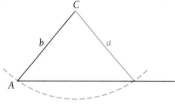

$a = b$, an arc of radius $a$ meets the base at just one point, other than $A$.

Case V:  One solution

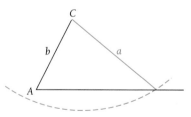

$a > b$, an arc of radius $a$ meets the base at just one point.

The following examples correspond to cases I, III, and V just described.

**Example 2**    *Case I: No solution.*    In triangle $ABC$, $a = 15$, $b = 25$, and $A = 47°$. Solve the triangle.

Solution    We look for $B$:

$$\frac{a}{\sin A} = \frac{b}{\sin B}$$

$$\frac{15}{\sin 47°} = \frac{25}{\sin B}$$

$$\sin B = \frac{25 \sin 47°}{15} \qquad \text{Solving for } \sin B$$

$$\sin B = 1.219.$$

Since there is no angle with a sine greater than 1, there is *no* solution.    ◀

**DO EXERCISE 2.**

**Example 3    *Case III: Two solutions.***    In triangle $ABC$, $a = 20$, $b = 15$, and $B = 30°$. Solve the triangle.

Solution    We look for $A$:

$$\frac{a}{\sin A} = \frac{b}{\sin B}$$

$$\frac{20}{\sin A} = \frac{15}{\sin 30°}$$

$$\sin A = \frac{20 \sin 30°}{15}$$

$$\sin A = 0.6667.$$

There are two angles less than 180° with a sine of 0.6667. They are 42° and 138°, to the nearest degree. This gives us two possible solutions.

*Possible Solution 1.*    If $A = 42°$, then

$$C = 180° - (30° + 42°) = 108°.$$

We now find $c$:

$$\frac{c}{\sin C} = \frac{b}{\sin B}$$

$$\frac{c}{\sin 108°} = \frac{15}{\sin 30°}$$

$$c = \frac{15 \sin 108°}{\sin 30°} \qquad \text{Solving for } c$$

$$c = 29.$$

These parts make a triangle, as shown; thus we have a solution.

*Possible Solution 2.*    If $A = 138°$, then

$$C = 180° - (30° + 138°) = 12°.$$

2. Solve this triangle.

$$a = 40, \quad b = 12, \quad B = 57°$$

3. In triangle $ABC$, $a = 25$, $b = 20$, and $B = 33°$. Solve the triangle.

We now find $c$:

$$\frac{c}{\sin C} = \frac{b}{\sin B}$$

$$\frac{c}{\sin 12°} = \frac{15}{\sin 30°}$$

$$c = \frac{15 \sin 12°}{\sin 30°} \qquad \text{Solving for } c$$

$$c = 6.$$

These parts make a triangle; thus we have a second solution.

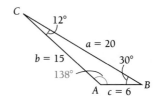

**DO EXERCISE 3.**

**Example 4** *Case V: One solution.*    In triangle $ABC$, $a = 25$, $b = 10$, and $A = 42°$. Solve the triangle.

Solution    We look for $B$:

$$\frac{b}{\sin B} = \frac{a}{\sin A}$$

$$\frac{10}{\sin B} = \frac{25}{\sin 42°}$$

$$\sin B = \frac{10 \sin 42°}{25} \qquad \text{Solving for } \sin B$$

$$\sin B = 0.2677.$$

Then $B = 16°$ or $B = 164°$, to the nearest degree. Since $a > b$, we know that there is only one solution. If we had not noticed this, we could tell it now. An angle of $164°$ cannot be an angle of this triangle because it already has an angle of $42°$ and these two angles would total more than $180°$. Thus $16°$ is the only possibility, and

$$C = 180° - (42° + 16°) = 122°.$$

We now find $c$:

$$\frac{c}{\sin C} = \frac{a}{\sin A}$$

$$\frac{c}{\sin 122°} = \frac{25}{\sin 42°}$$

$$c = \frac{25 \sin 122°}{\sin 42°} \qquad \text{Solving for } c$$

$$c = 32.$$

4. In triangle $ABC$, $b = 20$, $c = 10$, and $B = 38°$. Solve the triangle.

**DO EXERCISE 4.**

## ▶ The Area of a Triangle

The familiar formula for the area of a triangle, $A = \frac{1}{2}bh$, can be used only when $h$ is known. The method used to derive the law of sines can be used to derive an area formula that does not involve the height.

Consider a general triangle $ABC$ with area $K$.

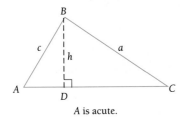

A is acute.

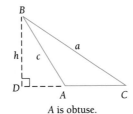

A is obtuse.

In each triangle $ADB$,

$$\frac{h}{c} = \sin A, \quad \text{or} \quad h = c \sin A.$$

Substituting into the formula $K = \frac{1}{2}bh$, we get

$$K = \frac{1}{2}b \cdot c \sin A.$$

Any pair of sides and the included angle could have been used. Thus we also have

$$K = \frac{1}{2}ab \sin C \quad \text{and} \quad K = \frac{1}{2}ac \sin B.$$

---

**THEOREM 2**        **Area of a Triangle**

The area $K$ of any triangle $ABC$ is one-half the product of the lengths of two sides and the sine of the included angle:

$$K = \frac{1}{2}bc \sin A = \frac{1}{2}ab \sin C = \frac{1}{2}ac \sin B.$$

---

**Example 5**   Find the area of triangle $ABC$ if $b = 10.1$ cm, $c = 6.8$ cm, and $A = 62.1°$.

**Solution**

$$
\begin{aligned}
K &= \tfrac{1}{2}bc \sin A && \text{Theorem 2}\\
&= \tfrac{1}{2}(10.1)(6.8)\sin 62.1° && \text{Substituting}\\
&= 30.3 \text{ cm}^2
\end{aligned}
$$

◀

**DO EXERCISE 5.**

5. Find the area of triangle $ABC$ if $a = 9.2$ in., $b = 8.0$ in., and $C = 37.6°$.

---

**EXERCISE SET** **8.1**

▲   Solve the triangle, if possible.

1. $A = 133°, B = 30°, b = 18$

2. $B = 120°, C = 30°, a = 16$

3. $B = 38°, C = 21°, b = 24$

4. $A = 131°, C = 23°, b = 10$

**5.** $A = 68.5°$, $C = 42.7°$, $c = 23.5$ cm

**6.** $B = 118.3°$, $C = 45.6°$, $b = 42.1$ ft

**7.** $c = 3$ mi, $B = 37.48°$, $C = 32.16°$

**8.** $a = 200$ m, $A = 32.76°$, $C = 21.97°$

**9.** $c = 6019$ km, $A = 16°56'$, $B = 59°23'$

**10.** $A = 120°18'$, $C = 27°46'$, $a = 16{,}435$ mi

**11.** $A = 129°32'$, $C = 18°28'$, $b = 1204$ in.

**12.** $B = 37°51'$, $C = 19°47'$, $b = 240.123$ mm

**B** Solve the triangle, if possible.

**13.** $A = 36.5°$, $a = 24$, $b = 34$

**14.** $C = 43.2°$, $c = 28$, $b = 27$

**15.** $A = 116.3°$, $a = 17.2$, $c = 13.5$

**16.** $A = 47.8°$, $a = 28.3$, $b = 18.2$

**17.** $C = 61°10'$, $c = 30.3$, $b = 24.2$

**18.** $B = 58°40'$, $a = 25.1$, $b = 32.6$

**19.** $a = 2345$ mi, $b = 2345$ mi, $A = 124.67°$

**20.** $b = 56.78$ yd, $c = 56.78$ yd, $C = 83.78°$

**21.** $a = 20.01$ cm, $b = 10.07$ cm, $A = 30.3°$

**22.** $A = 115°$, $c = 45.6$ yd, $a = 23.8$ yd

**23.** $a = 4000$ m, $b = 8000$ m, $A = 32°52'$

**24.** $a = 2$ cm, $b = 6$ cm, $A = 30°$

**25.** $b = 4.157$ km, $c = 3.446$ km, $C = 51°48'$

**26.** $b = 43.67$ mm, $c = 38.86$ mm, $C = 48.76°$

**27.** $A = 89°$, $a = 15.6$ in., $b = 18.4$ in.

**28.** $B = 36°$, $b = 8.0$ ft, $c = 10.0$ ft

**29.** $A = 41°50'$, $a = 90.0$ mi, $c = 110.0$ mi

**30.** $C = 46°30'$, $a = 56.2$ m, $c = 22.1$ m

**C** Find the area of the triangle.

**31.** $B = 42°$, $a = 7.2$ ft, $c = 3.4$ ft

**32.** $A = 17°12'$, $b = 10$ in., $c = 13$ in.

**33.** $C = 82°54'$, $a = 4$ yd, $b = 6$ yd

**34.** $C = 75.16°$, $a = 1.5$ m, $b = 2.1$ m

**35.** $B = 135.2°$, $a = 46.12$ ft, $c = 36.74$ ft

**36.** $A = 113°$, $b = 18.2$ cm, $c = 23.7$ cm

**A**, **B** Solve. Keep in mind the two types of bearing considered in Section 6.2.

**37.** Points $A$ and $B$ are on opposite sides of a lunar crater. Point $C$ is 50 m from $A$. The measure of $\angle BAC$ is determined to be 112° and the measure of $\angle ACB$ is determined to be 42°. What is the width of the crater?

**38.** A guy wire to a pole makes a 71° angle with level ground. At a point 25 ft farther from the pole than the guy wire, the angle of elevation of the top of the pole is 37°. How long is the guy wire?

**39.** A pole leans away from the sun at an angle of 7° to the vertical (see the figure). When the angle of elevation of the sun is 51°, the pole casts a shadow 47 ft long on level ground. How long is the pole?

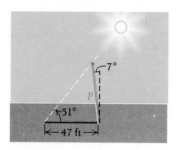

**40.** A vertical pole stands by a road that is inclined 10° to the horizontal. When the angle of elevation of the sun is 23°, the pole casts a shadow 38 ft long directly downhill along the road. How long is the pole?

**41.** A reconnaissance plane leaves its airport on the east coast of the United States and flies in a direction of 085°. Because of bad weather, it returns to another airport 230 km to the north of its home base. For the return trip, it flies in a direction of 283°. What was the total distance that it flew?

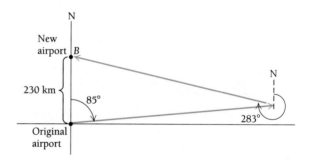

**42.** Lookout station $B$ is 10.2 km east of station $A$. The bearing of a fire from $A$ is S 10°40' W. The bearing of the fire from $B$ is S 31°20' W. How far is the fire from $A$? from $B$?

**43.** A ranger in fire tower $A$ spots a fire at a direction of 295°. A ranger in fire tower $B$, located 45 mi at a direction of 045° from tower $A$, spots the same fire at a direction of 255°. How far from tower $A$ is the fire? from tower $B$?

**44.** An airplane leaves airport $A$ and flies 200 km. At this time its direction from airport $B$, 250 km to the west of $A$, is 120°. How far is the airplane from $B$?

**45.** A boat leaves a lighthouse $A$ and sails 5.1 km. At this time it is sighted from lighthouse $B$, 7.2 km west of $A$. The bearing of the boat from $B$ is N 65°10' E. How far is the boat from $B$?

**46.** Mackinac Island is located 35 mi N 65°20' W of Cheboygan, Michigan, where the Coast Guard cutter Mackinaw is stationed. A freighter in distress radios the Coast Guard cutter for help. It radios their position as N 25°40' E of Mackinac Island and N 10°10' W of Cheboygan. How far is the freighter from Cheboygan?

●  SYNTHESIS

**47.** Prove each of the following formulas for a general triangle *ABC*.

**a)** $\dfrac{a+b}{c} = \dfrac{\cos\frac{1}{2}(A-B)}{\sin\frac{1}{2}C}$

**b)** $\dfrac{a-b}{c} = \dfrac{\sin\frac{1}{2}(A-B)}{\cos\frac{1}{2}C}$

**48.** Prove the following area formulas for a general triangle *ABC* with area represented by *K*.

$$K = \frac{a^2 \sin B \sin C}{2 \sin A} = \frac{c^2 \sin A \sin B}{2 \sin C}$$
$$= \frac{b^2 \sin C \sin A}{2 \sin B}$$

**49.** Prove that the area of a parallelogram is the product of two sides and the sine of the included angle.

**50.** Prove that the area of a quadrilateral is one-half the product of the lengths of its diagonals and the sine of the angle between the diagonals.

●  CHALLENGE

**51.** Consider the following triangle. Prove that

$$\frac{\sin\alpha}{z} + \frac{\sin\beta}{x} = \frac{\sin(\alpha+\beta)}{y}.$$

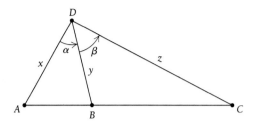

**52.** When two objects, such as ships, airplanes, or runners, move in straight-line paths, if the distance between them is decreasing and if the bearing from one of them to the other is constant, they will collide. ("Constant bearing means collision," as mariners put it.) Prove that this statement is true.

# 8.2
# THE LAW OF COSINES

The law of sines is used to solve triangles given a side and two angles (AAS and ASA) or given two sides and an angle opposite one of them (SSA). A second law, called the *law of cosines*, is needed to solve triangles given two sides and the included angle (SAS) or given three sides (SSS).

## The Law of Cosines

To derive this property, we consider any triangle *ABC* placed on a coordinate system. We position the origin at one of the vertices, say *C*, and the positive half of the *x*-axis along one of the sides, say *CB*. Let $(x, y)$ be the coordinates of vertex *A*. Point *B* has coordinates $(a, 0)$, and point *C* has coordinates $(0, 0)$.

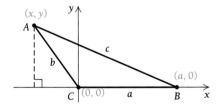

Then

$$\cos C = \frac{x}{b}, \quad \text{so} \quad x = b\cos C,$$

and

$$\sin C = \frac{y}{b}, \quad \text{so} \quad y = b \sin C.$$

Thus,

$$(x, y) = (b \cos C, b \sin C),$$

and point $A$ has coordinates $(b \cos C, b \sin C)$.

Next, we use the distance formula to determine $c^2$:

$$c^2 = (x - a)^2 + (y - 0)^2,$$

or

$$c^2 = (b \cos C - a)^2 + (b \sin C - 0)^2.$$

Now we multiply and simplify:

$$c^2 = b^2 \cos^2 C - 2ab \cos C + a^2 + b^2 \sin^2 C$$
$$= a^2 + b^2(\sin^2 C + \cos^2 C) - 2ab \cos C$$
$$= a^2 + b^2 - 2ab \cos C.$$

Had we placed the origin at one of the other vertices, we would have obtained

$$a^2 = b^2 + c^2 - 2bc \cos A$$

or

$$b^2 = a^2 + c^2 - 2ac \cos B.$$

These results can be summarized as follows.

---

**THEOREM 3**             **The Law of Cosines**

In any triangle $ABC$,

$$a^2 = b^2 + c^2 - 2bc \cos A,$$
$$b^2 = a^2 + c^2 - 2ac \cos B,$$

and

$$c^2 = a^2 + b^2 - 2ab \cos C.$$

(In any triangle, the square of a side is the sum of the squares of the other two sides, minus twice the product of those sides and the cosine of the included angle.)

---

CAUTION! Should the law of sines be applied after the law of cosines, one should be alert for the possibility of the ambiguous case.

## Solving Triangles (SAS)

When two sides of a triangle and the included angle are known, we can use the law of cosines to find the third side. The law of cosines or the law of sines can then be used to finish solving the triangle.

**Example 1**  In triangle $ABC$, $a = 24$, $c = 32$, and $B = 115°$. Solve the triangle.

**Solution**  We first sketch the triangle and label it.

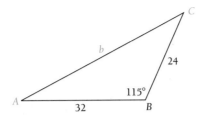

Next we find the third side. From the law of cosines,

$$b^2 = a^2 + c^2 - 2ac \cos B$$
$$b^2 = 24^2 + 32^2 - 2 \cdot 24 \cdot 32(\cos 115°)$$
$$b^2 = 576 + 1024 - 1536 \cos 115°$$
$$b^2 = 2249$$
$$b = 47.$$

We now have $a = 24$, $b = 47$, and $c = 32$. We need to find the other two angle measures. At this point, we can find them in two ways. One way uses the law of sines. The ambiguous case may arise, however, and we would have to be alert to this possibility. The advantage of using the law of cosines again is that if we solve for the cosine and its value is negative, then we know that the angle is obtuse. If the value of the cosine is positive, then the angle is acute. Thus we use the law of cosines to find a second angle.

Let us find angle $A$. We select the formula from the law of cosines that contains $\cos A$ and substitute:

$$a^2 = b^2 + c^2 - 2bc \cos A$$
$$24^2 = 47^2 + 32^2 - 2(47)(32) \cos A$$
$$576 = 2209 + 1024 - 3008 \cos A$$
$$-2657 = -3008 \cos A$$
$$\cos A = 0.8833.$$

Then $A = 28°$. The third angle is now easy to find:

$$C = 180° - (115° + 28°) = 37°.$$

Answers may vary depending on the order in which they are found. Had we found the measure of angle $C$ first in Example 1, the angle measures would have been $C = 39°$ and $A = 26°$. The answers at the back of the text were generated by considering alphabetical order. Variances in rounding also changes the answers. Had we used 47.4 for $b$ in Example 1, the angle measures would have been $A = 27°$ and $C = 38°$.  ◀

**DO EXERCISE 1.**

## ▶ Solving Triangles (SSS)

When all three sides of a triangle are known, the law of cosines can be used to solve the triangle.

1. In triangle $ABC$, $b = 18$, $c = 28$, and $A = 122°$. Solve the triangle.

**2.** In triangle $ABC$, $a = 25$, $b = 10$, and $c = 20$. Solve the triangle.

**Example 2**    In triangle $ABC$, $a = 18$, $b = 25$, and $c = 12$. Solve the triangle.

Solution    We sketch the triangle.

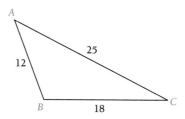

Let us find angle $B$. We select the formula from the law of cosines that contains $\cos B$; in other words, $b^2 = a^2 + c^2 - 2ac \cos B$. We substitute and solve for $\cos B$:

$$b^2 = a^2 + c^2 - 2ac \cos B$$
$$25^2 = 18^2 + 12^2 - 2(18)(12) \cos B$$
$$625 = 324 + 144 - 432 \cos B$$
$$157 = -432 \cos B$$
$$\cos B = -0.3634.$$

Then $B = 111.31°$. Similarly, we find angle $A$*:

$$a^2 = b^2 + c^2 - 2bc \cos A$$
$$18^2 = 25^2 + 12^2 - 2(25)(12) \cos A$$
$$324 = 625 + 144 - 600 \cos A$$
$$-445 = -600 \cos A$$
$$\cos A = 0.7417.$$

Thus, $A = 42.12°$. Then $C = 180° - (111.31° + 42.12°) = 26.57°$.    ◀

**DO EXERCISE 2.**

 ## Choosing the Appropriate Law

The following summarizes the situations in which to use the law of sines and the law of cosines.

---

To solve an oblique triangle:

**A)** Use the *law of sines* when:

    **1.** Two angles and any side of a triangle are known (AAS and ASA).

    **2.** Two sides of a triangle and an angle opposite one of them are known (SSA). Keep in mind that there may be no solution, one solution, or two solutions. The latter is known as the *ambiguous case.*

**B)** Use the *law of cosines* when:

    **3.** Two sides of a triangle and the included angle are known (SAS).

    **4.** All three sides of the triangle are known (SSS).

---

\* The law of sines could be used at this point, but be alert for the ambiguous case.

Let us practice determining which law to use.

**Example 3**   In triangle $ABC$, $a = 20$, $b = 15$, and $B = 30°$. Determine which law applies. Then solve the triangle.

Solution   We are given two sides and an angle opposite one of them. In this case, the law of sines applies. Thus there may be no solution, one solution, or two solutions (the ambiguous case). This is indeed the ambiguous case. The triangle is solved in Example 3 of Section 8.1.   ◀

**Example 4**   In triangle $ABC$, $a = 24$, $c = 32$, and $B = 115°$. Determine which law applies. Then solve the triangle.

Solution   We are given two sides and the included angle. In this case, the law of cosines applies. The triangle is solved in Example 1 of this section.   ◀

**DO EXERCISES 3–8.**

Determine which law applies. Then solve the triangle.

**3.** $a = 50, b = 20, c = 40$

**4.** $b = 40, c = 20, B = 38°$

**5.** $b = 54, c = 84, A = 122.4°$

**6.** $a = 75, b = 60, B = 33°$

**7.** $A = 43°35', a = 4.5, b = 6.0$

**8.** $b = 6, B = 57.1°, a = 20$

---

● **EXERCISE SET** **8.2**

**A** Solve the triangle, if possible.

**1.** $A = 30°, b = 12, c = 24$
**2.** $C = 60°, a = 15, b = 12$
**3.** $B = 133°, a = 12, c = 15$
**4.** $A = 116°, b = 31, c = 25$
**5.** $B = 72°40', c = 16$ m, $a = 78$ m
**6.** $A = 24°30', b = 68$ ft, $c = 14$ ft
**7.** $C = 22.28°, a = 25.4$ cm, $b = 73.8$ cm
**8.** $B = 72.66°, a = 23.78$ km, $c = 25.74$ km
**9.** $A = 96°13', b = 15.8$ yd, $c = 18.4$ yd
**10.** $C = 28°43', a = 6$ mm, $b = 9$ mm
**11.** $a = 60.12$ mi, $b = 40.23$ mi, $C = 48.7°$
**12.** $b = 10.2$ in., $c = 17.3$ in., $A = 53.456°$

**B** Solve the triangle, if possible.

**13.** $a = 12, b = 14, c = 20$
**14.** $a = 22, b = 22, c = 35$
**15.** $a = 16$ m, $b = 20$ m, $c = 32$ m
**16.** $a = 2.2$ cm, $b = 4.1$ cm, $c = 2.4$ cm
**17.** $a = 2$ ft, $b = 3$ ft, $c = 8$ ft
**18.** $a = 17$ yd, $b = 15.4$ yd, $c = 1.5$ yd
**19.** $a = 11.2$ cm, $b = 5.4$ cm, $c = 7$ cm
**20.** $a = 26.12$ km, $b = 21.34$ km, $c = 19.25$ km

**C** Determine which law applies. Then solve the triangle.

**21.** $A = 70°, B = 12°, b = 21.4$
**22.** $a = 15, c = 7, B = 62°$
**23.** $a = 3.3, b = 2.7, c = 2.8$
**24.** $a = 1.5, b = 2.5, A = 58°$
**25.** $a = 60, b = 40, C = 47°$
**26.** $a = 3.6, b = 6.2, c = 4.1$
**27.** $B = 110°30', C = 8°10', c = 0.912$
**28.** $B = 52°, C = 15°, b = 60.4$

**A, B**

**29.** A ship leaves a harbor and sails 15 nautical mi east. It then sails 18 nautical mi in a direction of S 27° E. How far is it, then, from the harbor and in what direction?

**30.** An airplane leaves an airport and flies west 147 km. It then flies 200 km in a direction of 220°. How far is it, then, from the airport and in what direction?

**31.** Two ships leave harbor at the same time. The first sails N 15° W at 25 knots (a knot is one nautical mile per hour). The second sails N 32° E at 20 knots. After 2 hr, how far apart are the ships?

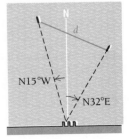

**32.** Two airplanes leave an airport at the same time. The first flies 150 km/h in a direction of 320°. The second flies 200 km/h in a direction of 200°. After 3 hr, how far apart are the planes?

**33.** A hill is inclined 5° to the horizontal. A 45-ft pole stands at the top of the hill. How long a rope will it take to reach from the top of the pole to a point 35 ft downhill from the base of the pole?

**34.** A hill is inclined 15° to the horizontal. A 40-ft pole stands at the top of the hill. How long a rope will it take to reach from the top of the pole to a point 68 ft downhill from the base of the pole?

**35.** A piece of wire 5.5 m long is bent into a triangular shape. One side is 1.5 m long and another is 2 m long. Find the angles of the triangle.

**36.** A triangular lot has sides of 120 ft, 150 ft, and 100 ft. Find the angles of the lot.

**37.** A slow-pitch softball diamond is a square 65 ft on a side. The pitcher's mound is 46 ft from home. How far is it from the pitcher's mound to first base?

**38.** A baseball diamond is a square 90 ft on a side. The pitcher's mound is 60.5 ft from home. How far does the pitcher have to run to cover first base?

**39.** The longer base of an isosceles trapezoid measures 14 ft. The nonparallel sides measure 10 ft, and the base angles measure 80°.

**a)** Find the length of a diagonal.
**b)** Find the area.

**40.** An isosceles triangle has a vertex angle of 38° and this angle is included by two sides, each measuring 20 ft. Find the area of the triangle.

**41.** A field in the shape of a parallelogram has sides that measure 50 yd and 70 yd. One angle of the field measures 78°. Find the area of the field.

**42.** An aircraft takes off to fly a 180-mi trip. After flying 75 mi, it is 10 mi off course. How much should the heading be corrected to then fly straight to the destination, assuming no wind correction?

**43.** A fence along one side of a triangular lot is 28 m long and makes angles of 25° and 58° with the sides. How long are the other two sides?

**44.** A guy wire to a pole makes a 63°40′ angle with level ground. At a point 8.2 m farther from the pole than the guy wire, the angle of elevation of the top of the pole is 42°30′. How long is the guy wire?

**45.** A triangular swimming pool has sides of length 40 ft, 35.3 ft, and 26.6 ft. Find the angles of the triangle.

**46.** A triangular swimming pool measures 44 ft on one side and 32.8 ft on another side. These sides form an angle that measures 40.8°. How long is the other side?

**47.** A pole leans away from the sun at an angle of 8° to the vertical. When the angle of elevation of the sun is 53°,

the pole casts a shadow 44 ft long on level ground. How long is the pole?

**48.** A hill is inclined 6° to the horizontal. A 42-ft pole stands at the top of the hill. How long a wire will it take to reach from the top of the pole to a point 35 ft downhill from the base of the pole?

**49.** Three circles are arranged as shown in the figure below. Find the length $PQ$.

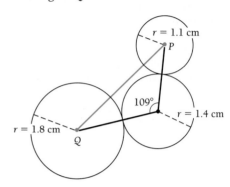

**50.** Three coins are arranged as shown in the figure below. Find the angle $\phi$.

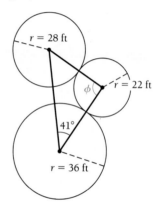

● **SYNTHESIS**

**51.** *Heron's formula.* If $a$, $b$, and $c$ are the lengths of the sides of a triangle, then the area $K$ of the triangle is given by

$$K = \sqrt{s(s-a)(s-b)(s-c)},$$

where $s = \frac{1}{2}(a + b + c)$. The number $s$ is called the **semiperimeter.** Prove Heron's formula. (*Hint:* Use the area formula $K = \frac{1}{2}bc \sin A$ developed in Section 8.1.)

**52.** Use Heron's formula to find the area of the triangular swimming pool described in Exercise 45.

**53.** The measures of two sides of a parallelogram are 50 and 60 in., while one diagonal is 90 in. How long is the other diagonal?

**54.** The Port Huron-to-Mackinac sailboat race covers a distance of 259 mi from Port Huron, Michigan, to Mackinac Island. During the race, the course is measured at a heading of 035° from Port Huron to Cove Island for a distance of 181 mi. From Cove Island, the boats head 295° to Mackinac Island. With a north wind, one boat started on a course of 320° and sailed for 85 mi. The navigator then determined that the boat could "tack" (change heading) and sail directly for Cove Island. How far is the boat from Cove Island, and what must its new heading be in order to reach Cove Island?

**55.** From the top of a hill 20 ft above the surface of a lake, a tree is sighted across the lake. The angle of elevation is 11°. From the same position, the top of the tree is seen by reflection in the lake. The angle of depression is 14°. What is the height of the tree? (*Hint:* The angle of incidence, $\theta$, equals the angle of reflection, $\alpha$.)

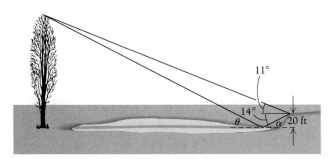

**56.** A bridge is being built across a canyon. The length of the bridge is 5045 ft. From the deepest point in the canyon, the angles of elevation of the ends of the bridge are 78° and 72°. How deep is the canyon?

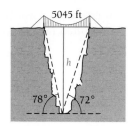

**57.** Find a formula for the area of an isosceles triangle in terms of the congruent sides and their included angle. Under what conditions will the area of a triangle with fixed congruent sides be maximum?

**58.** *Surveying.* In surveying, a series of bearings and distances is called a **transverse.** Measurements are taken as shown in the following figure.

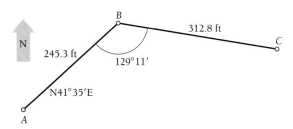

**a)** Compute the bearing *BC.*
**b)** Compute the distance and bearing *AC.*

● **CHALLENGE** _____

**59.** Show that in any triangle *ABC,*
$$a^2 + b^2 + c^2 = 2(bc \cos A + ac \cos B + ab \cos C).$$

**60.** Show that in any triangle *ABC,*
$$\frac{\cos A}{a} + \frac{\cos B}{b} + \frac{\cos C}{c} = \frac{a^2 + b^2 + c^2}{2abc}.$$

**61.** A reconnaissance plane patrolling at 5000 ft sights a submarine at bearing 35° at an angle of depression of 25°. A carrier is at bearing 105° and at an angle of depression of 60°. How far is the submarine from the carrier?

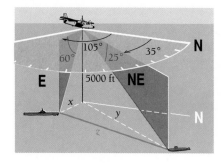

**62.** Two of NASA's tracking stations are located near the equator: One is in Ethiopia, at 40° east longitude; another is near Quito, Ecuador, at 78° west longitude. Assume that both stations, represented by *E* and *Q* in the figure below, are on the equator and that the radius of the earth is 6380 km. A satellite in orbit over the equator is observed at the same instant from both tracking stations. The angles of elevation above the horizon are 5° from Quito and 10° from Ethiopia. Find the distance of the satellite from earth at the instant of observation.*

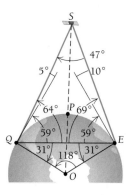

_____

* *Source: Space Mathematics,* National Aeronautics and Space Administration, pp. 104–106.

**OBJECTIVES**

You should be able to:

**A** Given two vectors, find their sum, or resultant.

**B** Solve applied problems involving finding sums of vectors.

**C** Resolve vectors into components.

**D** Add vectors using components.

**E** Change from rectangular to polar notation for vectors, and from polar to rectangular notation for vectors.

**F** Do simple manipulations in vector algebra.

# 8.3
# VECTORS AND APPLICATIONS

In many applications, there occur certain quantities in which a direction is specified. Any such quantity having a *direction* and a *magnitude* is called a **vector quantity,** or **vector.** Here are some examples of vector quantities.

**Displacement**    An object moves a certain distance in a certain direction.

A train travels 100 mi to the northeast.

A person takes 5 steps to the west.

A batter hits a ball 100 m along the left-field foul line.

**Velocity**    An object travels at a certain speed in a certain direction.

The wind is blowing 15 mph from the northwest.

An airplane is traveling 450 km/h in a direction of 243°.

**Force**    A push or pull is exerted on an object in a certain direction.

A force of 15 newtons (a newton is a unit of force used in physics, abbreviated N) is exerted downward on the handle of a jack.

A 25-lb upward force is required to lift a box.

A wagon is being pulled up a 30° incline, requiring an effort of 200 lb.

We represent vectors graphically by directed line segments, or arrows. The length is chosen, according to some scale, to represent the **magnitude of the vector,** and the direction of the arrow represents the **direction of the vector.** For example, if we let 1 cm represent 5 km/h, then a 15-km/h wind from the northwest would be represented by an arrow 3 cm long, as shown below.

## **A**  Vector Addition

To "add" vectors, we find a single vector that would have the same effect as the vectors combined. Suppose a person takes 4 steps east and then 3 steps north. He will then be 5 steps from the starting point in the direction shown. The **sum** of the two vectors is the vector 5 steps in magnitude and in the direction shown. The sum is also called the **resultant** of the two vectors.

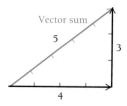

In general, if we have two vectors **a** and **b**, we can add them as in the above example.* That is, we place the tail of one arrow at the head of the other and then find the vector that forms the third side of a triangle. Or, we can place the tails of the arrows together, complete a parallelogram, and find the diagonal of the parallelogram.

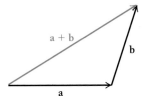 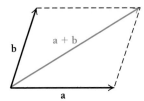

**Example 1** Forces of 15 newtons and 25 newtons act on an object at right angles to each other. Find their sum, or resultant, giving the angle that it makes with the larger force.

Solution

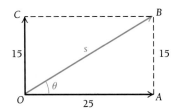

We make a drawing, this time a rectangle, using s for the length of vector **OB**. Since *OAB* is a right triangle, we have

$$\tan \theta = \frac{15}{25} = 0.6.$$

Thus $\theta$, the angle that the resultant makes with the larger force, is 31° to the nearest degree. Now, using the Pythagorean theorem, we have

$$s^2 = 15^2 + 25^2.$$

Thus, $s = \sqrt{15^2 + 25^2} = 29.2$, to the nearest tenth. The resultant **OB** has a magnitude of 29.2 N and makes an angle of 31° with the larger force. ◀

**DO EXERCISE 1.**

Vector subtraction, **a** − **b**, is defined as **a** + (−**b**), where −**b** is the vector found from **b** by heading in the opposite direction. We can show **a** − **b** as follows, though we will not give much attention to vector subtraction.

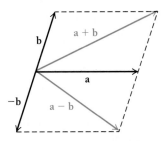

* It is general practice to denote vectors using boldface type; sometimes in handwriting, we write a half arrow over the letters, $\vec{a}$.

1. Two forces of 5.2 N and 14.6 N act at right angles to each other. Find the resultant, specifying, to the nearest degree, the angle that it makes with the larger force.

2. An airplane heads in a direction of 90° at a 140-km/h airspeed. The wind is 50 km/h from 210°. Find the direction and the speed of the airplane over the ground.

 **Applications**

**Example 2**   An airplane travels on a bearing of 100° at a 180-km/h airspeed while a wind is blowing 45 km/h from 220°. Find the speed of the airplane over the ground and the direction of its track over the ground.

**Solution**   We first make a drawing. The wind is represented by **OC** and the velocity vector of the airplane by **OA**. The resultant velocity is **v**, the sum of the two vectors. We denote the length of **v** by |**v**|.

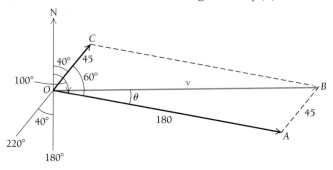

The measure of ∠*COA* is 60°, so ∠*CBA* = 60°. Now since the sum of all the angles of the parallelogram is 360° and ∠*OCB* and ∠*OAB* have the same measure, each must be 120°. By the law of cosines in △*OAB*, we have

$$|\mathbf{v}|^2 = 45^2 + 180^2 - 2 \cdot 45 \cdot 180 \cos 120°$$
$$= 42{,}525.$$

Thus, |**v**| is 206 km/h. By the law of sines in the same triangle,

$$\frac{45}{\sin \theta} = \frac{206}{\sin 120°},$$

or

$$\sin \theta = \frac{45 \sin 120°}{206}$$
$$= 0.1892.$$

Thus, $\theta = 11°$, to the nearest degree. The ground speed of the airplane is 206 km/h, and its track is in the direction of 100° − 11°, or 89°.  ◀

**DO EXERCISE 2.** _____

 **Components of Vectors**

Given a vector, it is often convenient to reverse the addition procedure, that is, to find two vectors whose sum is the given vector. Usually the two vectors we seek will be perpendicular. They are called **components** of the given vector, and the process of finding them is called **resolving** a vector into its components.

**Example 3**   A certain vector **a** has a magnitude of 130 and is inclined 40° with the horizontal. Resolve the vector into horizontal and vertical components.

**Solution**    We first make a drawing showing horizontal and vertical vectors whose sum is the given vector **a**. From △*OAB*, we see that

$$|\mathbf{h}| = 130 \cos 40° = 100$$

and

$$|\mathbf{v}| = 130 \sin 40° = 84.$$

These are the components we seek.    ◀

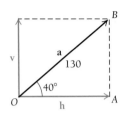

**Example 4**    An airplane is flying at 200 km/h in a direction of 305°. Find the westerly and northerly components of its velocity.

**Solution**    We first make a drawing showing westerly and northerly vectors whose sum is the given velocity. From △*OAB*, we see that

$$|\mathbf{n}| = 200 \cos 55° = 115 \text{ km/h}$$

and

$$|\mathbf{w}| = 200 \sin 55° = 164 \text{ km/h}.$$    ◀

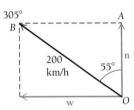

**DO EXERCISE 3.**

▷ **Adding Vectors Using Components**

Given a vector, we can resolve it into components. Given components of a vector, we can find the vector. We simply find the vector sum of the components.

**Example 5**    A vector **v** has a westerly component of 12 and a southerly component of 16. Find the vector.

**Solution**    We first make a drawing showing westerly and southerly components of vector **v**. From the drawing, we see that

$$\tan \theta = \tfrac{12}{16} = 0.75$$
$$\theta = 37°;$$

$$|\mathbf{v}| = \sqrt{12^2 + 16^2}$$
$$= 20.$$

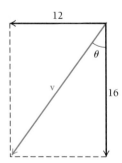

Thus the vector has a magnitude of 20 and a direction of S 37° W.    ◀

**DO EXERCISE 4.**

When we have two vectors resolved into components, we can add the two vectors by adding the components.

3. A vector of magnitude 100 points southeast. Resolve the vector into easterly and southerly components.

4. A wind has an easterly component (*from* the east) of 10 km/h and a southerly component (*from* the south) of 16 km/h. Find the magnitude and the direction of the wind.

**5.** One vector has components of 7 up and 9 to the left. A second vector has components of 3 down and 12 to the left. Find the sum, expressing it:

a) in terms of components;

b) giving magnitude and direction.

**Example 6**   A vector **v** has a westerly component of 3 and a northerly component of 5. A second vector **w** has an easterly component of 8 and a northerly component of 4. Find the components of the sum. Find the sum.

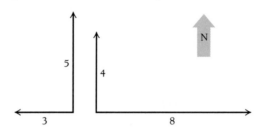

**Solution**   Adding the east–west components, we obtain 5 east. Adding the north–south components, we obtain 9 north. The components of **v** + **w** are 5 east and 9 north. To find **v** + **w**, we find both the direction and the magnitude:

$$\tan \theta = \frac{9}{5} = 1.8$$
$$\theta = 61°;$$
$$|\mathbf{v} + \mathbf{w}| = \sqrt{9^2 + 5^2}$$
$$= 10.3.$$

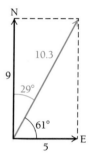

The sum $|\mathbf{v} + \mathbf{w}|$ has a magnitude of 10.3 and a direction of N 29° E.    ◀

**DO EXERCISE 5.** _____

## Analytic Representation of Vectors

If we place a vector on a coordinate system so that the tail of the vector is at the origin, we say that the vector is in **standard position.** Then if we know the coordinates of the other end of the vector, we know the vector. The coordinates will be the $x$-component and the $y$-component of the vector. Thus we can consider an ordered pair $(a, b)$ to be a vector. To emphasize that we are thinking of a vector, we usually write $\langle a, b \rangle$. When vectors are given in this form, it is easy to add them. We simply add the respective components.

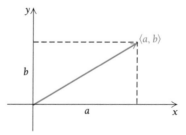

**Example 7**   Find **u** + **v**, where $\mathbf{u} = \langle 3, -7 \rangle$ and $\mathbf{v} = \langle 4, 2 \rangle$.

Solution    The sum is found by adding the *x*-components and then adding the *y*-components:

$$\mathbf{u} + \mathbf{v} = \langle 3, -7 \rangle + \langle 4, 2 \rangle = \langle 7, -5 \rangle.$$

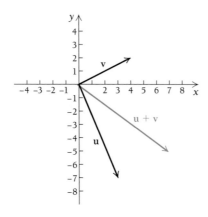

**DO EXERCISES 6 AND 7.**

Add the vectors, giving answers as ordered pairs.

6. $\langle 7, 2 \rangle$ and $\langle 5, -10 \rangle$

7. $\langle -2, 7 \rangle$ and $\langle 14, -3 \rangle$

### ▶ **Polar and Rectangular Notation**

Vectors can also be specified by giving their length and direction. When this is done, we say that we have **polar notation** for the vector. An example of polar notation is $(15, 260°)$.

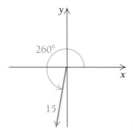

The angle is measured from the positive half of the *x*-axis counterclockwise. In Section 8.4, we consider a coordinate system (called a **polar coordinate system**) in which polar notation is very natural.

**Example 8**    Find polar notation for the vector **v**, where $\mathbf{v} = \langle 7, -2 \rangle$.

Solution    We see the reference angle $\phi$ in the figure:

$$\tan \phi = \frac{-2}{7} = -0.2857$$

$$\phi = -16°;$$

$$|\mathbf{v}| = \sqrt{(-2)^2 + 7^2}$$
$$= 7.3.$$

Thus polar notation for **v** is

$$(7.3, -16°), \quad \text{or} \quad (7.3, 344°).$$

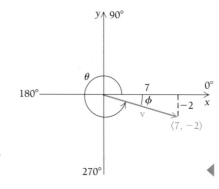

8. Find polar notation for the vector $\langle -8, 3 \rangle$.

**DO EXERCISE 8.**

9. Find rectangular notation for the vector $(15, 341°)$.

Given polar notation for a vector, we can also find the ordered-pair notation (**rectangular notation**).

**Example 9**  Find rectangular notation for the vector **w**, where **w** = $(14, 155°)$.

Solution    From the figure, we have

$$x = 14 \cos 155° = -12.7,$$
$$y = 14 \sin 155° = 5.9.$$

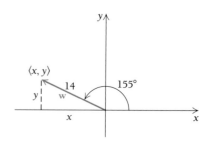

Thus the rectangular notation for **w** is $\langle -12.7, 5.9 \rangle$.

**DO EXERCISE 9.**

## F ▶ Properties of Vectors and Vector Algebra

With vectors represented in rectangular notation $\langle a, b \rangle$, it is easy to determine many of their properties. We consider the set of all ordered pairs of real numbers $\langle a, b \rangle$. This is the set of all vectors on a plane. Consider the sum of any two vectors:

$$\mathbf{u} = \langle a, b \rangle, \qquad\qquad \mathbf{v} = \langle c, d \rangle,$$
$$\mathbf{u} + \mathbf{v} = \langle a + c, b + d \rangle, \qquad \mathbf{v} + \mathbf{u} = \langle c + a, d + b \rangle.$$

We have just shown that addition of vectors is commutative. Other properties are also not difficult to show. The system of vectors is not a field, but it does have some of the familiar field properties discussed in Section 1.1.

---

**Properties of Vectors**

*Commutative law.*   Addition of vectors is commutative.

*Associative law.*   Addition of vectors is associative.

*Identity.*   There is an additive identity, the vector $\langle 0, 0 \rangle$.

*Inverses.*   Every vector **v**, given by $\mathbf{v} = \langle a, b \rangle$, has an opposite $-\mathbf{v}$, given by $-\mathbf{v} = \langle -a, -b \rangle$.

*Subtraction.*   We define $\mathbf{u} - \mathbf{v}$ to be the vector that when added to **v** gives **u**. It follows that if $\mathbf{u} = \langle a, b \rangle$ and $\mathbf{v} = \langle c, d \rangle$, then $\mathbf{u} - \mathbf{v} = \langle a - c, b - d \rangle$. Thus subtraction is always possible. Also, $\mathbf{u} - \mathbf{v} = \mathbf{u} + (-\mathbf{v})$.

*Length, or absolute value.*   If $\mathbf{u} = \langle a, b \rangle$, then it follows that $|\mathbf{u}| = \sqrt{a^2 + b^2}$.

---

A vector can be multiplied by a real number. A real number in this context is called a **scalar.** For example, $2\mathbf{v}$ is a vector in the same direction as **v** but twice as long. Analytically, we define scalar multiplication as follows.

---

**DEFINITION**    **Scalar Multiplication**

If $r$ is any real number and $\mathbf{v} = \langle a, b \rangle$, then $r\mathbf{v} = \langle ar, br \rangle$.

---

**Example 10**   Do the following calculations, where $\mathbf{u} = \langle 4, 3 \rangle$ and $\mathbf{v} = \langle -5, 8 \rangle$: (a) $\mathbf{u} + \mathbf{v}$; (b) $\mathbf{u} - \mathbf{v}$; (c) $3\mathbf{u} - 4\mathbf{v}$; (d) $|3\mathbf{u} - 4\mathbf{v}|$.

Solution

a)  $\mathbf{u} + \mathbf{v} = \langle 4, 3 \rangle + \langle -5, 8 \rangle = \langle -1, 11 \rangle$

b)  $\mathbf{u} - \mathbf{v} = \mathbf{u} + (-\mathbf{v}) = \langle 4, 3 \rangle + \langle 5, -8 \rangle = \langle 9, -5 \rangle$

c)  $3\mathbf{u} - 4\mathbf{v} = 3\langle 4, 3 \rangle - 4\langle -5, 8 \rangle = \langle 12, 9 \rangle - \langle -20, 32 \rangle$
    $= \langle 32, -23 \rangle$

d)  $|3\mathbf{u} - 4\mathbf{v}| = \sqrt{32^2 + (-23)^2} = \sqrt{1553} = 39.4$    ◀

**DO EXERCISE 10.**

10.  For $\mathbf{u} = \langle -3, 5 \rangle$ and $\mathbf{v} = \langle 4, 2 \rangle$, find each of the following.

   a)  $\mathbf{u} + \mathbf{v}$
   b)  $\mathbf{v} - \mathbf{u}$
   c)  $5\mathbf{u} - 2\mathbf{v}$
   d)  $|2\mathbf{u} + 3\mathbf{v}|$

---

● **EXERCISE SET** **8.3**

  Magnitudes of vectors **a** and **b** and the angle between the vectors $\theta$ are given. Find the resultant, to the nearest tenth, giving the direction by specifying to the nearest degree the angle that it makes with the vector **a**.

1.  $|\mathbf{a}| = 45$, $|\mathbf{b}| = 35$, $\theta = 90°$
2.  $|\mathbf{a}| = 54$, $|\mathbf{b}| = 43$, $\theta = 90°$
3.  $|\mathbf{a}| = 10$, $|\mathbf{b}| = 12$, $\theta = 67°$
4.  $|\mathbf{a}| = 25$, $|\mathbf{b}| = 30$, $\theta = 75°$
5.  $|\mathbf{a}| = 20$, $|\mathbf{b}| = 20$, $\theta = 117°$
6.  $|\mathbf{a}| = 30$, $|\mathbf{b}| = 30$, $\theta = 123°$
7.  $|\mathbf{a}| = 23$, $|\mathbf{b}| = 47$, $\theta = 27°$
8.  $|\mathbf{a}| = 32$, $|\mathbf{b}| = 74$, $\theta = 72°$

**B**

9.  Two forces of 5.8 N and 12.3 N act on an object at right angles. Find the magnitude of the resultant and the angle that it makes with the smaller force.

10.  Two forces of 32 N and 45 N act on an object at right angles. Find the magnitude of the resultant and the angle that it makes with the smaller force.

11.  Forces of 420 N and 300 N act on an object. The angle between the forces is 50°. Find the resultant, giving the angle that it makes with the larger force.

12.  Forces of 410 N and 600 N act on an object. The angle between the forces is 47°. Find the resultant, giving the angle that it makes with the smaller force.

13.  A balloon is rising 12 ft/sec while a wind is blowing 18 ft/sec. Find the speed of the balloon and the angle that it makes with the horizontal.

14.  A balloon is rising 10 ft/sec while a wind is blowing 5 ft/sec. Find the speed of the balloon and the angle that it makes with the horizontal.

15.  A boat heads 35°, propelled by a force of 750 lb. A wind from 320° exerts a force of 150 lb on the boat. How large is the resultant force, and in what direction is the boat moving?

16.  A boat heads 220°, propelled by a force of 650 lb. A wind from 080° exerts a force of 100 lb on the boat. How large is the resultant force, and in what direction is the boat moving?

17.  A ship sails N 80° E for 120 nautical mi, then S 20° W for 200 nautical mi. How far is it then from the starting point, and in what direction?

18.  An airplane flies 032° for 210 km, then 280° for 170 km. How far is it, then, from the starting point, and in what direction?

19.  A motorboat has a speed of 15 km/h. It crosses a river whose current has a speed of 3 km/h. In order to cross the river at right angles, in what direction should the boat be pointed?

20.  An airplane has an airspeed of 150 km/h. It is to make a flight in a direction of 080° while there is a 25-km/h wind from 350°. What should the airplane's heading be?

▲, ⬤

21. A vector **u** with magnitude 150 is inclined to the right and upward 52° from the horizontal. Find the horizontal and vertical components of **u**.

22. A vector **u** with magnitude 170 is inclined to the right and downward 63° from the horizontal. Find the horizontal and vertical components of **u**.

23. An airplane is flying 220° at 250 km/h. Find the magnitude of the southerly and westerly components of its velocity **v**.

24. A wind is blowing from 310° at 25 mph. Find the magnitude of the southerly and easterly components of the wind velocity **v**.

▲, ⬤

25. A force **f** has a westerly component of 25 N and a southerly component of 35 N. Find the magnitude and the direction of **f**.

26. A force **f** has a component of 65 N upward and a component of 90 N to the left. Find the magnitude and the direction of **f**.

27. Vector **u** has a westerly component of 15 and a northerly component of 22. Vector **v** has a northerly component of 6 and an easterly component of 8. Find:
    a) the components of **u** + **v**;
    b) the magnitude and the direction of **u** + **v**.

28. Vector **u** has a component of 18 up and a component of 12 to the left. Vector **v** has a component of 5 down and a component of 35 to the right. Find:
    a) the components of **u** + **v**;
    b) the magnitude and the direction of **u** + **v**.

▷  In Exercises 29–42, rectangular notation for vectors is given.

Add, writing rectangular notation for the answer.

29. $\langle 3, 7 \rangle + \langle 2, 9 \rangle$

30. $\langle 5, -7 \rangle + \langle -3, 2 \rangle$

31. $\langle 17, 7.6 \rangle + \langle -12.2, 6.1 \rangle$

32. $\langle -15.2, 37.1 \rangle + \langle 7.9, -17.8 \rangle$

33. $\langle -650, -750 \rangle + \langle -12, 324 \rangle$

34. $\langle -354, -973 \rangle + \langle -75, 256 \rangle$

▷, ⬤   Find polar notation.

35. $\langle 3, 4 \rangle$          36. $\langle 4, 3 \rangle$

37. $\langle 10, -15 \rangle$       38. $\langle 17, -10 \rangle$

39. $\langle -3, -4 \rangle$        40. $\langle -4, -3 \rangle$

41. $\langle -10, 15 \rangle$       42. $\langle -17, 10 \rangle$

Find rectangular notation.

43. $(4, 30°)$          44. $(8, 60°)$

45. $(10, 235°)$        46. $(15, 210°)$

47. $(20, 330°)$        48. $(20, 200°)$

49. $(100, -45°)$       50. $(150, -60°)$

⬤, ⬤   Do the calculations for the following vectors:
**u** = $\langle 3, 4 \rangle$, **v** = $\langle 5, 12 \rangle$, and **w** = $\langle -6, 8 \rangle$.

51. $3\mathbf{u} + 2\mathbf{v}$          52. $3\mathbf{u} - 2\mathbf{v}$

53. $(\mathbf{u} + \mathbf{v}) - \mathbf{w}$          54. $\mathbf{u} - (\mathbf{v} + \mathbf{w})$

55. $|\mathbf{u}| + |\mathbf{v}|$          56. $|\mathbf{u}| - |\mathbf{v}|$

57. $|\mathbf{u} + \mathbf{v}|$          58. $|\mathbf{u} - \mathbf{v}|$

59. $2|\mathbf{u} + \mathbf{v}|$          60. $2|\mathbf{u}| + 2|\mathbf{v}|$

⬤   SYNTHESIS

61. If PQ is any vector, what is PQ + QP?

62. The **inner product** of vectors **u** · **v** is a scalar, defined as follows: **u** · **v** = |**u**| |**v**| cos θ, where θ is the angle between the vectors. Show that vectors **u** and **v** are perpendicular if and only if **u** · **v** = 0.

⬤   CHALLENGE

63. Show that scalar multiplication is distributive over vector addition.

64. Let **u** = $\langle 3, 4 \rangle$. Find a vector that has the same direction as **u** but length 1.

65. Prove that for any vectors **u** and **v**, $|\mathbf{u} + \mathbf{v}| \leq |\mathbf{u}| + |\mathbf{v}|$.

---

**OBJECTIVES**

You should be able to:

A  Plot points given their polar coordinates, and determine polar coordinates of points on a graph.

B  Convert from rectangular to polar equations and from polar to rectangular equations.

C  Graph polar equations.

# 8.4
# POLAR COORDINATES

A  In graphing, we locate a point with an ordered pair of numbers $(a, b)$. We can consider any such ordered pair to be a vector. From our work with vectors, we know that we could also locate a point with a vector, given a length and a direction. When we use rectangular notation to locate points, we describe the coordinate system as **rectangular.**\* When we use polar notation, we describe the coordinate system as **polar.** As this diagram shows, any point has **rectangular coordinates** $(x, y)$ and **polar coordinates** $(r, \theta)$. On a polar graph, the origin is called the **pole** and the positive half of the x-axis is called the **polar axis.**

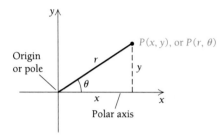

To plot points on a polar graph:

1. Locate $\theta$.
2. Move a distance $r$ from the pole in that direction.
3. If $r$ is negative, move in the opposite direction.

Polar graph paper, shown below, facilitates plotting. Point B illustrates that $\theta$ may be in radians. Point E illustrates that the polar coordinates of a point are not unique.

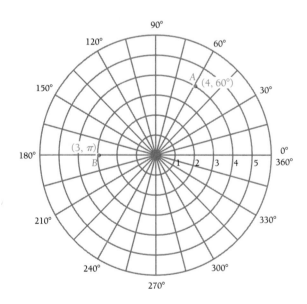

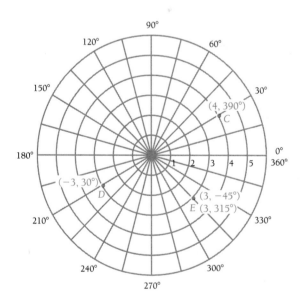

**DO EXERCISES 1–4 ON THE FOLLOWING PAGE.**

\* Also called *Cartesian*, after the French mathematician René Descartes.

1. Plot each of the following points.

   a) $(3, 60°)$       b) $(0, 10°)$

   c) $(-5, 120°)$   d) $(5, -60°)$

   e) $(2, 3\pi/2)$

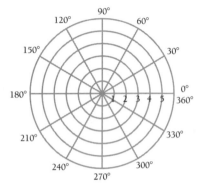

2. Find polar coordinates of each of these points. Give two answers for each.

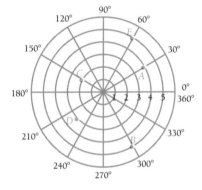

3. Find the polar coordinates of each of these points. (See Example 8 of Section 8.3.)

   a) $(3, 3)$          b) $(0, -4)$

   c) $(-3, 3\sqrt{3})$   d) $(2\sqrt{3}, -2)$

4. Find the rectangular coordinates of each of these points. (See Example 9 of Section 8.3.)

   a) $(5, 30°)$

   b) $(10, \pi/3)$

   c) $(-5, 45°)$

   d) $(-8, -5\pi/6)$

## Polar and Rectangular Equations

Some curves have simpler equations in polar coordinates than in rectangular coordinates. For others, the reverse is true. Applying the definitions of sine and cosine,

$$\frac{x}{r} = \cos \theta \quad \text{and} \quad \frac{y}{r} = \sin \theta,$$

we obtain the following relationships between the two kinds of coordinates. These allow us to convert an equation from rectangular to polar coordinates:

$$x = r \cos \theta, \qquad y = r \sin \theta.$$

**Example 1**    Convert to a polar equation: $x^2 + y^2 = 25$.

Solution

$$x^2 + y^2 = 25$$
$$(r \cos \theta)^2 + (r \sin \theta)^2 = 25 \qquad \text{Substituting for } x \text{ and } y$$
$$r^2 \cos^2 \theta + r^2 \sin^2 \theta = 25$$
$$r^2(\cos^2 \theta + \sin^2 \theta) = 25$$
$$r^2 = 25$$

◀

Example 1 illustrates that the polar equation of a circle centered at the origin is much simpler than the rectangular equation.

**Example 2**    Convert to a polar equation: $2x - y = 5$.

Solution

$$2x - y = 5$$
$$2(r \cos \theta) - (r \sin \theta) = 5$$
$$r(2 \cos \theta - \sin \theta) = 5$$

◀

**DO EXERCISES 5 AND 6 ON THE FOLLOWING PAGE.**

The relationships that we need to convert from polar equations to rectangular equations can be determined easily from the triangle. They are as follows:

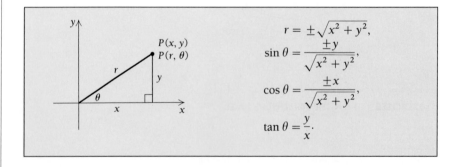

To convert to rectangular notation, we substitute as needed from either of the two lists, choosing the + or the − signs as appropriate in each case.

**Example 3**   Convert to a rectangular (Cartesian) equation: $r = 4$.

**Solution**

$$r = 4$$
$$+\sqrt{x^2 + y^2} = 4 \qquad \text{Substituting for } r$$
$$x^2 + y^2 = 16 \qquad \text{Squaring} \qquad \blacktriangleleft$$

In squaring, as we did in Example 3, we must be careful not to introduce solutions of the equation that are not already present. We did not, because the graph is a circle of radius 4 centered at the origin, in both cases.

**Example 4**   Convert to a rectangular equation: $r \cos \theta = 6$.

**Solution**   From our first list, we know that $x = r \cos \theta$, so we have $x = 6$. $\blacktriangleleft$

**Example 5**   Convert to a rectangular equation: $r = 2 \cos \theta + 3 \sin \theta$.

**Solution**   We first multiply on both sides by $r$:

$$r^2 = 2r \cos \theta + 3r \sin \theta$$
$$x^2 + y^2 = 2x + 3y. \qquad \text{Substituting} \qquad \blacktriangleleft$$

**DO EXERCISES 7–9.**

 **Graphing Polar Equations**

To graph a polar equation, we usually make a table of values, choosing values of $\theta$ and calculating corresponding values of $r$. We plot the points and then complete the graph, as in the rectangular case. A difference occurs in the case of a polar equation, because as $\theta$ increases sufficiently, points may, in some cases, begin to repeat and the curve will be traced again and again. If such a point is reached, the curve is complete.

**Example 6**   Graph: $r = \cos \theta$.

**Solution**   We first make a table of values.

| $\theta$ | 0° | 30° | 45° | 60° | 90° | 120° | 135° | 150° | 180° |
|---|---|---|---|---|---|---|---|---|---|
| $r$ | 1 | 0.866 | 0.707 | 0.5 | 0 | −0.5 | −0.707 | −0.866 | −1 |

Convert to a polar equation.

5. $2x + 5y = 9$

6. $x^2 + y^2 + 8x = 0$

Convert to a Cartesian equation.

7. $r = 7$

8. $r \sin \theta = 5$

9. $r - 3 \cos \theta = 5 \sin \theta$

**10.** Graph: $r = 1 - \sin\theta$.

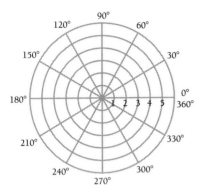

We plot these points and note that the last point is the same as the first.

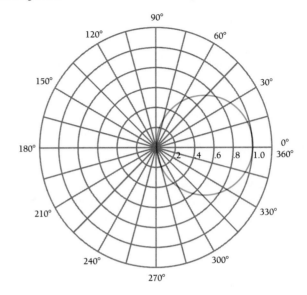

We try another point, $(-0.866, 210°)$, and find that it has already been plotted. Since the repeating has started, we plot no more points, but draw the graph, which is a circle as shown. ◀

**DO EXERCISE 10.**

**Example 7**   Graph: $r = 4 \sin 3\theta$.

---

   **TECHNOLOGY CONNECTION**

Some graphers allow the graphing of polar equations. Check the documentation for your grapher. Usually the equation must first be written in the form $r = f(\theta)$. When drawing such a graph, it is necessary to decide on not only the best viewing box but also the range of values for $\theta$. Typically, we begin with a range of 0 to $2\pi$ for $\theta$ for radians, or 0 to 360 for degrees. If the graph does not appear to be complete, these limits are extended, usually in multiples of $2\pi$ radians or 360°. (The letter $T$ may be used instead of $\theta$, if the grapher can't use Greek letters.) Because most polar graphs are curved, it is important that you choose a viewing box in which there is no distortion in either the horizontal or vertical direction.

For example, to graph the polar coordinate equation $r - 7\cos\theta = 7$, the first step is to solve for $r$, rewriting the equation in the form $r = 7 + 7\cos\theta$. If we let the values of $\theta$ increase from 0 to 360° in steps of 1° and use a viewing box of $[-15, 15] \times [-10, 10]$, the following graph will be drawn.

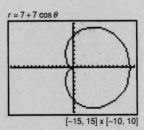

$r = 7 + 7\cos\theta$

$[-15, 15] \times [-10, 10]$

Use a grapher to draw the graphs of each of the following polar coordinate equations. Use a viewing box of $[-15, 15] \times [-10, 10]$.

**TC 1.** $r = 8 \sin 4\theta$

**TC 2.** $r = 10 \tan \theta \sin \theta$

**TC 3.** $r = 4 \sin 2\theta - 8$

**TC 4.** $r = 7 \cos 3\theta + 4 \sin 2\theta$

Solution   We first make a table of values.

| $\theta$ | 0° | 15° | 30° | 45° | 60° | 75° | 90° |
|---|---|---|---|---|---|---|---|
| $r$ | 0 | 2.83 | 4 | 2.83 | 0 | −2.83 | −4 |

| $\theta$ | 105° | 120° | 135° | 150° | 165° | 180° |
|---|---|---|---|---|---|---|
| $r$ | −2.83 | 0 | 2.83 | 4 | 2.83 | 0 |

We plot these points and sketch the curve. We then try another point, (−2.83, 195°), and find that it has already been plotted. We plot no more points since the repeating has begun.

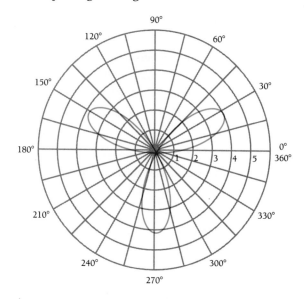

This curve is often referred to as a **three-leafed rose.**

**DO EXERCISE 11.**

11. Graph: $r = -2 \sin 3\theta$.

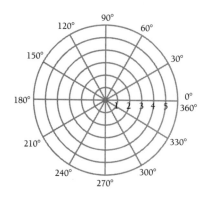

● **EXERCISE SET** **8.4**

A   Using polar coordinates, graph each of the following points.

**1.** (4, 30°)
**2.** (5, 45°)
**3.** (0, 37°)
**4.** (0, 48°)
**5.** (−6, 150°)
**6.** (−5, 135°)
**7.** (−8, 210°)
**8.** (−5, 270°)
**9.** (3, −30°)
**10.** (6, −45°)
**11.** (7, −315°)
**12.** (4, −270°)
**13.** (−3, −30°)
**14.** (−6, −45°)
**15.** (−3.2, 27°)
**16.** (−6.8, 34°)
**17.** $\left(6, \dfrac{\pi}{4}\right)$
**18.** $\left(5, \dfrac{\pi}{6}\right)$
**19.** $\left(4, \dfrac{3\pi}{2}\right)$
**20.** $\left(3, \dfrac{3\pi}{4}\right)$

**21.** $\left(-6, \dfrac{\pi}{4}\right)$

**22.** $\left(-5, \dfrac{\pi}{6}\right)$

**23.** $\left(-4, -\dfrac{3\pi}{2}\right)$

**24.** $\left(-3, -\dfrac{3\pi}{4}\right)$

**B**    Find the polar coordinates of the point.

**25.** $(4, 4)$

**26.** $(5, 5)$

**27.** $(0, 5)$

**28.** $(0, -3)$

**29.** $(4, 0)$

**30.** $(-5, 0)$

**31.** $(3, 3\sqrt{3})$

**32.** $(-3, -3\sqrt{3})$

**33.** $(\sqrt{3}, 1)$

**34.** $(-\sqrt{3}, 1)$

**35.** $(3\sqrt{3}, 3)$

**36.** $(4\sqrt{3}, -4)$

Find the Cartesian coordinates of the point.

**37.** $(4, 45°)$

**38.** $(5, 60°)$

**39.** $(0, 23°)$

**40.** $(0, -34°)$

**41.** $(-3, 45°)$

**42.** $(-5, 30°)$

**43.** $(6, -60°)$

**44.** $(3, -120°)$

**45.** $\left(10, \dfrac{\pi}{6}\right)$

**46.** $\left(12, \dfrac{3\pi}{4}\right)$

**47.** $\left(-5, \dfrac{5\pi}{6}\right)$

**48.** $\left(-6, \dfrac{3\pi}{4}\right)$

Convert to a polar equation.

**49.** $3x + 4y = 5$

**50.** $5x + 3y = 4$

**51.** $x = 5$

**52.** $y = 4$

**53.** $x^2 + y^2 = 36$

**54.** $x^2 + y^2 = 16$

**55.** $x^2 - 4y^2 = 4$

**56.** $x^2 - 5y^2 = 5$

Convert to a rectangular equation.

**57.** $r = 5$

**58.** $r = 8$

**59.** $\theta = \dfrac{\pi}{4}$

**60.** $\theta = \dfrac{3\pi}{4}$

**61.** $r \sin \theta = 2$

**62.** $r \cos \theta = 5$

**63.** $r = 4 \cos \theta$

**64.** $r = -3 \sin \theta$

**65.** $r - r \sin \theta = 2$

**66.** $r + r \cos \theta = 3$

**67.** $r - 2 \cos \theta = 3 \sin \theta$

**68.** $r + 5 \sin \theta = 7 \cos \theta$

**C**    Graph.

**69.** $r = 4 \cos \theta$

**70.** $r = 4 \sin \theta$

**71.** $r = 1 - \cos \theta$ (Cardioid)

**72.** $r = 1 + \sin \theta$ (Cardioid)

**73.** $r = \sin 2\theta$ (Four-leafed rose)

**74.** $r = 3 \cos 2\theta$ (Four-leafed rose)

**75.** $r = 2 \cos 3\theta$ (Three-leafed rose)

**76.** $r = \sin 3\theta$ (Three-leafed rose)

**77.** $r \cos \theta = 4$

**78.** $r \sin \theta = 6$

**79.** $r = \dfrac{5}{1 + \cos \theta}$ (Limaçon)

**80.** $r = \dfrac{3}{1 + \sin \theta}$ (Limaçon)

**81.** $r = \theta$ (Spiral of Archimedes)

**82.** $r = 3\theta$ (Spiral of Archimedes)

**83.** $r^2 = \sin 2\theta$ (Lemniscate)

**84.** $r^2 = 4 \cos 2\theta$ (Lemniscate)

**85.** $r = e^{\theta/10}$ or $\ln r = \theta/10$ (Logarithmic spiral)

**86.** $r = 10^{2\theta}$ or $\log r = 2\theta$ (Logarithmic spiral)

● CHALLENGE _____

**87.** Convert to a rectangular equation:
$$r = \sec^2 \frac{\theta}{2}.$$

**88.** The center of a regular hexagon is at the origin, and one vertex is the point $(4, 0°)$. Find the coordinates of the other vertices.

◪ TECHNOLOGY CONNECTION _____

Use a grapher to graph each of the following.

**89.** $r = \sin \theta \tan \theta$ (Cissoid)

**90.** $r = \cos 2\theta \sec \theta$ (Strophoid)

**91.** $r = 2 \cos 2\theta - 1$ (Bow tie)

**92.** $r = \cos 2\theta - 2$ (Peanut)

**93.** $r = \frac{1}{4} \tan^2 \theta \sec \theta$ (Semicubical parabola)

**94.** $r = \sin 2\theta + \cos \theta$ (Twisted sister)

# 8.5

## COMPLEX NUMBERS

## Graphical Representation and Polar Notation

### A   Graphs

The real numbers can be graphed on a line. Complex numbers are graphed on a plane. We graph a complex number $a + bi$ in the same way that we graph an ordered pair of real numbers $(a, b)$. However, in place of an $x$-axis, we have a **real axis,** and in place of a $y$-axis, we have an **imaginary axis.**

**Examples**   Graph each of the following.

1.  $3 + 2i$

2.  $-4 + 5i$

3.  $-5 - 4i$

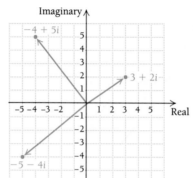

Horizontal distances correspond to the real part of a number. Vertical distances correspond to the imaginary part. Graphs of the numbers are shown above as vectors. The horizontal component of the vector for $a + bi$ is $a$ and the vertical component is $b$. Vectors are sometimes used to study complex numbers.

**DO EXERCISE 1.**

Adding complex numbers is like adding vectors using components. For example, to add $3 + 2i$ and $5 + 4i$, we add the real parts and the imaginary parts to obtain $8 + 6i$. Graphically, then, the sum of two complex numbers looks like a vector sum. It is the diagonal of a parallelogram.

**Example 4**   Show graphically $2 + 2i$ and $3 - i$. Show also their sum.

**Solution**

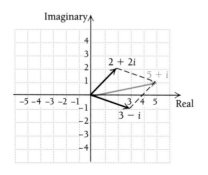

1.  Graph each of the following complex numbers.

    a)  $5 - 3i$

    b)  $-3 + 4i$

    c)  $-5 - 2i$

    d)  $5 + 5i$

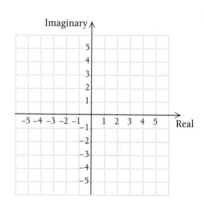

2. Graph each pair of complex numbers and then graph their sum.

a) $2 + 3i$,   $1 - 5i$

b) $-5 + 2i$,   $-1 - 4i$

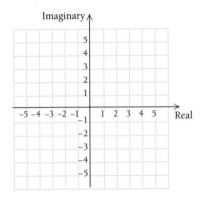

3. Find the absolute value.

a) $|4 - 3i|$

b) $|-12 - 5i|$

**DO EXERCISE 2.**

##  Polar Notation for Complex Numbers

We can locate endpoints of arrows as shown in the preceding examples or by giving the length of the arrow and the angle that it makes with the positive half of the real axis, as we did in Section 8.4. We now develop **polar notation** for complex numbers using this idea. From the figure below, we see that the length of the vector is $\sqrt{a^2 + b^2}$. Note that this quantity is a real number. It is called the **absolute value** of $a + bi$.

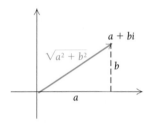

| **DEFINITION** | **Absolute Value of a Complex Number** |
| --- | --- |

The *absolute value of a complex number* $a + bi$ is denoted $|a + bi|$ and is defined to be $\sqrt{a^2 + b^2}$.

**Example 5**   Find $|3 + 4i|$.

Solution
$$|3 + 4i| = \sqrt{3^2 + 4^2} = \sqrt{9 + 16} = 5$$

**DO EXERCISE 3.**

Now let us consider any complex number $a + bi$. Suppose that its absolute value is $r$. Let us also suppose that the angle that the vector makes with the real axis is $\theta$. As this figure shows, we have $a/r = \cos\theta$ and $b/r = \sin\theta$, so

$$a = r\cos\theta \quad \text{and} \quad b = r\sin\theta.$$

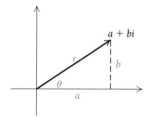

Thus,

$$a + bi = r\cos\theta + (r\sin\theta)i \quad \text{Substituting}$$
$$= r(\cos\theta + i\sin\theta).$$

This is polar notation for $a + bi$. The angle $\theta$ is called the **argument.**

Polar notation for complex numbers is also called **trigonometric notation,** and is often abbreviated $r \operatorname{cis} \theta$.

---

### Polar Notation for Complex Numbers

$$a + bi = r(\cos \theta + i \sin \theta) = r \operatorname{cis} \theta$$

(Polar notation for complex numbers is also called trigonometric notation.)

---

### ▷ B , ▷ C   **Change of Notation**

To change from polar notation to **binomial,** or **rectangular,** notation $a + bi$, we proceed as in Section 8.4, recalling that $a = r \cos \theta$ and $b = r \sin \theta$.

**Example 6**   Write binomial notation for $2(\cos 120° + i \sin 120°)$.

**Solution**   Rewriting, we have

$$2(\cos 120° + i \sin 120°) = 2 \cos 120° + (2 \sin 120°)i.$$

Thus,

$$a = 2 \cos 120° = -1,$$
$$b = 2 \sin 120° = \sqrt{3},$$

and

$$2(\cos 120° + i \sin 120°) = -1 + i\sqrt{3}. \qquad \blacktriangleleft$$

**Example 7**   Write binomial notation for $\sqrt{8} \operatorname{cis} \dfrac{7\pi}{4}$.

**Solution**

$$\sqrt{8} \operatorname{cis} \frac{7\pi}{4} = \sqrt{8}\left(\cos \frac{7\pi}{4} + i \sin \frac{7\pi}{4}\right),$$

$$a = \sqrt{8} \cos \frac{7\pi}{4} = \sqrt{8} \cdot \frac{1}{\sqrt{2}} = 2,$$

$$b = \sqrt{8} \sin \frac{7\pi}{4} = \sqrt{8} \cdot \frac{-1}{\sqrt{2}} = -2$$

Thus,

$$\sqrt{8} \operatorname{cis} \frac{7\pi}{4} = 2 - 2i. \qquad \blacktriangleleft$$

**DO EXERCISES 4 AND 5.**

To change from binomial notation to polar notation, we remember that $r = \sqrt{a^2 + b^2}$ and that $\theta$ is an angle for which $\sin \theta = b/r$ and $\cos \theta = a/r$.

**Example 8**   Find polar notation for $1 + i$.

---

4. Write rectangular notation for

$$\sqrt{2}(\cos 315° + i \sin 315°).$$

5. Write binomial notation for

$$2 \operatorname{cis}\left(-\frac{\pi}{6}\right).$$

Write polar notation.

6. $1 - i$

**Solution**    We note that $a = 1$ and $b = 1$. Then

$$r = \sqrt{1^2 + 1^2} = \sqrt{2},$$

$$\sin \theta = \frac{1}{\sqrt{2}}, \quad \text{or} \quad \frac{\sqrt{2}}{2},$$

and

$$\cos \theta = \frac{1}{\sqrt{2}}, \quad \text{or} \quad \frac{\sqrt{2}}{2}.$$

Thus, $\theta = \pi/4$, or $45°$, and we have

$$1 + i = \sqrt{2} \operatorname{cis} \frac{\pi}{4}, \quad \text{or} \quad 1 + i = \sqrt{2} \operatorname{cis} 45°. \quad \blacktriangleleft$$

**Example 9**    Find polar notation for $\sqrt{3} - i$.

**Solution**    We note that $a = \sqrt{3}$ and $b = -1$. Then

$$r = \sqrt{(\sqrt{3})^2 + (-1)^2} = 2,$$

$$\sin \theta = -\frac{1}{2}, \quad \text{and} \quad \cos \theta = \frac{\sqrt{3}}{2}.$$

Thus, $\theta = 11\pi/6$, or $330°$, and we have

$$\sqrt{3} - i = 2 \operatorname{cis} \frac{11\pi}{6}, \quad \text{or} \quad 2 \operatorname{cis} 330°. \quad \blacktriangleleft$$

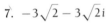

7. $-3\sqrt{2} - 3\sqrt{2}i$

In changing to polar notation, note that there are many angles satisfying the given conditions. We ordinarily choose the smallest positive angle.

**DO EXERCISES 6 AND 7.**

## ▷ Multiplication and Division with Polar Notation

Multiplication of complex numbers is somewhat easier to do with polar notation than with rectangular notation. We simply multiply the absolute values and add the arguments. Let us state this more formally and prove it.

---

**THEOREM 4**

For any complex numbers $r_1 \operatorname{cis} \theta_1$ and $r_2 \operatorname{cis} \theta_2$,

$$(r_1 \operatorname{cis} \theta_1)(r_2 \operatorname{cis} \theta_2) = r_1 \cdot r_2 \operatorname{cis} (\theta_1 + \theta_2).$$

---

**Proof.**    Let us first multiply $a_1 + b_1 i$ by $a_2 + b_2 i$:

$$(a_1 + b_1 i)(a_2 + b_2 i) = (a_1 a_2 - b_1 b_2) + (a_2 b_1 + a_1 b_2)i.$$

Recall that

$$a_1 = r_1 \cos \theta_1, \qquad b_1 = r_1 \sin \theta_1,$$

and

$$a_2 = r_2 \cos \theta_2, \qquad b_2 = r_2 \sin \theta_2.$$

We substitute these in the product above, to obtain

$$r_1(\cos\theta_1 + i\sin\theta_1) \cdot r_2(\cos\theta_2 + i\sin\theta_2)$$
$$= (r_1 r_2 \cos\theta_1 \cos\theta_2 - r_1 r_2 \sin\theta_1 \sin\theta_2)$$
$$+ (r_1 r_2 \sin\theta_1 \cos\theta_2 + r_1 r_2 \cos\theta_1 \sin\theta_2)i.$$

This simplifies to

$$r_1 r_2(\cos\theta_1 \cos\theta_2 - \sin\theta_1 \sin\theta_2) + r_1 r_2(\sin\theta_1 \cos\theta_2 + \cos\theta_1 \sin\theta_2)i.$$

Now, using identities for sums of angles, we simplify, obtaining

$$r_1 r_2 \cos(\theta_1 + \theta_2) + r_1 r_2 \sin(\theta_1 + \theta_2)i,$$

or

$$r_1 r_2 \operatorname{cis}(\theta_1 + \theta_2),$$

which was to be shown.

To divide complex numbers, we do the reverse of the above. We state that fact, but omit the proof.

### THEOREM 5

For any complex numbers $r_1 \operatorname{cis}\theta_1$ and $r_2 \operatorname{cis}\theta_2$ ($r_2 \neq 0$),

$$\frac{r_1 \operatorname{cis}\theta_1}{r_2 \operatorname{cis}\theta_2} = \frac{r_1}{r_2}\operatorname{cis}(\theta_1 - \theta_2).$$

**Example 10**   Find the product of $3\operatorname{cis}40°$ and $7\operatorname{cis}20°$.

Solution
$$3\operatorname{cis}40° \cdot 7\operatorname{cis}20° = 3 \cdot 7\operatorname{cis}(40° + 20°) \quad \text{Using Theorem 4}$$
$$= 21\operatorname{cis}60°$$

**Example 11**   Find the product of $2\operatorname{cis}\pi$ and $3\operatorname{cis}\left(-\frac{\pi}{2}\right)$.

Solution
$$2\operatorname{cis}\pi \cdot 3\operatorname{cis}\left(-\frac{\pi}{2}\right) = 2 \cdot 3\operatorname{cis}\left(\pi - \frac{\pi}{2}\right)$$
$$= 6\operatorname{cis}\frac{\pi}{2}$$

**Example 12**   Convert to polar notation and multiply: $(1 + i)(\sqrt{3} - i)$.

Solution   We first find polar notation (see Examples 8 and 9):
$$1 + i = \sqrt{2}\operatorname{cis}45°, \quad \sqrt{3} - i = 2\operatorname{cis}330°.$$
We then multiply, using Theorem 4:
$$(\sqrt{2}\operatorname{cis}45°)(2\operatorname{cis}330°) = 2 \cdot \sqrt{2}\operatorname{cis}375°, \quad \text{or } 2\sqrt{2}\operatorname{cis}15°.$$

Multiply.

**8.** $5 \text{ cis } 25° \cdot 4 \text{ cis } 30°$

**9.** $8 \text{ cis } \pi \cdot \dfrac{1}{2} \text{ cis } \dfrac{\pi}{4}$

**10.** Convert to polar notation and multiply:

$$(1 + i)(2 + 2i).$$

**11.** Divide:

$$10 \text{ cis } \dfrac{\pi}{2} \div 5 \text{ cis } \dfrac{\pi}{4}.$$

**12.** Convert to polar notation and divide:

$$\dfrac{\sqrt{3} - i}{1 + i}.$$

**Example 13**  Divide $2 \text{ cis } \pi$ by $4 \text{ cis } \dfrac{\pi}{2}$.

**Solution**

$$\dfrac{2 \text{ cis } \pi}{4 \text{ cis } \dfrac{\pi}{2}} = \dfrac{2}{4} \text{ cis } \left( \pi - \dfrac{\pi}{2} \right) \qquad \text{Using Theorem 5}$$

$$= \dfrac{1}{2} \text{ cis } \dfrac{\pi}{2}$$

**Example 14**  Convert to polar notation and divide: $(1 + i)/(1 - i)$.

**Solution**  We first convert to polar notation:

$$1 + i = \sqrt{2} \text{ cis } 45°, \qquad \text{See Example 8.}$$
$$1 - i = \sqrt{2} \text{ cis } 315°.$$

We now divide, using Theorem 5:

$$\dfrac{1 + i}{1 - i} = \dfrac{\sqrt{2} \text{ cis } 45°}{\sqrt{2} \text{ cis } 315°} = 1 \cdot \text{cis } (45° - 315°)$$

$$= 1 \cdot \text{cis } (-270°), \quad \text{or cis } 90°.$$

**DO EXERCISES 8–12.**

---

## ● EXERCISE SET 8.5

**A**  Graph each pair of complex numbers and then graph their sum.

**1.** $3 + 2i, \quad 2 - 5i$
**2.** $4 + 3i, \quad 3 - 4i$
**3.** $-5 + 3i, \quad -2 - 3i$
**4.** $-4 + 2i, \quad -3 - 4i$
**5.** $2 - 3i, \quad -5 + 4i$
**6.** $3 - 2i, \quad -5 + 5i$
**7.** $-2 - 5i, \quad 5 + 3i$
**8.** $-3 - 4i, \quad 6 + 3i$

**B**  Find rectangular notation.

**9.** $3(\cos 30° + i \sin 30°)$
**10.** $6(\cos 150° + i \sin 150°)$
**11.** $10 \text{ cis } 270°$
**12.** $5 \text{ cis } (-60°)$
**13.** $\sqrt{8} \left( \cos \dfrac{\pi}{4} + i \sin \dfrac{\pi}{4} \right)$
**14.** $5 \left( \cos \dfrac{\pi}{3} + i \sin \dfrac{\pi}{3} \right)$
**15.** $\sqrt{8} \text{ cis } \dfrac{5\pi}{4}$
**16.** $\sqrt{8} \text{ cis } \left( -\dfrac{\pi}{4} \right)$

**C**  Find polar notation.

**17.** $1 - i$
**18.** $\sqrt{3} + i$
**19.** $10\sqrt{3} - 10i$
**20.** $-10\sqrt{3} + 10i$
**21.** $-5$
**22.** $-5i$

**D**  Convert to polar notation and then multiply or divide.

**23.** $(1 - i)(2 + 2i)$
**24.** $(1 + i\sqrt{3})(1 + i)$
**25.** $(2\sqrt{3} + 2i)(2i)$
**26.** $(3\sqrt{3} - 3i)(2i)$
**27.** $\dfrac{1 - i}{1 + i}$
**28.** $\dfrac{1 - i}{\sqrt{3} - i}$
**29.** $\dfrac{2\sqrt{3} - 2i}{1 + \sqrt{3}i}$
**30.** $\dfrac{3 - 3\sqrt{3}i}{\sqrt{3} - i}$

**●  SYNTHESIS**

**31.** Show that for any complex number $z$,
$$|z| = |-z|.$$
(*Hint:* Let $z = a + bi$.)

**32.** Show that for any complex number $z$,
$$|z| = |\bar{z}|.$$
(*Hint:* Let $z = a + bi$.)

**33.** Show that for any complex number $z$,
$$|z\bar{z}| = |z^2|.$$

**34.** Show that for any complex number $z$,
$$|z^2| = |z|^2.$$

**35.** Show that for any complex numbers $z$ and $w$,

$$|z \cdot w| = |z| \cdot |w|.$$

(*Hint:* Let $z = r_1 \operatorname{cis} \theta_1$ and $w = r_2 \operatorname{cis} \theta_2$.)

**36.** Show that for any complex number $z$ and any nonzero, complex number $w$,

$$\left| \frac{z}{w} \right| = \frac{|z|}{|w|}. \qquad \text{(Use the hint for Exercise 35.)}$$

**37.** On a complex plane, graph $|z| = 1$.

**38.** On a complex plane, graph $z + \bar{z} = 3$.

● CHALLENGE _____

**39.** Find polar notation for $(\cos \theta + i \sin \theta)^{-1}$.

**40.** Compute $(5 \operatorname{cis} 20°)^3$.

# 8.6

## COMPLEX NUMBERS: DeMOIVRE'S THEOREM AND $n$th ROOTS OF COMPLEX NUMBERS

 **Powers of Complex Numbers**

An important theorem about powers and roots of complex numbers is named for the French mathematician Abraham DeMoivre (1667–1754). Let us consider a number $r \operatorname{cis} \theta$ and its square:

$$(r \operatorname{cis} \theta)^2 = (r \operatorname{cis} \theta)(r \operatorname{cis} \theta) = r \cdot r \operatorname{cis} (\theta + \theta)$$
$$= r^2 \operatorname{cis} 2\theta.$$

Similarly, we see that

$$(r \operatorname{cis} \theta)^3 = r \cdot r \cdot r \operatorname{cis} (\theta + \theta + \theta)$$
$$= r^3 \operatorname{cis} 3\theta.$$

DeMoivre's theorem is the generalization of these results.

| THEOREM 6 | DeMoivre's Theorem |
| --- | --- |

For any complex number $r \operatorname{cis} \theta$ and any natural number $n$,

$$(r \operatorname{cis} \theta)^n = r^n \operatorname{cis} n\theta,$$

or

$$[r(\cos \theta + i \sin \theta)]^n = r^n[\cos n\theta + i \sin n\theta].$$

**OBJECTIVES**

You should be able to:

A Use DeMoivre's theorem to raise complex numbers to powers.

B Find the $n$th roots of a complex number.

*Example 1*  Find $(1 + i)^9$.

*Solution*  We first find polar notation: $1 + i = \sqrt{2} \operatorname{cis} 45°$. Then

$$(1 + i)^9 = (\sqrt{2} \operatorname{cis} 45°)^9 = \sqrt{2}^9 \operatorname{cis} (9 \cdot 45°)$$
$$= 2^{9/2} \operatorname{cis} 405°$$
$$= 16\sqrt{2} \operatorname{cis} 45° \qquad \text{405° has the same terminal side as 45°.}$$
$$= 16\sqrt{2}(\cos 45° + i \sin 45°)$$
$$= 16\sqrt{2}\left(\frac{\sqrt{2}}{2} + i\frac{\sqrt{2}}{2}\right)$$
$$= 16 + 16i. \qquad \blacktriangleleft$$

1. Find $(1 - i)^{10}$.

**Example 2**    Find $(\sqrt{3} - i)^{10}$.

Solution    We first find polar notation: $\sqrt{3} - i = 2 \text{ cis } 330°$. Then

$$(\sqrt{3} - i)^{10} = (2 \text{ cis } 330°)^{10}$$
$$= 2^{10} \text{ cis } (10 \cdot 330°)$$
$$= 1024 \text{ cis } 3300°$$
$$= 1024 \text{ cis } 60°$$
$$= 1024[\cos 60° + i \sin 60°]$$
$$= 1024\left[\frac{1}{2} + i\frac{\sqrt{3}}{2}\right]$$
$$= 512 + 512\sqrt{3}i.$$
◀

**DO EXERCISES 1 AND 2.**

 ## Roots of Complex Numbers

As we will see, every nonzero complex number has two square roots. A number has three cube roots, four fourth roots, and so on. In general, a nonzero complex number has $n$ different $n$th roots. They can be found by the formula that we now state and prove.

2. Find $(\sqrt{3} + i)^{4}$.

---

**THEOREM 7**

The $n$th roots of a complex number $r \text{ cis } \theta$, $r \neq 0$, are given by

$$r^{1/n} \text{ cis } \left(\frac{\theta}{n} + k \cdot \frac{360°}{n}\right),$$

where $k = 0, 1, 2, \ldots, n - 1$.

---

Using DeMoivre's theorem, we show that this formula gives us $n$ different roots. We take the expression for the $n$th roots and raise it to the $n$th power, to show that we get $r \text{ cis } \theta$:

$$\left[r^{1/n} \text{ cis } \left(\frac{\theta}{n} + k \cdot \frac{360°}{n}\right)\right]^{n} = (r^{1/n})^{n} \text{ cis } \left(\frac{\theta}{n} \cdot n + k \cdot n \cdot \frac{360°}{n}\right)$$
$$= r \text{ cis } (\theta + k \cdot 360°)$$
$$= r \text{ cis } \theta.$$

Thus we know that the formula gives us $n$th roots for any natural number $k$. Next, we show that there are at least $n$ different roots. To see this, consider substituting 0, 1, 2, and so on, for $k$. When $k = n$, the cycle begins to repeat, but from 0 to $n - 1$, the angles obtained and their sines and cosines are all different. There cannot be more than $n$ different $n$th roots. That fact follows from the **fundamental theorem of algebra**, considered in the theory of polynomial functions.

**Example 3**    Find the square roots of $2 + 2\sqrt{3}i$.

Solution   We first find polar notation: $2 + 2\sqrt{3}i = 4 \operatorname{cis} 60°$. Then using Theorem 7 gives us

$$(4 \operatorname{cis} 60°)^{1/2} = 4^{1/2} \operatorname{cis}\left(\frac{60°}{2} + k \cdot \frac{360°}{2}\right), \quad k = 0, 1$$

$$= 2 \operatorname{cis}\left(30° + k \cdot \frac{360°}{2}\right), \quad k = 0, 1.$$

Thus the roots are $2 \operatorname{cis} 30°$ and $2 \operatorname{cis} 210°$, or

$$\sqrt{3} + i \quad \text{and} \quad -\sqrt{3} - i. \qquad \blacktriangleleft$$

**DO EXERCISES 3 AND 4.**

In Example 3, it should be noted that the two square roots of the number are opposites of each other. The same is true of the square roots of any complex number. To see this, let us find the square roots of any complex number $r \operatorname{cis} \theta$:

$$(r \operatorname{cis} \theta)^{1/2} = r^{1/2} \operatorname{cis}\left(\frac{\theta}{2} + k \cdot \frac{360°}{2}\right), \quad k = 0, 1$$

$$= r^{1/2} \operatorname{cis} \frac{\theta}{2} \quad \text{or} \quad r^{1/2} \operatorname{cis}\left(\frac{\theta}{2} + 180°\right).$$

Now let us look at the two numbers on a graph. They lie on a line, so if one number has binomial notation $a + bi$, the other has binomial notation $-a - bi$. Thus their sum is 0 and they are opposites of each other.

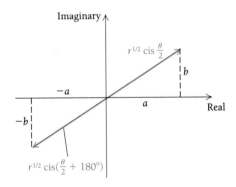

**Example 4**   Find the cube roots of 1. Then locate them on a graph.

Solution   We have

$$1 = 1 \operatorname{cis} 0° \qquad \text{Polar notation}$$

$$(1 \operatorname{cis} 0°)^{1/3} = 1^{1/3} \operatorname{cis}\left(\frac{0°}{3} + k \cdot \frac{360°}{3}\right), \quad k = 0, 1, 2.$$

The roots are $1 \operatorname{cis} 0°$, $1 \operatorname{cis} 120°$, and $1 \operatorname{cis} 240°$, or

$$1, \quad -\frac{1}{2} + \frac{\sqrt{3}}{2}i, \quad \text{and} \quad -\frac{1}{2} - \frac{\sqrt{3}}{2}i.$$

Find the square roots.

3. $2i$

4. $10i$

**5.** Find and graph the cube roots of $-1$.

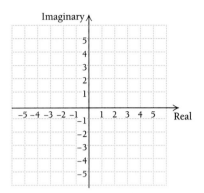

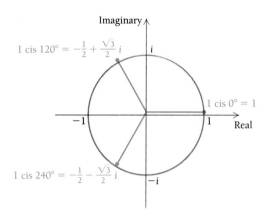

Note in Example 4 that the graphs of the cube roots lie equally spaced about a circle. The same is true of the *n*th roots of any complex number.

**DO EXERCISE 5.**

 **TECHNOLOGY CONNECTION**

If your grapher has a PARAMETRIC mode, you can use it to graphically find all the roots of a number. Parametric equations define *x* and *y* in terms of a third variable, usually *t*. For example, a parametric function is defined by the pair of equations $x = t^2$ and $y = 3t$. To begin the graph of this function, we select a value of *t*, substitute that value into the two equations to find the corresponding values for *x* and *y*, and then plot the point $(x, y)$. This is repeated with many different values of *t*, producing many ordered pairs $(x, y)$ that are then plotted. By drawing a smooth curve through these points, we obtain a graph of the equation.

To find the fifth roots of the number 8, we do the following:

- Set the mode to DEGREE and PARAMETRIC.
- Set the range for *t*:

$$t_{min} = 0, \quad t_{max} = 360,$$
$$t_{step} = 72 \quad \text{(this is because } 360/5 = 72\text{).}$$

- Set the viewing rectangle:

$$[-4, 4] \times [-3, 3].$$

- Set the parametric equations as follows:

$$x_t = (8 \wedge (1/5)) \cos t,$$
$$y_t = (8 \wedge (1/5)) \sin t.$$

- After the graph has been drawn, use the TRACE feature to see the fifth roots of 8.

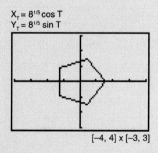

$$X_T = 8^{1/5} \cos T$$
$$Y_T = 8^{1/5} \sin T$$

$[-4, 4] \times [-3, 3]$

Use a grapher to find each of the following graphically.

**TC 1.** The fourth roots of 5    **TC 2.** The cube roots of $-10$
**TC 3.** The 10th roots of 1    **TC 4.** The 15th roots of 8

## ●EXERCISE SET 8.6

**A** Raise the number to the power and write polar notation for the answer.

**1.** $\left(2\operatorname{cis}\dfrac{\pi}{3}\right)^3$      **2.** $\left(3\operatorname{cis}\dfrac{\pi}{2}\right)^4$

**3.** $\left(2\operatorname{cis}\dfrac{\pi}{6}\right)^6$      **4.** $\left(2\operatorname{cis}\dfrac{\pi}{5}\right)^5$

**5.** $(1+i)^6$      **6.** $(1-i)^6$

Raise the number to the power and write rectangular notation for the answer.

**7.** $(2\operatorname{cis}240°)^4$      **8.** $(2\operatorname{cis}120°)^4$

**9.** $(1+\sqrt{3}i)^4$      **10.** $(-\sqrt{3}+i)^6$

**11.** $\left(\dfrac{1}{\sqrt{2}}+\dfrac{1}{\sqrt{2}}i\right)^{10}$      **12.** $\left(\dfrac{1}{\sqrt{2}}-\dfrac{1}{\sqrt{2}}i\right)^{12}$

**13.** $\left(\dfrac{\sqrt{3}}{2}+\dfrac{1}{2}i\right)^{12}$      **14.** $\left(\dfrac{\sqrt{3}}{2}+\dfrac{1}{2}i\right)^{14}$

**B** Solve, that is, find the square roots or the cube roots.

**15.** $x^2=i$      **16.** $x^2=-i$

**17.** $x^2=2\sqrt{2}-2\sqrt{2}i$      **18.** $x^2=2\sqrt{2}+2\sqrt{2}i$

**19.** $x^2=-1+\sqrt{3}i$      **20.** $x^2=-\sqrt{3}-i$

**21.** $x^3=i$      **22.** ▤ $x^3=68.4321$

**23.** Find and graph the fourth roots of 16.

**24.** Find and graph the fourth roots of $i$.

**25.** Find and graph the fifth roots of $-1$.

**26.** Find and graph the sixth roots of 1.

**27.** Find the fourth roots of $-8+i8\sqrt{3}$.

**28.** Find the cube roots of $-64i$.

**29.** Find the cube roots of $2\sqrt{3}-2i$.

**30.** Find the cube roots of $1-i\sqrt{3}$.

Find all the complex solutions of the equation.

**31.** $x^3=1$      **32.** $x^5-1=0$

**33.** $x^5+1=0$      **34.** $x^4+i=0$

**35.** $x^5+\sqrt{3}+i=0$      **36.** $x^3+(4-4\sqrt{3}i)=0$

**37.** $x^4+81=0$      **38.** $x^6+1=0$

● SYNTHESIS

Solve. Evaluate square roots.

**39.** $x^2+(1-i)x+i=0$

**40.** $3x^2+(1+2i)x+1-i=0$

▤ TECHNOLOGY CONNECTION

Use a grapher to find each of the following.

**41.** The fourth roots of 12      **42.** The sixth roots of $-1$

**43.** The tenth roots of 8      **44.** The ninth roots of $-4$

## SUMMARY AND REVIEW 8

● **TERMS TO KNOW**

Angle of elevation, p. 505
Angle of depression, p. 505
Bearing, p. 506
Oblique triangle, p. 511
Law of sines, p. 513
Ambiguous case, p. 514
Law of cosines, p. 520
Vector, p. 526
Magnitude of a vector, p. 526

Direction of a vector, p. 526
Vector addition, p. 526
Resultant, p. 526
Components of vectors, p. 528
Polar coordinates, p. 535
Rectangular coordinates, p. 535
Polar equation, p. 536
Rectangular equation, p. 536

Absolute value of a complex number, p. 542
Polar notation for complex numbers, p. 543
Rectangular notation for complex numbers, p. 543
DeMoivre's theorem, p. 547
Roots of complex numbers, p. 548

● **REVIEW EXERCISES**

1. Solve triangle $ABC$, where $a = 25$, $b = 20$, and $B = 35°$.

2. Solve triangle $ABC$, where $B = 118.3°$, $C = 27.5°$, and $b = 0.974$.

3. In triangle $ABC$, $a = 3.7$, $c = 4.9$, and $B = 135°13'$. Find $b$.

4. In an isosceles triangle, the base angles each measure $52.3°$ and the base is 513 cm long. Find the lengths of the other two sides.

5. A triangular flower garden has sides of lengths 11 ft, 9 ft, and 6 ft. Find the angles of the garden, to the nearest tenth of a degree.

6. A parallelogram has sides of lengths 3.21 cm and 7.85 cm. One of its angles measures $147°$. Find the area of the parallelogram.

7. A car is moving north at 45 mph. A ball is thrown from the car at 37 mph in a direction of N 45° E. Find the speed and the direction of the ball over the ground.

8. Find the area of triangle $ABC$ if $b = 9.8$ m, $c = 7.3$ m, and $A = 67.3°$.

9. Vector **u** has components of 12 lb up and 18 lb to the right. Vector **v** has components of 35 lb down and 35 lb to the left. Find the components of **u** + **v** and the magnitude and the direction of **u** + **v**.

10. Find rectangular notation for the vector $(20, -200°)$.

11. Find polar notation for the vector $\langle -2, 3 \rangle$.

12. Do the following calculations for $\mathbf{u} = \langle 4, 3 \rangle$ and $\mathbf{v} = \langle -3, 4 \rangle$.

    **a)** $4\mathbf{u} - 3\mathbf{v}$
    **b)** $|\mathbf{u} + \mathbf{v}|$

Find the polar coordinates.

13. $(0, -4)$

14. $(\sqrt{3}, -1)$

Find the Cartesian coordinates.

15. $(-4, 60°)$

16. $\left( 4, \dfrac{2\pi}{3} \right)$

17. Convert to a polar equation:
    $$x^2 + y^2 + 2x - 3y = 0.$$

18. Convert to a rectangular equation:
    $$r = -4 \sin \theta.$$

Graph.

19. $r = \cos 2\theta$

20. $r = 2\theta$

21. $r = 3(1 - \sin \theta)$

22. Graph the pair of complex numbers $-3 - 2i$ and $4 + 7i$ and their sum.

23. Find rectangular notation for $2(\cos 135° + i \sin 135°)$.

24. Find polar notation for $1 + i$.

25. Find the product of $7 \text{ cis } 18°$ and $10 \text{ cis } 32°$.

26. Convert to polar notation and then divide:
    $$\frac{1 + i}{\sqrt{3} + i}.$$

27. Find $(1 - i)^5$ and write polar notation for the answer.

28. Find $(3 \text{ cis } 120°)^4$ and write rectangular notation for the answer.

29. Find the cube roots of $1 + i$.

30. Solve $x^3 - 27 = 0$.

31. Find the square roots of $4i$.

● **SYNTHESIS**

32. A parallelogram has sides of lengths 3.42 and 6.97. Its area is 18.4. Find the sizes of its angles.

33. Let $\mathbf{u} = \langle 12, 5 \rangle$. Find a vector that has the same direction as **u** but length 3.

34. Convert to a rectangular equation:
    $$r = \csc^2 \frac{\theta}{2}.$$

● **THINKING AND WRITING**

1. Summarize how you can algebraically tell when solving triangles whether there is no solution, one solution, or two solutions.

2. Explain why the law of sines cannot be used to find the first angle when solving a triangle given three sides.

3. We considered five cases of solving triangles given two sides and an angle opposite one of them. Discuss the relationship between side $a$ and the height $h$ in each.

# CHAPTER TEST 8

1. Solve triangle $ABC$, where $B = 46°$, $C = 19°$, and $b = 70.2$.

2. In triangle $ABC$, $b = 11$, $c = 13$, and $A = 58°$. Find $a$.

3. A triangular flower bed measures 8 ft on one side, and that side makes angles of $38°40'$ and $50°$ with the other sides. How long are the other two sides?

4. Find the area of triangle $ABC$ if $a = 3.1$ cm, $c = 2.7$ cm, and $B = 52.5°$.

5. Two forces of 20 N and 11 N act on an object at right angles. Find the magnitude of the resultant and the angle that it makes with the smaller force.

6. Find rectangular notation for the vector $(25, 150°)$.

7. Find polar notation for the vector $\langle -10, 6 \rangle$.

8. Do the following calculations for $\mathbf{u} = \langle 2, -6 \rangle$ and $\mathbf{v} = \langle 3, 8 \rangle$.

   **a)** $\mathbf{u} - \mathbf{v}$
   **b)** $|2\mathbf{u} + \mathbf{v}|$

9. Vector $\mathbf{u}$ has an easterly component of 10 and a northerly component of 15. Vector $\mathbf{v}$ has a southerly component of 8 and a westerly component of 15. Find:

   **a)** the components of $\mathbf{u} + \mathbf{v}$;
   **b)** the magnitude and the direction of $\mathbf{u} + \mathbf{v}$.

10. Find Cartesian coordinates of the point $(-3, 30°)$.

11. Convert $r = 25$ to a rectangular equation.

12. Solve $x^2 = 8i$.

13. Graph the pair of complex numbers $2 - 7i$ and $-4 + 2i$ and their sum.

14. Find rectangular notation for $3 \operatorname{cis}(-30°)$.

15. Find polar notation for $-1 + \sqrt{3}\,i$.

16. Find the product of $2 \operatorname{cis} 50°$ and $5 \operatorname{cis} 40°$.

17. Divide $3 \operatorname{cis} 120°$ by $6 \operatorname{cis} 40°$.

18. Find $(1 + i)^6$. Write polar notation for the answer.

19. Find the cube roots of $-8$.

Graph.

20. $r = 4 \cos 3\theta$

21. $r = \dfrac{2}{1 + \sin \theta}$

● SYNTHESIS _____

22. Let $\mathbf{u} = (-2, 7)$. Find a vector that has the same direction as $\mathbf{u}$ but has length 2.

# Answers

Margin Exercise 1(a)   **3.** 20 m   **4.** 127.3 ft
**5.** 67.2211°   **6.** 37°27′   **7.** 0.2588   **8.** 0.4163
**9.** 0.2588   **10.** 1.8040   **11.** 0.4710   **12.** 0.2301
**13.** 1.7535   **14.** 24.80°   **15.** 50°   **16.** 45.50°
**17.** sin 15° = 0.2588, cos 15° = 0.9659, tan 15° = 0.2679,
cot 15° = 3.732, sec 15° = 1.035, csc 15° = 3.864

**18.** $\sin\theta = \dfrac{\sqrt{15}}{8}$; $\tan\theta = \dfrac{\sqrt{15}}{7}$; $\sec\theta = \dfrac{8}{7}$; $\csc\theta = \dfrac{8}{\sqrt{15}}$,

or $\dfrac{8\sqrt{15}}{15}$; $\cot\theta = \dfrac{7}{\sqrt{15}}$, or $\dfrac{7\sqrt{15}}{15}$   **19.** 42°

**20.** $a = 14.7$, $b = 13.4$   **21.** $A = 41.79°$, $a = 6.4$, $c = 9.7$
**22.** $B = 52.2°$, $a = 6.69$, $c = 10.91$

**Exercise Set 6.1, pp. 365–367**

**1.** $\sin\theta = \dfrac{8}{17}$; $\cos\theta = \dfrac{15}{17}$; $\tan\theta = \dfrac{8}{15}$; $\cot\theta = \dfrac{15}{8}$;

$\sec\theta = \dfrac{17}{15}$; $\csc\theta = \dfrac{17}{8}$

**3.** $\sin\theta = \dfrac{3}{h}$; $\cos\theta = \dfrac{7}{h}$;

$\tan\theta = \dfrac{3}{7}$; $\cot\theta = \dfrac{7}{3}$; $\sec\theta = \dfrac{h}{7}$; $\csc\theta = \dfrac{h}{3}$

**5.** $\sin\theta = 0.8788$; $\cos\theta = 0.4771$; $\tan\theta = 1.8419$;
$\cot\theta = 0.5429$; $\sec\theta = 2.0958$; $\csc\theta = 1.1379$
**7.** 62.4 m
**9.** sin 30° = 0.5, cos 30° = 0.866, tan 30° = 0.577,
cot 30° = 1.732, sec 30° = 1.155, csc 30° = 2
**11.** sin 30° = 0.5, cos 30° = 0.866, tan 30° = 0.577,
cot 30° = 1.732, sec 30° = 1.155, csc 30° = 2   **13.** 8°36′
**15.** 72°15′   **17.** 46°23′   **19.** 67°50′   **21.** 87°20′44″
**23.** 48°1′30″   **25.** 9.75°   **27.** 35.83°   **29.** 80.55°
**31.** 3.03°   **33.** 19.7897°   **35.** 31.9653°   **37.** 0.9511
**39.** 0.0454   **41.** 0.8857   **43.** 0.9632   **45.** 2.0627
**47.** 1.0171   **49.** 30.83°   **51.** 49.37°   **53.** 72.46°
**55.** 9.17°
**57.** sin 25° = 0.4226, cos 25° = 0.9063, tan 25° = 0.4663,
cot 25° = 2.145, sec 25° = 1.103, csc 25° = 2.366

**59.** $\cos\theta = \dfrac{7}{25}$, $\tan\theta = \dfrac{24}{7}$, $\cot\theta = \dfrac{7}{24}$, $\sec\theta = \dfrac{25}{7}$,

$\csc\theta = \dfrac{25}{24}$

**61.** $\sin\phi = \dfrac{2}{\sqrt{5}}$, or $\dfrac{2\sqrt{5}}{5}$; $\cos\phi = \dfrac{1}{\sqrt{5}}$, or $\dfrac{\sqrt{5}}{5}$; $\cot\phi = \dfrac{1}{2}$;

$\sec\phi = \sqrt{5}$; $\csc\phi = \dfrac{\sqrt{5}}{2}$   **63.** $B = 60°$, $a = 3$, $b = 5$

**65.** $A = 22.7°$, $a = 53$, $c = 137$
**67.** $B = 47°38′$, $b = 25.4$, $c = 34.4$
**69.** $B = 53°48′$, $b = 37.2$, $c = 46.1$
**71.** $A = 77.35°$, $a = 437.1$, $c = 448.0$
**73.** $B = 72.6°$, $a = 4.3$, $c = 14.3$
**75.** $A = 66°48′$, $b = 150$, $c = 381$
**77.** $B = 42.42°$, $a = 35.7$, $b = 32.6$
**79.** $A = 7.7°$, $a = 0.132$, $b = 0.973$
**81.** $A = 34.05°$, $B = 55.95°$, $c = 22.3$
**83.** $A = 53.13°$, $B = 36.87°$, $b = 12.0$

**CHAPTER 6**

**Margin Exercises, Section 6.1**

**1. (a)** $\sin\theta = \dfrac{12}{13}$, $\cos\theta = \dfrac{5}{13}$, $\tan\theta = \dfrac{12}{5}$, $\cot\theta = \dfrac{5}{12}$,

$\sec\theta = \dfrac{13}{5}$, $\csc\theta = \dfrac{13}{12}$; **(b)** $\sin\phi = \dfrac{5}{13}$, $\cos\phi = \dfrac{12}{13}$,

$\tan\phi = \dfrac{5}{12}$, $\cot\phi = \dfrac{12}{5}$, $\sec\phi = \dfrac{13}{12}$, $\csc\phi = \dfrac{13}{5}$

**2. (a)** $\sin\theta = \dfrac{12}{13}$, $\cos\theta = \dfrac{5}{13}$, $\tan\theta = \dfrac{12}{5}$; same values as

**85.** $A = 62.44°$, $B = 27.56°$, $a = 3.56$   **87.** 3.3
**89.** Area $= \frac{1}{2}b \cdot a$, where $a = c \sin A$, so Area $= \frac{1}{2}bc \sin A$.

**Margin Exercises, Section 6.2**
**1. (a)** 76 m; **(b)** 142 m   **2.** 28.55°   **3.** 35.7 ft
**4.** 240.3 ft   **5.** 23.1 km   **6.** 15 mi

**Exercise Set 6.2, pp. 371–373**
**1.** 48.0 ft   **3.** 239 ft   **5.** 1.72°   **7.** 30.22°   **9.** 3.5 mi
**11.** 17,067 ft   **13.** 24 km   **15.** 25.9 cm   **17.** $\frac{25}{3}$ cm
**19.** 23 ft   **21.** 8 km   **23.** 3.4 km
**25.** $d = \sqrt{8000h + h^2}$, where $d$ and $h$ are in miles; 38.9 miles
**27.** Cut so that $\theta = 79.38°$   **29.** 27°

**Margin Exercises, Section 6.3**
**1. (a)** 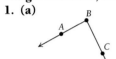   **(b)**

**2. (a)** 360°; **(b)** 180°; **(c)** 90°; **(d)** 45°; **(e)** 60°; **(f)** 30°
**3. (a)** I; **(b)** III; **(c)** IV; **(d)** III; **(e)** I; **(f)** IV; **(g)** I
**4.** Positive: 405°, 765°; negative: −315°, −675°. Answers
may vary.   **5. (a)** Obtuse; **(b)** right; **(c)** acute; **(d)** straight
**6.** Complement: 4°47′22″; supplement: 94°47′22″
**7.** Complement: 55.11°; supplement: 145.11°

**8.** $\sin\theta = -\frac{3}{5}$; $\cos\theta = -\frac{4}{5}$; $\tan\theta = \frac{3}{4}$; $\cot\theta = \frac{4}{3}$;

$\sec\theta = -\frac{5}{4}$; $\csc\theta = -\frac{5}{3}$

**9.** $\sin\theta = \frac{\sqrt{3}}{2}$; $\cos\theta = -\frac{1}{2}$; $\tan\theta = -\sqrt{3}$; $\cot\theta = -\frac{1}{\sqrt{3}}$,

or $-\frac{\sqrt{3}}{3}$; $\sec\theta = -2$; $\csc\theta = \frac{2}{\sqrt{3}}$, or $\frac{2\sqrt{3}}{3}$

**10.** $\sin\theta = -\frac{2}{\sqrt{5}}$, or $-\frac{2\sqrt{5}}{5}$; $\cos\theta = -\frac{1}{\sqrt{5}}$, or $-\frac{\sqrt{5}}{5}$;

$\tan\theta = 2$
**11.** $\sin\theta$ and $\csc\theta$: +, +, −, −; $\cos\theta$ and $\sec\theta$: +,
−, −, +; $\tan\theta$ and $\cot\theta$: +, −, +, −
**12.** The cosine and secant are positive; the other four
function values are negative.
**13.** The sine and cosecant are positive; the other four
function values are negative.
**14.** The cosine and secant are positive; the other four
function values are negative.
**15.** $\sin 270° = -1$; $\cos 270° = 0$; $\tan 270°$: undefined;
$\cot 270° = 0$; $\sec 270°$: undefined; $\csc 270° = -1$

**16.** $\sin 120° = \frac{\sqrt{3}}{2}$; $\cos 120° = -\frac{1}{2}$; $\tan 120° = -\sqrt{3}$;

$\cot 120° = -\frac{1}{\sqrt{3}}$, or $-\frac{\sqrt{3}}{3}$; $\sec 120° = -2$;

$\csc 120° = \frac{2}{\sqrt{3}}$, or $\frac{2\sqrt{3}}{3}$

**17.** $\sin 570° = -\frac{1}{2}$; $\cos 570° = -\frac{\sqrt{3}}{2}$; $\tan 570° = \frac{1}{\sqrt{3}}$,

or $\frac{\sqrt{3}}{3}$; $\cot 570° = \sqrt{3}$; $\sec 570° = -\frac{2}{\sqrt{3}}$, or $-\frac{2\sqrt{3}}{3}$;

$\csc 570° = -2$

**18.** $\sin(-945°) = \frac{\sqrt{2}}{2}$; $\cos(-945°) = -\frac{\sqrt{2}}{2}$;

$\tan(-945°) = -1$; $\cot(-945°) = -1$;

$\sec(-945°) = -\frac{2}{\sqrt{2}}$, or $-\sqrt{2}$; $\csc(-945°) = \frac{2}{\sqrt{2}}$, or $\sqrt{2}$

**19.** East: 75 km; south 130 km   **20.** −0.7547
**21.** 0.3057   **22.** −0.8973   **23.** −2.171   **24.** 0.9999
**25.** 9.230   **26.** −0.0523   **27.** −0.0816   **28.** 304.83°
**29.** 111.98°   **30.** 193.67°   **31.** 213.7°
**32.** $\sin\theta = \frac{1}{\sqrt{10}}$, or $\frac{\sqrt{10}}{10}$; $\cos\theta = -\frac{3}{\sqrt{10}}$, or $-\frac{3\sqrt{10}}{10}$;

$\tan\theta = -\frac{1}{3}$; $\sec\theta = -\frac{\sqrt{10}}{3}$; $\csc\theta = \sqrt{10}$

**33.** $\sin\theta = -\frac{\sqrt{7}}{4}$; $\tan\theta = -\frac{\sqrt{7}}{3}$; $\cot\theta = -\frac{3}{\sqrt{7}}$, or $-\frac{3\sqrt{7}}{7}$;

$\sec\theta = \frac{4}{3}$; $\csc\theta = -\frac{4}{\sqrt{7}}$, or $-\frac{4\sqrt{7}}{7}$

**Exercise Set 6.3, pp. 389–390**
**1.** I   **3.** III   **5.** II   **7.** I   **9.** III   **11.** II
**13.** Positive: 418°, 778°; negative: −302°, −662°. Answers
may vary.
**15.** Positive: 240°, 600°; negative: −480°, −840°. Answers
may vary.
**17.** 32°37′; 122°37′   **19.** 16°14′49″; 106°14′49″
**21.** 22.69°; 112.69°   **23.** 78.7656°; 168.7656°

**25.** $\sin\theta = \frac{5}{13}$; $\cos\theta = -\frac{12}{13}$; $\tan\theta = -\frac{5}{12}$; $\cot\theta = -\frac{12}{5}$;

$\sec\theta = -\frac{13}{12}$; $\csc\theta = \frac{13}{5}$

**27.** $\sin\theta = -\frac{3}{4}$; $\cos\theta = -\frac{\sqrt{7}}{4}$; $\tan\theta = \frac{3}{\sqrt{7}}$, or $\frac{3\sqrt{7}}{7}$;

$\cot\theta = \frac{\sqrt{7}}{3}$; $\sec\theta = -\frac{4}{\sqrt{7}}$, or $-\frac{4\sqrt{7}}{7}$; $\csc\theta = -\frac{4}{3}$

**29.** $\sin\theta = -\frac{2}{\sqrt{13}}$, or $-\frac{2\sqrt{13}}{13}$; $\cos\theta = \frac{3}{\sqrt{13}}$, or $\frac{3\sqrt{13}}{13}$;

$\tan\theta = -\frac{2}{3}$

**31.** $\sin \theta = \dfrac{5}{\sqrt{41}}$, or $\dfrac{5\sqrt{41}}{41}$; $\cos \theta = \dfrac{4}{\sqrt{41}}$, or $\dfrac{4\sqrt{41}}{41}$;

$\tan \theta = \dfrac{5}{4}$

**33.** The cosine and secant are positive; the other four function values are negative.

**35.** The sine and cosecant are positive; the other four function values are negative.

**37.** The sine and cosecant are positive; the other four function values are negative.

**39.** The sine and cosecant are positive; the other four function values are negative.

**41.** 0   **43.** $\sqrt{3}$   **45.** $\dfrac{2}{\sqrt{2}}$, or $\sqrt{2}$   **47.** Does not exist

**49.** $\dfrac{1}{2}$   **51.** $-\dfrac{2}{\sqrt{3}}$, or $-\dfrac{2\sqrt{3}}{3}$   **53.** Does not exist

**55.** $-\dfrac{2}{\sqrt{2}}$, or $-\sqrt{2}$   **57.** $\sqrt{3}$   **59.** 2   **61.** $-1$

**63.** $-\sqrt{3}$   **65.** $-\dfrac{\sqrt{2}}{2}$   **67.** $-1$   **69.** $-1$   **71.** $-\dfrac{1}{2}$

**73.** 1   **75.** $\dfrac{2}{\sqrt{2}}$, or $\sqrt{2}$   **77.** $\dfrac{\sqrt{2}}{2}$   **79.** $-\dfrac{1}{\sqrt{3}}$, or $-\dfrac{\sqrt{3}}{3}$

**81.** $-\dfrac{2}{\sqrt{2}}$, or $-\sqrt{2}$   **83.** $-1$

**85.** $\sin 319° = -0.6561$, $\cos 319° = 0.7547$, $\tan 319° = -0.8693$, $\cot 319° = -1.1504$, $\sec 319° = 1.3250$, $\csc 319° = -1.5242$

**87.** $\sin 115° = 0.9063$, $\cos 115° = -0.4226$, $\tan 115° = -2.1445$, $\cot 115° = -0.4663$, $\sec 115° = -2.3663$, $\csc 115° = 1.1034$   **89.** 109 km

**91.** $-2.122$   **93.** $-1.194$   **95.** $0.5937$   **97.** $0.1882$

**99.** $-3.351$   **101.** $-0.6947$   **103.** $275.38°$

**105.** $200.15°$   **107.** $193.82°$   **109.** $162.51°$

**111.** $\cos \theta = -\dfrac{2\sqrt{2}}{3}$, $\tan \theta = \dfrac{\sqrt{2}}{4}$, $\cot \theta = 2\sqrt{2}$,

$\sec \theta = -\dfrac{3\sqrt{2}}{4}$, $\csc \theta = -3$

**113.** $\sin \theta = -\dfrac{4}{5}$, $\tan \theta = -\dfrac{4}{3}$, $\cot \theta = -\dfrac{3}{4}$, $\sec \theta = \dfrac{5}{3}$,

$\csc \theta = -\dfrac{5}{4}$

**115.** $\sin \theta = -\dfrac{\sqrt{5}}{5}$, $\cos \theta = \dfrac{2\sqrt{5}}{5}$, $\tan \theta = -\dfrac{1}{2}$,

$\sec \theta = \dfrac{\sqrt{5}}{2}$, $\csc \theta = -\sqrt{5}$   **117.** 38.25 in.

**Margin Exercises, Section 6.4**

**1. (a)** $\dfrac{\pi}{2}$; **(b)** $\dfrac{3\pi}{4}$; **(c)** $\dfrac{3\pi}{2}$   **2. (a)** $\dfrac{\pi}{4}$; **(b)** $\dfrac{7\pi}{4}$; **(c)** $\dfrac{5\pi}{4}$

**3. (a)** $\dfrac{\pi}{6}$; **(b)** $\dfrac{4\pi}{3}$; **(c)** $\dfrac{11\pi}{6}$

**4.**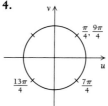

**5.**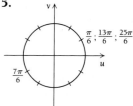

**6.**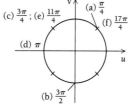

**7.** $N\left(\dfrac{\sqrt{3}}{2}, -\dfrac{1}{2}\right),$

$P\left(-\dfrac{\sqrt{3}}{2}, -\dfrac{1}{2}\right),$

$R\left(-\dfrac{\sqrt{3}}{2}, \dfrac{1}{2}\right)$

**8. (a)** $\left(\dfrac{3}{5}, \dfrac{4}{5}\right)$; **(b)** $\left(-\dfrac{3}{5}, -\dfrac{4}{5}\right)$; **(c)** $\left(-\dfrac{3}{5}, \dfrac{4}{5}\right)$;

**(d)** yes, by symmetry

**9. (a)** $\left(-\dfrac{\sqrt{35}}{6}, \dfrac{1}{6}\right)$; **(b)** $\left(\dfrac{\sqrt{35}}{6}, -\dfrac{1}{6}\right)$;

**(c)** $\left(\dfrac{\sqrt{35}}{6}, \dfrac{1}{6}\right)$; **(d)** yes, by symmetry

**10. (a)** III; **(b)** I; **(c)** IV; **(d)** III

**11. (a)** $\dfrac{5\pi}{4}$; **(b)** $\dfrac{7\pi}{4}$; **(c)** $-4\pi$

**12. (a)** 1.26; **(b)** 5.23; **(c)** $-5.50$

**13. (a)** $240°$; **(b)** $450°$; **(c)** $-144°$   **14.** 57.6 m   **15.** 6

**16.** $\dfrac{3}{2}$   **17.** 377 cm/sec   **18.** $r$ in cm, $\omega$ in radians/sec

**19.** km/yr   **20.** 7.2 radians/sec   **21.** 70.4 radians, or 11.21 revolutions

**Exercise Set 6.4, pp. 402–405**

**1.**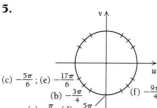

**3.**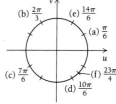

**5.**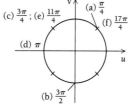

**7.** $M: \dfrac{2\pi}{3}, -\dfrac{4\pi}{3}$; $N: \dfrac{5\pi}{6}$,

$-\dfrac{7\pi}{6}$; $P: \dfrac{5\pi}{4}, -\dfrac{3\pi}{4}$; $Q: \dfrac{11\pi}{6}$,

$-\dfrac{\pi}{6}$

**9. (a)** $\left(-\dfrac{\sqrt{2}}{2}, -\dfrac{\sqrt{2}}{2}\right)$; **(b)** $\left(\dfrac{\sqrt{2}}{2}, \dfrac{\sqrt{2}}{2}\right)$; **(c)** $\left(\dfrac{\sqrt{2}}{2}, -\dfrac{\sqrt{2}}{2}\right)$

**11.** (a) $\left(\dfrac{2}{3}, -\dfrac{\sqrt{5}}{3}\right)$; (b) $\left(-\dfrac{2}{3}, \dfrac{\sqrt{5}}{3}\right)$; (c) $\left(-\dfrac{2}{3}, -\dfrac{\sqrt{5}}{3}\right)$

**13.** $\left(\dfrac{\sqrt{3}}{2}, -\dfrac{1}{2}\right)$   **15.** $\left(\dfrac{3}{4}, \dfrac{\sqrt{7}}{4}\right)$   **17.** $\dfrac{\pi}{6}$   **19.** $\dfrac{\pi}{3}$   **21.** $\dfrac{5\pi}{12}$

**23.** $0.2095\pi$   **25.** $1.1922\pi$   **27.** $2.093$   **29.** $5.582$

**31.** $3.489$   **33.** $2.0550$   **35.** $0.0236$   **37.** $57.32°$

**39.** $1440°$   **41.** $135°$   **43.** $74.69°$   **45.** $135.6°$

**47.** $30° = \dfrac{\pi}{6}$, $60° = \dfrac{\pi}{3}$, $135° = \dfrac{3\pi}{4}$, $180° = \pi$, $225° = \dfrac{5\pi}{4}$,

$270° = \dfrac{3\pi}{2}$, $315° = \dfrac{7\pi}{4}$

**49.** 1.1 radians, $63°$   **51.** 5.233 radians   **53.** 16 m

**55.** 3150 cm/min   **57.** 52.3 cm/sec   **59.** 1047 mph

**61.** 68.8 radians/sec   **63.** 10 mph   **65.** 75.4 ft

**67.** 377 radians   **69.** M: $\dfrac{8\pi}{3}$; N: $\dfrac{17\pi}{6}$; P: $\dfrac{13\pi}{4}$; Q: $\dfrac{23\pi}{6}$

**71.** $\pm\dfrac{2\sqrt{2}}{3}$   **73.** (a) 53.33; (b) 170; (c) 25; (d) 142.86

**75.** 111.7 km, 69.8 mi   **77.** $\dfrac{1}{30}$ radian

**79.** 2.093 radians/sec

**81.** (a) 395.35 rpm/sec, angular acc $= \dfrac{\triangle\omega}{t}$;

(b) 41.4 radians/sec$^2$   **83.** $37.5°$

**85.** 1.15 statute miles

**Margin Exercises, Section 6.5**

**1.** $\dfrac{\sqrt{2}}{2}$   **2.** $\dfrac{\sqrt{3}}{3}$   **3.** $-\dfrac{\sqrt{3}}{2}$   **4.** 1   **5.** Undefined

**6.** 0.3640   **7.** 0.8367   **8.** $-0.2768$   **9.** 0.6549

**10.**

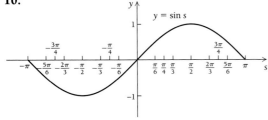

**11.** Yes   **12.** The set of all real numbers

**13.** The set of real numbers from $-1$ to 1, inclusive

**14.** (a) $(-a, -b)$; (b) $(-a, -b)$

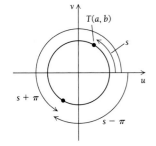

**15.**

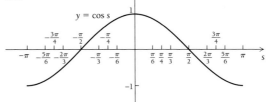

**16.** Yes   **17.** The set of all real numbers

**18.** The set of all real numbers from $-1$ to 1, inclusive

**19.** $\cos y = d$, $\sin y = e$   **20.** $\tan s = \dfrac{v}{u} = \dfrac{\sin s}{\cos s}$

**21.** $\cot s = \dfrac{u}{v} = \dfrac{\cos s}{\sin s}$

**22.** 0, not defined, $-1$, not defined

**23.** $-1, -1, -\sqrt{2}, \sqrt{2}$   **24.** $\dfrac{\sqrt{3}}{3}, \sqrt{3}, \dfrac{2\sqrt{3}}{3}, 2$

**25.** $-\sqrt{3}, -\dfrac{\sqrt{3}}{3}, 2, -\dfrac{2\sqrt{3}}{3}$

**26.** Not defined, not defined, not defined

**27.** $-\dfrac{\pi}{2}, \dfrac{3\pi}{2}, -\dfrac{3\pi}{2}, \dfrac{5\pi}{2}, -\dfrac{5\pi}{2}$, etc.

**28.**

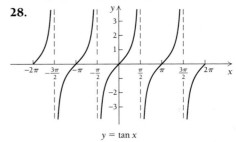

**29.** $\pi$

**30.** The set of all real numbers except $\dfrac{\pi}{2} + k\pi$, $k$ any integer

**31.** The set of all real numbers   **32.** Odd

**33.** $0, \pi, -\pi, 2\pi, -2\pi$, etc.

**34.**

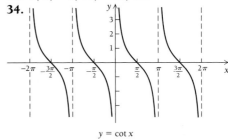

**35.** $\pi$

**36.** The set of all real numbers except $k\pi$, $k$ any integer

**37.** The set of all real numbers

**38.** Positive: I, III; negative: II, IV

**39.** Odd

**40.** (a) $\dfrac{\pi}{2}, -\dfrac{\pi}{2}, \dfrac{3\pi}{2}, -\dfrac{3\pi}{2}$, etc.; (b) $0, \pi, -\pi, 2\pi, -2\pi$, etc.

**41.**

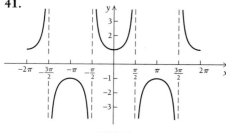

$y = \sec x$

**42.**

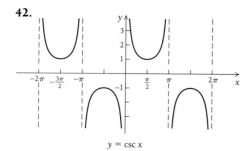

$y = \csc x$

**43.** $2\pi$; $2\pi$
**44.** The set of real numbers except $\pi/2 + k\pi$, $k$ any integer; the set of real numbers except $k\pi$, $k$ any integer
**45.** Its range consists of all real numbers 1 and greater, in addition to all real numbers $-1$ and less; its range is the same as that of the secant function.
**46.** Positive: I, IV; negative: II, III
**47.** Positive: I, II; negative: III, IV    **48.** Even    **49.** Odd

**50.**

| Function | I | II | III | IV |
|----------|---|----|----|----|
| sin, csc | + | + | − | − |
| cos, sec | + | − | − | + |
| tan, cot | + | − | + | − |

**51.** $\tan s = \dfrac{\sin s}{\cos s} = \dfrac{1}{\dfrac{\cos s}{\sin s}} = \dfrac{1}{\cot s}$    **52.** $\cos s = \dfrac{1}{\dfrac{1}{\cos s}} = \dfrac{1}{\sec s}$

**53.** $\sin^2 s + \cos^2 s = 1$    **54.** $\sin^2 x = 1 - \cos^2 x$,

$\dfrac{\sin^2 s}{\cos^2 s} + \dfrac{\cos^2 s}{\cos^2 s} = \dfrac{1}{\cos^2 s}$    $\sin x = \pm\sqrt{1 - \cos^2 x}$

$\tan^2 s + 1 = \sec^2 s$

**55.** $\cos s = -\dfrac{\sqrt{55}}{8}$; $\tan s = \dfrac{3}{\sqrt{55}}$, or $\dfrac{3\sqrt{55}}{55}$; $\cot s = \dfrac{\sqrt{55}}{3}$;

$\sec s = -\dfrac{8}{\sqrt{55}}$, or $-\dfrac{8\sqrt{55}}{55}$; $\csc s = -\dfrac{8}{3}$

**56.** $\sin s = \dfrac{3}{\sqrt{10}}$, or $\dfrac{3\sqrt{10}}{10}$; $\cos s = -\dfrac{1}{\sqrt{10}}$, or $-\dfrac{\sqrt{10}}{10}$;

$\cot s = -\dfrac{1}{3}$; $\sec s = -\sqrt{10}$; $\csc s = \dfrac{\sqrt{10}}{3}$

**Technology Connection, Section 6.5**
**TC1.**
$y = \cos^2 x \sin x$

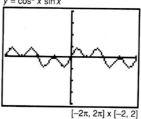

$[-2\pi, 2\pi] \times [-2, 2]$

**TC2.**
$y = (\cos x + 1)(\tan x)$

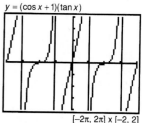

$[-2\pi, 2\pi] \times [-2, 2]$

**TC3.**
$y = \csc x + \cot x$

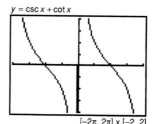

$[-2\pi, 2\pi] \times [-2, 2]$

**TC4.**
$y = \sec x \cot x$

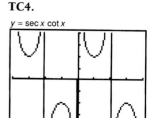

$[-2\pi, 2\pi] \times [-2, 2]$

**Exercise Set 6.5, pp. 425–427**

**1.** 0    **3.** $-1$    **5.** $\dfrac{1}{2}$    **7.** 1    **9.** $-1$    **11.** $-1$

**13.** $-\dfrac{\sqrt{2}}{2}$    **15.** $\sqrt{3}$    **17.** $-0.6435$    **19.** $-0.8391$

**21.** 0    **23.** 1.236    **25.** 0.6801    **27.** $-4.494$    **29.** $-1$
**31.** $-0.7470$    **33.** $\cos x$    **35.** $-\sin x$    **37.** $-\cos x$
**39.** $\cos x$    **41.** $-\cos x$
**43.** **(a)** See Margin Exercise 10;
**(b)**

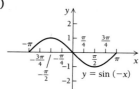

$y = \sin(-x)$

**(c)** same as **(b)**; **(d)** the same
**45.** **(a)** See Margin Exercise 10;
**(b)**

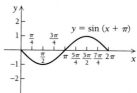

$y = \sin(x + \pi)$

**(c)** same as **(b)**; **(d)** the same
**47.** **(a)** See Margin Exercise 15;
**(b)**

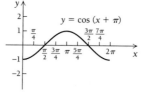

$y = \cos(x + \pi)$

**(c)** same as (b); **(d)** the same

**49.** 1   **51.** $\dfrac{\sqrt{3}}{3}$   **53.** $\sqrt{2}$   **55.** Undefined   **57.** $-\sqrt{3}$

**59.** $\sqrt{2}$   **61.** $-\dfrac{\sqrt{3}}{3}$   **63.** cos, sec   **65.** sin, cos, sec, csc

**67.** Positive: I, III; negative: II, IV
**69.** Positive: I, IV; negative: II, III
**71.**

|      | $\pi/16$ | $\pi/8$ | $\pi/6$ |
|------|---------|---------|---------|
| sin  | 0.19509 | 0.38268 | 0.50000 |
| cos  | 0.98079 | 0.92388 | 0.86603 |
| tan  | 0.19891 | 0.41421 | 0.57735 |
| cot  | 5.02737 | 2.41424 | 1.73206 |
| sec  | 1.01959 | 1.08239 | 1.15469 |
| csc  | 5.12584 | 2.61315 | 2.00000 |

|      | $\pi/4$ | $3\pi/8$ | $7\pi/16$ |
|------|---------|----------|-----------|
| sin  | 0.70711 | 0.92388 | 0.98079 |
| cos  | 0.70711 | 0.38268 | 0.19509 |
| tan  | 1.00000 | 2.41424 | 5.02737 |
| cot  | 1.00000 | 0.41421 | 0.19891 |
| sec  | 1.41421 | 2.61315 | 5.12584 |
| csc  | 1.41421 | 1.08239 | 1.01959 |

**73.** $\sin s = \dfrac{2\sqrt{2}}{3}$; $\tan s = 2\sqrt{2}$; $\cot s = \dfrac{1}{2\sqrt{2}}$, or $\dfrac{\sqrt{2}}{4}$;

$\sec s = 3$; $\csc s = \dfrac{3}{2\sqrt{2}}$, or $\dfrac{3\sqrt{2}}{4}$

**75.** $\sin s = -\dfrac{3}{\sqrt{10}}$, or $-\dfrac{3\sqrt{10}}{10}$; $\cos s = -\dfrac{1}{\sqrt{10}}$,

or $-\dfrac{\sqrt{10}}{10}$; $\cot s = \dfrac{1}{3}$; $\sec s = -\sqrt{10}$; $\csc s = -\dfrac{\sqrt{10}}{3}$

**77.** $\sin s = \dfrac{4}{5}$; $\cos s = -\dfrac{3}{5}$; $\tan s = -\dfrac{4}{3}$; $\cot s = -\dfrac{3}{4}$;

$\csc s = \dfrac{5}{4}$

**79.** $\cos s = -\dfrac{\sqrt{21}}{5}$; $\tan s = \dfrac{2}{\sqrt{21}}$, or $\dfrac{2\sqrt{21}}{21}$;

$\cot s = \dfrac{\sqrt{21}}{2}$; $\sec s = -\dfrac{5}{\sqrt{21}}$, or $-\dfrac{5\sqrt{21}}{21}$; $\csc s = -\dfrac{5}{2}$

**81. (a)** $\dfrac{\pi}{2} + 2k\pi$, $k$ any integer; **(b)** $\dfrac{3\pi}{2} + 2k\pi$, $k$ any integer

**83.** $x = k\pi$, $k$ any integer
**85.** $f \circ g(x) = \cos^2 x + 2\cos x$, $g \circ f(x) = \cos(x^2 + 2x)$

**87.**

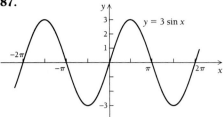

**89.** The domain is the set of all real numbers. The range is the set of all real numbers from 0 to 1 inclusive. The period is $\pi$. The amplitude is $\frac{1}{2}$.
**91.** The domain consists of the intervals

$$\left[-\frac{\pi}{2} + 2k\pi, \frac{\pi}{2} + 2k\pi\right], k \text{ any integer.}$$

**93.** The domain is the set of all real numbers except

$\dfrac{\pi}{2} + k\pi$ for any integer $k$.

**95.** The limit of $(\sin\theta)/\theta$ as $\theta$ approaches 0 is 1.
**97.** The graph of $\sec(x - \pi)$ is like that of $\sec x$, moved $\pi$ units to the right. The graph of $-\sec x$ is that of $\sec x$ reflected across the $x$-axis. The graphs are identical.
**99.** If the graph of $\tan x$ were reflected across the $y$-axis and then translated to the right a distance of $\pi/2$, the graph of $\cot x$ would be obtained. There are other ways to describe the relation.
**101.** The sine and tangent functions; the cosine and cotangent functions
**103.**

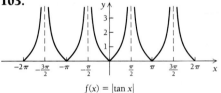

$f(x) = |\tan x|$

**105.**

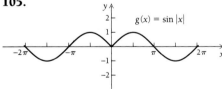

**107. (a)** $\triangle OPA \sim \triangle ODB$   **(b)** $\triangle OPA \sim \triangle ODB$

Thus, $\dfrac{AP}{OA} = \dfrac{BD}{OB}$     $\dfrac{OD}{OP} = \dfrac{OB}{OA}$

$\dfrac{\sin\theta}{\cos\theta} = \dfrac{BD}{1}$     $\dfrac{OD}{1} = \dfrac{1}{\cos\theta}$

$\tan\theta = BD$     $OD = \sec\theta$

**(c)** $\triangle OAP \sim \triangle ECO$

$$\frac{OE}{PO} = \frac{CO}{AP}$$

$$\frac{OE}{1} = \frac{1}{\sin\theta}$$

$$OE = \csc\theta$$

**(d)** $\triangle OAP \sim \triangle ECO$

$$\frac{CE}{AO} = \frac{CO}{AP}$$

$$\frac{CE}{\cos\theta} = \frac{1}{\sin\theta}$$

$$CE = \frac{\cos\theta}{\sin\theta}$$

$$CE = \cot\theta$$

**109.** It is true that $\cos\theta \leq \sec\theta$ for all $\theta$ in the intervals $\left(-\frac{\pi}{2} + k\cdot 2\pi, \frac{\pi}{2} + k\cdot 2\pi\right)$, $k$ any integer.

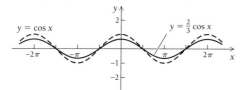

**Margin Exercises, Section 6.6**

**1.**
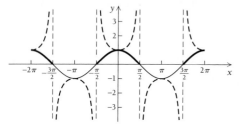

**2.** The amplitude is $\frac{2}{3}$.
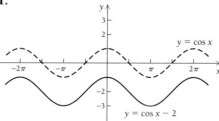

**3.** The amplitude is 2.
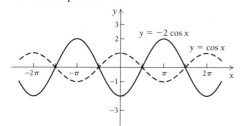

**4.** The period is $4\pi$.
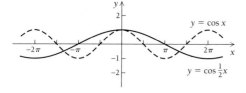

**5.** The period is $\pi$.
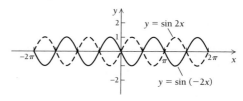

**6.**
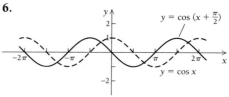

**7.**
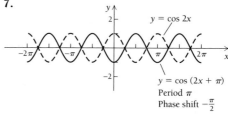
Period $\pi$
Phase shift $-\frac{\pi}{2}$

**8.**
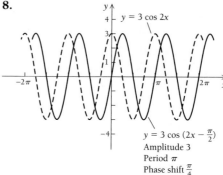
Amplitude 3
Period $\pi$
Phase shift $\frac{\pi}{4}$

**9.**
$y = -2\sin(\pi x) + 2$
Amplitude 2
Period 2
Phase shift 0
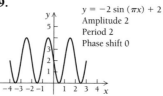

**10.**
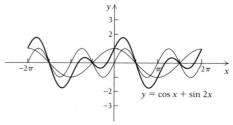

**Technology Connection, Section 6.6**

**TC 1.**

$f(x) = \sin x + 4$

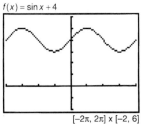

$[-2\pi, 2\pi] \times [-2, 6]$

**TC 2.**

$f(x) = \cos x - 2$

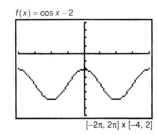

$[-2\pi, 2\pi] \times [-4, 2]$

**TC 11.**

$f(x) = \cos\left(-\frac{1}{4}x\right)$

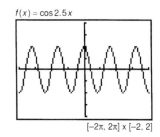

$[-2\pi, 2\pi] \times [-2, 2]$

**TC 12.**

$f(x) = \cos 2.5x$

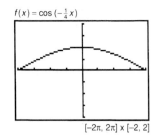

$[-2\pi, 2\pi] \times [-2, 2]$

**TC 3.**

$f(x) = \cos x + 3.5$

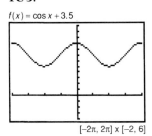

$[-2\pi, 2\pi] \times [-2, 6]$

**TC 4.**

$f(x) = \sin x - 0.6$

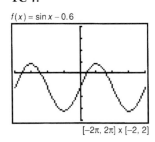

$[-2\pi, 2\pi] \times [-2, 2]$

**TC 13.**

$f(x) = \cos\left(x - \frac{\pi}{4}\right)$

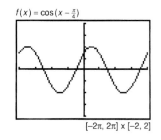

$[-2\pi, 2\pi] \times [-2, 2]$

**TC 14.**

$f(x) = \cos(x + 135°)$

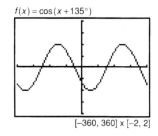

$[-360, 360] \times [-2, 2]$

**TC 5.**

$f(x) = 5 \sin x$

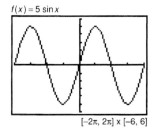

$[-2\pi, 2\pi] \times [-6, 6]$

**TC 6.**

$f(x) = -5 \sin x$

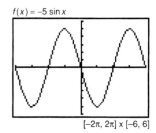

$[-2\pi, 2\pi] \times [-6, 6]$

**TC 15.**

$f(x) = \sin\left(x + \frac{3\pi}{2}\right)$

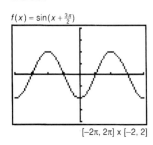

$[-2\pi, 2\pi] \times [-2, 2]$

**TC 16.**

$f(x) = \sin(x - 45°)$

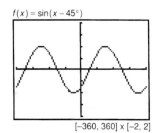

$[-360, 360] \times [-2, 2]$

**TC 7.**

$f(x) = \frac{1}{2} \cos x$

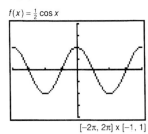

$[-2\pi, 2\pi] \times [-1, 1]$

**TC 8.**

$f(x) = -\frac{1}{2} \cos x$

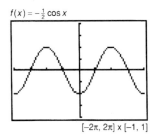

$[-2\pi, 2\pi] \times [-1, 1]$

**TC 17.**

$f(x) = 12.5 \cos(5x - 2.5\pi) - 3$

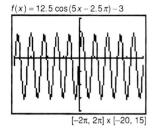

$[-2\pi, 2\pi] \times [-20, 15]$

**TC 18.**

$f(x) = -20 \sin(-3x - 2\pi) + 1.5$

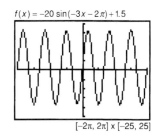

$[-2\pi, 2\pi] \times [-25, 25]$

**TC 9.**

$f(x) = \sin 4x$

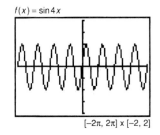

$[-2\pi, 2\pi] \times [-2, 2]$

**TC 10.**

$f(x) = \sin(-4x)$

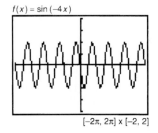

$[-2\pi, 2\pi] \times [-2, 2]$

**TC 19.**

$f(x) = 0.25 \cos\left(4x + \frac{\pi}{3}\right) + 3.5$

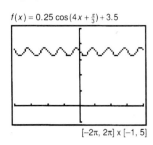

$[-2\pi, 2\pi] \times [-1, 5]$

**TC 20.**

$f(x) = 5 \cos x + \cos 5x$

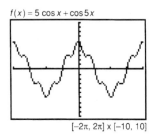

$[-2\pi, 2\pi] \times [-10, 10]$

**TC 21.**

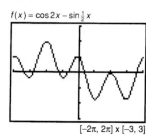

$f(x) = \cos 2x - \sin \frac{1}{2}x$

$[-2\pi, 2\pi] \times [-3, 3]$

**TC 22.**

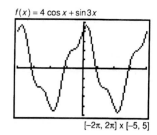

$f(x) = 4\cos x + \sin 3x$

$[-2\pi, 2\pi] \times [-5, 5]$

**Exercise Set 6.6, pp. 440–441**

**1.** Amplitude: 1, period: $2\pi$, phase shift: 0;

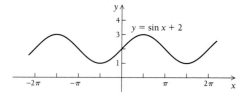

$y = \sin x + 2$

**3.** Amplitude: $\frac{1}{2}$, period: $2\pi$, phase shift: 0;

$y = \frac{1}{2}\sin x$

**5.** Amplitude: 2, period: $2\pi$, phase shift: 0;

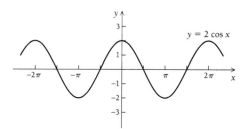

$y = 2\cos x$

**7.** Amplitude: $\frac{1}{2}$, period: $2\pi$, phase shift: 0;

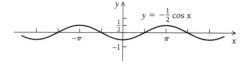

$y = -\frac{1}{2}\cos x$

**9.** Amplitude: 1, period: $\pi$, phase shift: 0;

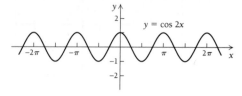

$y = \cos 2x$

**11.** Amplitude: 1, period: $\pi$, phase shift: 0;

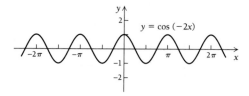

$y = \cos(-2x)$

**13.** Amplitude: 1; period: $4\pi$, phase shift: 0;

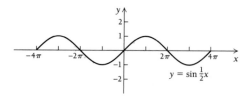

$y = \sin \frac{1}{2}x$

**15.** Amplitude: 1, period: $4\pi$, phase shift: 0;

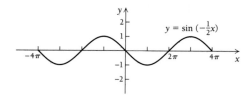

$y = \sin\left(-\frac{1}{2}x\right)$

**17.** Amplitude: 1, period: $\pi$, phase shift: $\frac{\pi}{2}$;

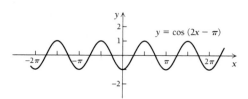

$y = \cos(2x - \pi)$

**19.** Amplitude: 2, period: $4\pi$, phase shift: $\pi$;

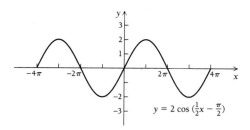

$y = 2\cos\left(\frac{1}{2}x - \frac{\pi}{2}\right)$

**21.** Amplitude: 3, period: $\frac{\pi}{2}$, phase shift: $\frac{\pi}{4}$;

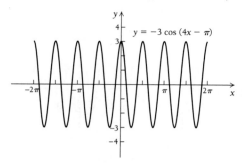

$y = -3\cos(4x - \pi)$

**23.** Amplitude: 3, period: 2, phase shift: $\dfrac{3}{\pi}$;

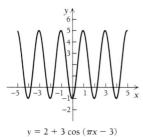

$$y = 2 + 3\cos(\pi x - 3)$$

**25.** Amplitude, 3; period, $\dfrac{2\pi}{3}$; phase shift, $\dfrac{\pi}{6}$

**27.** Amplitude, 5; period, $\dfrac{\pi}{2}$; phase shift, $-\dfrac{\pi}{12}$

**29.** Amplitude, $\dfrac{1}{2}$; period, 1; phase shift, $-\dfrac{1}{2}$

**31.**

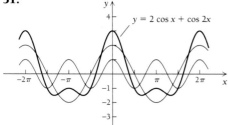

$y = 2\cos x + \cos 2x$

**33.**

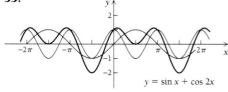

$y = \sin x + \cos 2x$

**35.**

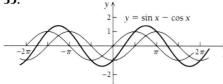

$y = \sin x - \cos x$

**37.**

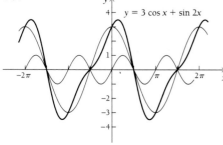

$y = 3\cos x + \sin 2x$

**39. (a)**            **(b)** 104.6°, 98.6°

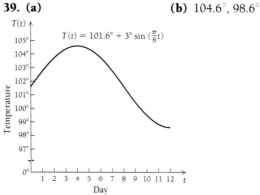

$T(t) = 101.6° + 3°\sin\left(\dfrac{\pi}{8}t\right)$

**41. (a)**            **(b)** Amplitude: 3000, period: 90, phase shift: 10

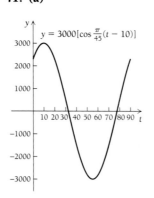

$y = 3000\left[\cos\dfrac{\pi}{45}(t - 10)\right]$

**43.** $y = 3 - 8.6\sin\left[\dfrac{2\pi}{5}(x + 11)\right]$; answers may vary.

**45.** $y = 2 + 16\cos\left[6\left(x + \dfrac{2}{\pi}\right)\right]$

**47.**

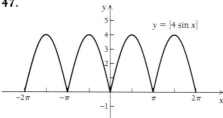

$y = |4\sin x|$

**49.**

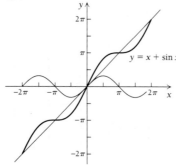

$y = x + \sin x$

**51.**

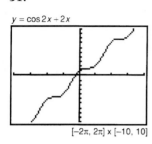

$y = \cos 2x + 2x$

$[-2\pi, 2\pi] \times [-10, 10]$

**53.**

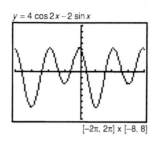

$y = 4 \cos 2x - 2 \sin x$

$[-2\pi, 2\pi] \times [-8, 8]$

**TC 3.**

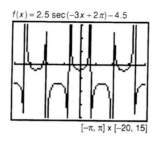

$f(x) = 2.5 \sec(-3x + 2\pi) - 4.5$

$[-\pi, \pi] \times [-20, 15]$

**TC 4.**

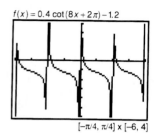

$f(x) = 0.4 \cot(8x + 2\pi) - 1.2$

$[-\pi/4, \pi/4] \times [-6, 4]$

**Margin Exercises, Section 6.7**

**1.**

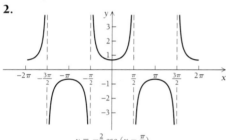

$y = 2 \cot\left(x + \frac{\pi}{4}\right)$

**2.**

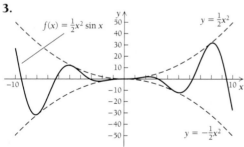

$y = -\frac{2}{3} \csc\left(x - \frac{\pi}{2}\right)$

**3.**

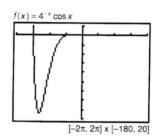

$f(x) = \frac{1}{2}x^2 \sin x$  $y = \frac{1}{2}x^2$  $y = -\frac{1}{2}x^2$

**TC 5.**

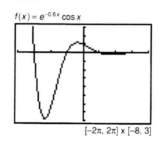

$f(x) = e^{-0.6x} \cos x$

$[-2\pi, 2\pi] \times [-8, 3]$

**TC 6.**

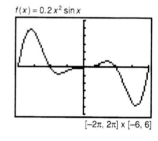

$f(x) = 0.2 x^2 \sin x$

$[-2\pi, 2\pi] \times [-6, 6]$

**TC 7.**

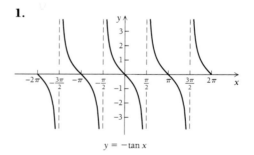

$f(x) = 4^{-x} \cos x$

$[-2\pi, 2\pi] \times [-180, 20]$

**Exercise Set 6.7, p. 446**

**1.**

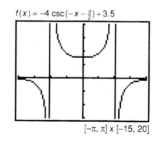

$y = -\tan x$

**Technology Connection, Section 6.7**

**TC 1.**

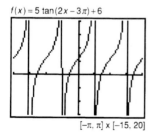

$f(x) = 5 \tan(2x - 3\pi) + 6$

$[-\pi, \pi] \times [-15, 20]$

**TC 2.**

$f(x) = -4 \csc\left(-x - \frac{\pi}{2}\right) + 3.5$

$[-\pi, \pi] \times [-15, 20]$

**3.**

$y = -\csc x$

**5.**

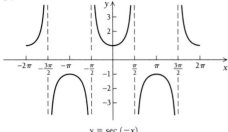

$y = \sec(-x)$

**7.**

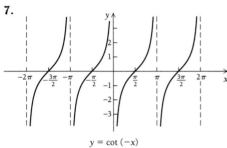

$y = \cot(-x)$

**9.**

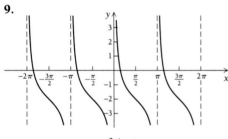

$y = -2 + \cot x$

**11.**

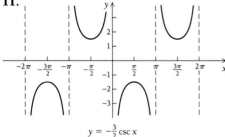

$y = -\frac{3}{2}\csc x$

**13.**

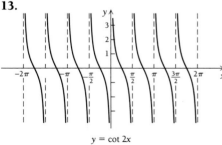

$y = \cot 2x$

**15.**

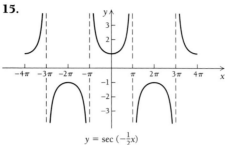

$y = \sec\left(-\frac{1}{2}x\right)$

**17.**

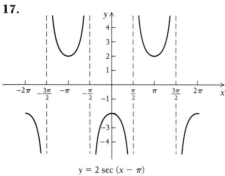

$y = 2\sec(x - \pi)$

**19.**

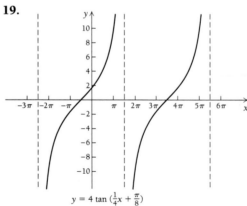

$y = 4\tan\left(\frac{1}{4}x + \frac{\pi}{8}\right)$

**21.**

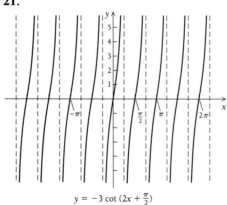

$y = -3\cot\left(2x + \frac{\pi}{2}\right)$

**23.**

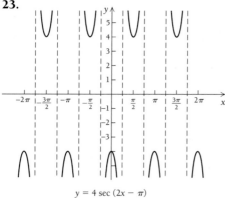

$$y = 4 \sec (2x - \pi)$$

**25.**

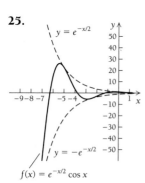

$$f(x) = e^{-x/2} \cos x$$

**27.**

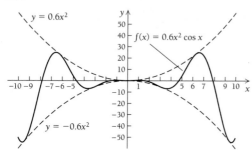

**29.**

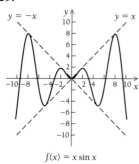

$$f(x) = x \sin x$$

**31.**

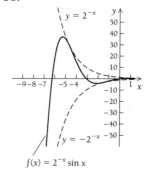

$$f(x) = 2^{-x} \sin x$$

**33.**

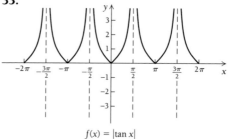

$$f(x) = |\tan x|$$

**35.**

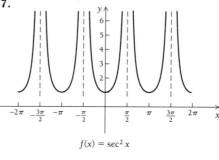

$$f(x) = (\ln x)(\sin x)$$

**37.**

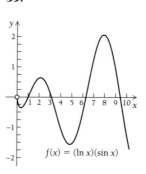

$$f(x) = \sec^2 x$$

**39. (a)**        **(b)** 4 inches

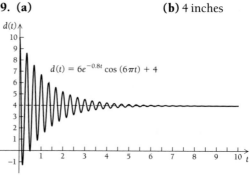

$$d(t) = 6e^{-0.8t} \cos (6\pi t) + 4$$

**41. (a)**

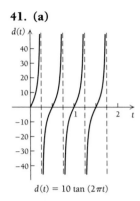

$$d(t) = 10 \tan (2\pi t)$$

**(b)** The function is undefined when the beam is parallel to the wall.
**43.** $-3.14, 3.14$    **45.** $-6.28, 6.28$    **47.** $-0.84, -0.41$

**Review Exercises: Chapter 6, pp. 447–448**

**1.** [6.1] $\sin \theta = \dfrac{3}{\sqrt{73}}$, or $\dfrac{3\sqrt{73}}{73}$; $\cos \theta = \dfrac{8}{\sqrt{73}}$, or $\dfrac{8\sqrt{73}}{73}$;

$\tan \theta = \dfrac{3}{8}$; $\cot \theta = \dfrac{8}{3}$; $\sec \theta = \dfrac{\sqrt{73}}{8}$; $\csc \theta = \dfrac{\sqrt{73}}{3}$

**2.** [6.1] $22°12'$    **3.** [6.1] $47.5575°$
**4.** [6.1] $\sin (90° - \theta) = 0.7314$, $\cos (90° - \theta) = 0.6820$, $\tan (90° - \theta) = 1.0724$, $\cot (90° - \theta) = 0.9325$, $\sec (90° - \theta) = 1.4663$, $\csc (90° - \theta) = 1.3673$
**5.** [6.1] $A = 58.09°$, $B = 31.91°$, $b = 4.5$
**6.** [6.1] $A = 38.83°$, $b = 37.9$, $c = 48.6$    **7.** [6.2] 1748 cm
**8.** [6.2] 14 ft
**9.** [6.3] Positive: $425°$, $785°$; negative: $-295°$, $-655°$

**10.** [6.3] $\sin \theta = \dfrac{3}{\sqrt{13}}$, or $\dfrac{3\sqrt{13}}{13}$; $\cos \theta = -\dfrac{2}{\sqrt{13}}$,

or $-\dfrac{2\sqrt{13}}{13}$; $\tan \theta = -\dfrac{3}{2}$; $\cot \theta = -\dfrac{2}{3}$; $\sec \theta = -\dfrac{\sqrt{13}}{2}$;

$\csc \theta = \dfrac{\sqrt{13}}{3}$    **11.** [6.3] $\dfrac{\sqrt{2}}{2}$    **12.** [6.3] 1

**13.** [6.3] $\sqrt{3}$    **14.** [6.3] $-\dfrac{\sqrt{3}}{2}$    **15.** [6.3] Undefined

**16.** [6.3] $-\dfrac{2\sqrt{3}}{3}$    **17.** [6.3] 2.1842    **18.** [6.3] $-0.5835$

**19.** [6.3] 1.1305    **20.** [6.3] $-0.2826$    **21.** [6.3] $-1.5350$
**22.** [6.3] 2.4586    **23.** [6.3] $205.26°$    **24.** [6.3] $47.18°$

**25.** [6.3] $\sin \theta = -\dfrac{2}{3}$; $\cos \theta = -\dfrac{\sqrt{5}}{3}$;

$\cot \theta = \dfrac{\sqrt{5}}{2}$; $\sec \theta = -\dfrac{3\sqrt{5}}{5}$; $\csc \theta = -\dfrac{3}{2}$

**26.** [6.4]

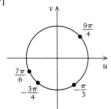

**27.** [6.4] $u$-axis: $\left(\dfrac{3}{5}, \dfrac{4}{5}\right)$; $v$-axis: $\left(-\dfrac{3}{5}, -\dfrac{4}{5}\right)$; origin: $\left(-\dfrac{3}{5}, \dfrac{4}{5}\right)$

**28.** [6.4] I, $0.483\pi$, 1.52    **29.** [6.4] II, $0.806\pi$, 2.53
**30.** [6.4] IV, $-0.167\pi$, $-0.524$    **31.** [6.4] $270°$

**32.** [6.4] $171.89°$    **33.** [6.4] $\dfrac{7\pi}{4}$, or 5.5 cm

**34.** [6.4] 2.25, $129°$    **35.** [6.4] 1131 cm/min
**36.** [6.4] 146,215 radians/hr    **37.** [6.5] $-1$    **38.** [6.5] 1

**39.** [6.5] $-\dfrac{\sqrt{3}}{2}$    **40.** [6.5] $\dfrac{1}{2}$    **4.1** [6.5] $\dfrac{\sqrt{3}}{3}$    **42.** [6.5] $-1$

**43.** [6.5] $-0.9056$    **44.** [6.5] 0.9218
**45.** [6.5] Does not exist    **46.** [6.5] 4.3813
**47.** [6.5] $-6.1685$    **48.** [6.5] 0.8090    **49.** [6.5] 1
**50.** [6.5] $\csc^2 s$
**51.** [6.5]

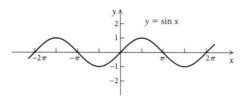

**52.** [6.5]

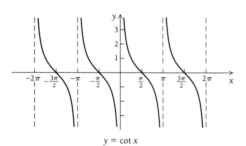

$y = \cot x$

**53.** [6.5]

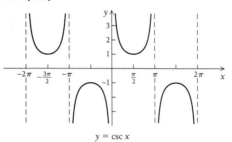

$y = \csc x$

**54.** [6.5] Period: $2\pi$; range: all reals from $-1$ to 1, inclusive
**55.** [6.5] Amplitude: 1; domain: all reals
**56.** [6.5] Period: $2\pi$; range: all real numbers 1 and greater, in addition to all real numbers $-1$ and less
**57.** [6.6]

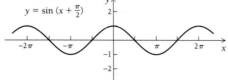

**58.** [6.6]

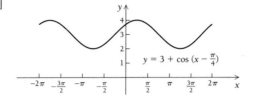

$y = 3 + \cos\left(x - \frac{\pi}{4}\right)$

**59.** [6.6] $\frac{\pi}{4}$    **60.** [6.6] $2\pi$

**61.** [6.6]

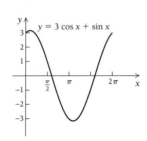

$y = 3\cos x + \sin x$

**62** [6.7]

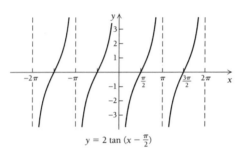

$y = 2\tan\left(x - \frac{\pi}{2}\right)$

**63** [6.7]

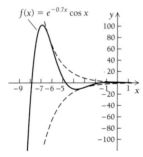

$f(x) = e^{-0.7x}\cos x$

**64.** [6.5] No, $\sin x = \frac{7}{5}$, but sines are never greater than 1.

**65.** [6.5] All values
**66.** [6.7] Domain: set of all reals; range $[-3, 3]$; period: $4\pi$

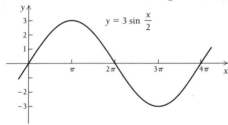

$y = 3\sin\frac{x}{2}$

**67.** [6.5] The domain consists of the intervals

$\left(-\frac{\pi}{2} + 2k\pi, \frac{\pi}{2} + 2k\pi\right)$, $k$ any integer

**68.** [6.3] $\cos x = -0.7890$; $\tan x = -0.7787$;
$\cot x = -1.2842$; $\sec x = -1.2674$; $\csc x = 1.6276$

**Test: Chapter 6, pp. 448–449**
**1.** [6.1] $19°15'$    **2.** [6.1] $72.8°$
**3.** [6.1] $\sin(90° - \theta) = 0.8910$, $\cos(90° - \theta) = 0.4540$,
$\tan(90° - \theta) = 1.963$, $\cot(90° - \theta) = 0.5095$,
$\sec(90° - \theta) = 2.203$, $\csc(90° - \theta) = 1.122$
**4.** [6.1] $B = 67.6°$, $a = 7.6$, $c = 19.9$    **5.** [6.1] $B = 66.87°$,
$A = 23.13°$, $b = 26$    **6.** [6.2] $35°$

**7.** [6.3] $\sin\theta = \frac{5}{\sqrt{34}}$, or $\frac{5\sqrt{34}}{34}$; $\cos\theta = -\frac{3}{\sqrt{34}}$, or $-\frac{3\sqrt{34}}{34}$;

$\tan\theta = -\frac{5}{3}$; $\cot\theta = -\frac{3}{5}$; $\sec\theta = -\frac{\sqrt{34}}{3}$; $\csc\theta = \frac{\sqrt{34}}{5}$

**8.** [6.5] $\frac{\sqrt{2}}{2}$    **9.** [6.3] $\frac{\sqrt{3}}{2}$    **10.** [6.3] 0    **11.** [6.3] 2

**12.** [6.3] $-0.830949$    **13.** [6.5] 2.80201
**14.** [6.5] 0.614213    **15.** [6.3] 1.27396    **16.** [6.5] $-3.73281$
**17.** [6.3] 0.781286    **18.** [6.3] $228.16°$

**19.** [6.3] $\cos\theta = -\frac{\sqrt{15}}{4}$, $\tan\theta = -\frac{\sqrt{15}}{15}$,

$\cot\theta = -\sqrt{15}$, $\sec\theta = -\frac{4\sqrt{15}}{15}$, $\csc\theta = 4$

**20.** [6.4] III, $-\frac{5\pi}{6}$, $-2.62$    **21.** [6.4] II, $\frac{13\pi}{20}$, 2.04

**22.** [6.4] III, $\frac{5\pi}{4}$, 3.93    **23.** [6.4] $315°$    **24.** [6.4] $114.59°$

**25.** [6.4] 10.47 cm    **26.** [6.4] $\frac{1}{2}$, $29°$    **27.** [6.4] 2010

**28.** [6.4] 1250    **29.** [6.5] $\sec^2 s$    **30.** [6.5] 1
**31.** [6.5]

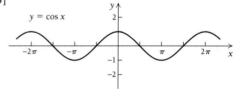

$y = \cos x$

**32.** [6.5] All reals from $-1$ to 1 inclusive    **33.** [6.5] $2\pi$
**34.** [6.5] All reals    **35.** [6.5] 1
**36.** [6.5]

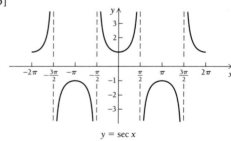

$y = \sec x$

**37.** [6.5] $2\pi$

**38.** [6.5] All reals except $\frac{\pi}{2} + k\pi$    **39.** [6.5] II, IV

**40.** [6.6]

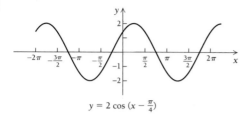

$$y = 2\cos\left(x - \frac{\pi}{4}\right)$$

**41.** [6.6] $2\pi$　**42.** [6.6] $\dfrac{\pi}{4}$

**43.** [6.6]

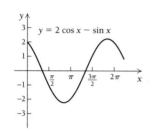

$y = 2\cos x - \sin x$

**44** [6.1], [6.4] **(a)** 0.99518; **(b)** 0.99518　**45.** [6.4] 132.5°

## CHAPTER 7

### Margin Exercises, Section 7.1

**1.** $\cos x + 1$　**2.** $\sin x$　**3.** $\sec\theta$　**4.** $\dfrac{1 + \sin x}{1 - \cot x}$

**5.** $\dfrac{5\sin x - \cos x}{\sin^2 x - \cos^2 x}$　**6.** $\dfrac{2}{\cos^2 x\,(\cot x - 2)}$

**7.** $\tan x \sin x\sqrt{\sin x}$　**8.** $\dfrac{\cos x}{\sqrt{3\cos x}}$

**9.** $\sqrt{4 - x^2} = 2\cos\theta$; $\cos\theta = \dfrac{\sqrt{4 - x^2}}{2}$; $\tan\theta = \dfrac{x}{\sqrt{4 - x^2}}$

### Technology Connection, Section 7.1
**TC 1.** A, B, C　**TC 2.** A, B, C, E

### Exercise Set 7.1, pp. 456–458
**1.** $\sin^2 x - \cos^2 x$　**3.** $\sin x - \sec x$　**5.** $\sin y + \cos y$
**7.** $\cot x - \tan x$　**9.** $1 - 2\sin y\cos y$
**11.** $2\tan x + \sec^2 x$　**13.** $\sin^2 y + \csc^2 y - 2$
**15.** $\cos^3 x - \sec^3 x$　**17.** $\cot^3 x - \tan^3 x$
**19.** $\cos^2 x$　**21.** $\cos x\,(\sin x + \cos x)$
**23.** $(\sin\theta - \cos\theta)(\sin\theta + \cos\theta)$　**25.** $\sin x\,(\sec x + 1)$
**27.** $(\sin x + \cos x)(\sin x - \cos x)$　**29.** $3(\cot y + 1)^2$
**31.** $(\csc^2 x + 5)(\csc x + 1)(\csc x - 1)$
**33.** $(\sin y + 3)(\sin^2 y - 3\sin y + 9)$
**35.** $(\sin v - \csc v)(\sin^2 v + 1 + \csc^2 v)$　**37.** $\tan x$
**39.** $\dfrac{2\cos^2 x}{9\sin x}$　**41.** $\cos x - 1$　**43.** $\cos\alpha + 1$
**45.** $\dfrac{2\tan x + 1}{3\sin x + 1}$　**47.** $\sec\theta$　**49.** 1

**51.** $\dfrac{1}{2\cos x}$, or $\dfrac{1}{2}\sec x$　**53.** $\dfrac{3\tan x}{\cos x - \sin x}$　**55.** $\dfrac{1}{3}\cot\gamma$

**57.** $\dfrac{1 - 2\sin y + 2\cos y}{\sin^2 y - \cos^2 y}$　**59.** $-1$　**61.** $\dfrac{5(\sin x - 3)}{3}$

**63.** $\sin x\cos x$　**65.** $\sqrt{\sin y}\,(\sin y + \cos y)$
**67.** $\sin x + \cos x$　**69.** $1 - \sin y$

**71.** $\sin x\,(\sqrt{2} + \sqrt{\cos x})$　**73.** $\dfrac{\sqrt{\sin x\cos x}}{\cos x}$　**75.** $\dfrac{\sqrt{\cos x}}{\cot x}$

**77.** $\dfrac{\sqrt{2}\cot x}{2}$　**79.** $\dfrac{\cos x}{1 - \sin x}$　**81.** $\dfrac{\sin x}{\sqrt{\cos x\sin x}}$

**83.** $\dfrac{\sin x}{\sqrt{\cos x}}$　**85.** $\dfrac{\cot x}{\sqrt{2}}$　**87.** $\dfrac{1 + \sin x}{\cos x}$

**89.** $\cos\theta = \dfrac{\sqrt{a^2 - x^2}}{a}$; $\tan\theta = \dfrac{x}{\sqrt{a^2 - x^2}}$

**91.** $\sin\theta = \dfrac{\sqrt{x^2 - 9}}{x}$; $\cos\theta = \dfrac{3}{x}$　**93.** $\sin\theta\tan\theta$

**95.** Let $x = \dfrac{\pi}{4}$. Then $\left(\sin\dfrac{\pi}{4} + \cos\dfrac{\pi}{4}\right)^2 = \left(\dfrac{\sqrt{2}}{2} + \dfrac{\sqrt{2}}{2}\right)^2 = $

$(\sqrt{2})^2 = 2$, and $\sin^2\dfrac{\pi}{4} + \cos^2\dfrac{\pi}{4} = \left(\dfrac{\sqrt{2}}{2}\right)^2 + \left(\dfrac{\sqrt{2}}{2}\right)^2 = $

$\dfrac{2}{4} + \dfrac{2}{4} = 1$.

**97.** Let $x = \dfrac{\pi}{6}$ and $y = \dfrac{\pi}{4}$. Then $\dfrac{\sin\dfrac{\pi}{6}}{\sin\dfrac{\pi}{4}} = \dfrac{\dfrac{1}{2}}{\dfrac{\sqrt{2}}{2}} = \dfrac{1}{\sqrt{2}}$, and

$\dfrac{x}{y} = \dfrac{\dfrac{\pi}{6}}{\dfrac{\pi}{4}} = \dfrac{2}{3}$.

**99.** Let $\alpha = \dfrac{\pi}{3}$ and $\beta = \dfrac{2\pi}{3}$. Then $\cos\left(\dfrac{\pi}{3} + \dfrac{2\pi}{3}\right) = $

$\cos\pi = -1$, and $\cos\dfrac{\pi}{3} + \cos\dfrac{2\pi}{3} = \dfrac{1}{2} + \left(-\dfrac{1}{2}\right) = 0$.

**101.** Let $\theta = \dfrac{\pi}{2}$. Then $\cos\left(2\cdot\dfrac{\pi}{2}\right) = \cos\pi = -1$,

and $2\cos\dfrac{\pi}{2} = 2\cdot 0 = 0$.

**103.** Let $\theta = \dfrac{\pi}{2}$. Then $\ln\left(\sin\dfrac{\pi}{2}\right) = \ln(1) = 0$, and

$\sin\left(\ln\dfrac{\pi}{2}\right) = \sin(0.451583) = 0.436390$.

### Margin Exercises, Section 7.2

**1.** $\dfrac{1}{2}$　**2.** $\dfrac{-\sqrt{6} + \sqrt{2}}{4}$, or $\dfrac{\sqrt{2}}{4}(1 - \sqrt{3})$　**3.** $-\dfrac{\sqrt{2} + \sqrt{6}}{4}$

**4.** $\dfrac{\sqrt{2} + \sqrt{6}}{4}$, or $\dfrac{\sqrt{2}}{4}(1 + \sqrt{3})$　**5.** $\sin(\alpha - \beta)$

**6.** $\tan(\alpha - \beta) = \dfrac{\sin(\alpha - \beta)}{\cos(\alpha - \beta)}$

$$= \frac{\sin \alpha \cos \beta - \cos \alpha \sin \beta}{\cos \alpha \cos \beta + \sin \alpha \sin \beta} \cdot \dfrac{\dfrac{1}{\cos \alpha \cos \beta}}{\dfrac{1}{\cos \alpha \cos \beta}}$$

$$= \frac{\dfrac{\sin \alpha \cos \beta}{\cos \alpha \cos \beta} - \dfrac{\cos \alpha \sin \beta}{\cos \alpha \cos \beta}}{\dfrac{\cos \alpha \cos \beta}{\cos \alpha \cos \beta} + \dfrac{\sin \alpha \sin \beta}{\cos \alpha \cos \beta}}$$

$$= \frac{\tan \alpha - \tan \beta}{1 + \tan \alpha \tan \beta}$$

**7.** $\dfrac{3 + \sqrt{3}}{3 - \sqrt{3}}$   **8.** $\cos \dfrac{7\pi}{12}$, or $\cos\left(-\dfrac{7\pi}{12}\right)$   **9.** $\cos 25°$

**10.** $\sin \dfrac{\pi}{6}$, or $\dfrac{1}{2}$   **11.** $\dfrac{3 + \sqrt{105}}{16}$

**Exercise Set 7.2, pp. 463–465**

**1.** $\dfrac{\sqrt{6} + \sqrt{2}}{4}$   **3.** $\dfrac{\sqrt{6} - \sqrt{2}}{4}$   **5.** $\dfrac{\sqrt{6} + \sqrt{2}}{4}$   **7.** $\dfrac{3 + \sqrt{3}}{3 - \sqrt{3}}$

**9.** $\sin 59°$, or 0.8572   **11.** $\tan 52°$, or 1.280   **13.** 1
**15.** 0   **17.** Undefined   **19.** 0.8448   **21.** $2 \sin \alpha \cos \beta$
**23.** $2 \cos \alpha \cos \beta$   **25.** $\cos u$   **27.** 0°   **29.** 135°
**31.** 22.83°

**33.** $\dfrac{\cos(x + h) - \cos x}{h} = \dfrac{\cos x \cos h - \sin x \sin h - \cos x}{h}$

$$= \frac{\cos x \cos h - \cos x}{h} - \frac{\sin x \sin h}{h}$$

$$= \cos x \left(\frac{\cos h - 1}{h}\right) - \sin x \left(\frac{\sin h}{h}\right)$$

**35.** $\dfrac{1}{7}$   **37.** $\dfrac{1}{6}$   **39.** $2 \sin \theta \cos \theta$

**41.** $\cot(\alpha + \beta) = \dfrac{\cos(\alpha + \beta)}{\sin(\alpha + \beta)}$

$$= \frac{\cos \alpha \cos \beta - \sin \alpha \sin \beta}{\sin \alpha \cos \beta + \cos \alpha \sin \beta} \cdot \dfrac{\dfrac{1}{\sin \alpha \sin \beta}}{\dfrac{1}{\sin \alpha \sin \beta}}$$

$$= \frac{\dfrac{\cos \alpha \cos \beta}{\sin \alpha \sin \beta} - 1}{\dfrac{\sin \alpha \cos \beta}{\sin \alpha \sin \beta} + \dfrac{\cos \alpha \sin \beta}{\sin \alpha \sin \beta}} = \frac{\cot \alpha \cot \beta - 1}{\cot \alpha + \cot \beta}$$

**43.** $\sin\left(x + \dfrac{3\pi}{2}\right) = \sin x \cos \dfrac{3\pi}{2} + \cos x \sin \dfrac{3\pi}{2}$

$$= (\sin x)(0) + (\cos x)(-1) = -\cos x$$

**45.** $\cos\left(x + \dfrac{3\pi}{2}\right) = \cos x \cos \dfrac{3\pi}{2} - \sin x \sin \dfrac{3\pi}{2}$

$$= (\cos x)(0) - (\sin x)(-1) = \sin x$$

**47.** $\tan\left(x + \dfrac{\pi}{4}\right) = \dfrac{\tan x + \tan \dfrac{\pi}{4}}{1 - \tan x \tan \dfrac{\pi}{4}} = \dfrac{1 + \tan x}{1 - \tan x}$

**49.** $\dfrac{\tan \alpha + \tan \beta}{1 + \tan \alpha \tan \beta} = \dfrac{\dfrac{\sin \alpha}{\cos \alpha} + \dfrac{\sin \beta}{\cos \beta}}{1 + \dfrac{\sin \alpha}{\cos \alpha} \cdot \dfrac{\sin \beta}{\cos \beta}}$

$$= \frac{\dfrac{\sin \alpha \cos \beta + \cos \alpha \sin \beta}{\cos \alpha \cos \beta}}{\dfrac{\cos \alpha \cos \beta + \sin \alpha \sin \beta}{\cos \alpha \cos \beta}}$$

$$= \frac{\sin \alpha \cos \beta + \cos \alpha \sin \beta}{\cos \alpha \cos \beta + \sin \alpha \sin \beta}$$

$$= \frac{\sin(\alpha + \beta)}{\cos(\alpha - \beta)}$$

**51.** $\sin(\alpha + \beta) + \sin(\alpha - \beta) = \sin \alpha \cos \beta + \cos \alpha \sin \beta +$ $\sin \alpha \cos \beta - \cos \alpha \sin \beta = 2 \sin \alpha \cos \beta$
**53.** $\cos(\alpha + \beta) = \cos \alpha \cos \beta - \sin \alpha \sin \beta$

$$= \cos \alpha \cos \beta - \cos\left(\frac{\pi}{2} - \alpha\right) \cos\left(\frac{\pi}{2} - \beta\right)$$

**Margin Exercises, Section 7.3**

**1.** $\cot\left(\dfrac{\pi}{2} - \theta\right) = \dfrac{\cos\left(\dfrac{\pi}{2} - \theta\right)}{\sin\left(\dfrac{\pi}{2} - \theta\right)} = \dfrac{\sin \theta}{\cos \theta} = \tan \theta$

**2.** $\sec\left(\dfrac{\pi}{2} - \theta\right) = \dfrac{1}{\cos\left(\dfrac{\pi}{2} - \theta\right)} = \dfrac{1}{\sin \theta} = \csc \theta$

**3.** $\csc\left(\dfrac{\pi}{2} - \theta\right) = \dfrac{1}{\sin\left(\dfrac{\pi}{2} - \theta\right)} = \dfrac{1}{\cos \theta} = \sec \theta$

**4.** $\sin 15° = 0.2588$; $\cos 15° = 0.9659$; $\tan 15° = 0.2679$; $\cot 15° = 3.732$; $\sec 15° = 1.035$; $\csc 15° = 3.864$
**5. (a)** and **(b)**

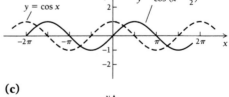

**(c)**

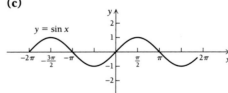

**(d)** the same; **(e)** $\cos\left(x - \dfrac{\pi}{2}\right) = \sin x$

**6.** $\cos\left(x - \dfrac{\pi}{2}\right) = \cos x \cos\dfrac{\pi}{2} + \sin x \sin\dfrac{\pi}{2}$

$\qquad\qquad = \cos x \cdot 0 + \sin x \cdot 1 = \sin x$

**7.** $\sin\left(x - \dfrac{\pi}{2}\right) = \sin x \cos\dfrac{\pi}{2} - \cos x \sin\dfrac{\pi}{2}$

$\qquad\qquad = \sin x \cdot 0 - \cos x \cdot 1 = -\cos x$

**8. (a)** Translate the graph of $y = \cos x$ to obtain the graph

of $y = \cos\left(x + \dfrac{\pi}{2}\right)$. Then graph $y = \sin x$ and reflect the

graph to obtain $y = -\sin x$.

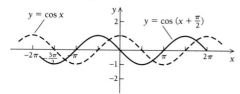

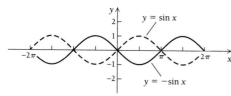

The graphs of $y = \cos\left(x + \dfrac{\pi}{2}\right)$ and $y = -\sin x$ are the

same. Thus, $\cos\left(x + \dfrac{\pi}{2}\right) = -\sin x$.

**(b)** $\cos\left(x + \dfrac{\pi}{2}\right) = \cos x \cos\dfrac{\pi}{2} - \sin x \sin\dfrac{\pi}{2}$

$\qquad\qquad = \cos x \cdot 0 - \sin x \cdot 1 = -\sin x$

**9.** $\cot\left(x + \dfrac{\pi}{2}\right) = \dfrac{\cos\left(x + \dfrac{\pi}{2}\right)}{\sin\left(x + \dfrac{\pi}{2}\right)} = \dfrac{-\sin x}{\cos x} = -\tan x$

**10.** $\csc\left(x - \dfrac{\pi}{2}\right) = \dfrac{1}{\sin\left(x - \dfrac{\pi}{2}\right)} = \dfrac{1}{-\cos x} = -\sec x$

**11.** $\sin(x + \pi) = -\sin x$    **12.** $\tan(\theta - 450°) = -\cot\theta$

**Technology Connection, Section 7.3**
**TC 1.** No    **TC 2.** Yes    **TC 3.** Yes    **TC 4.** Yes

**Exercise Set 7.3, pp. 470–471**
**1.** $\sin 25° = 0.4226$; $\cos 25° = 0.9063$; $\tan 25° = 0.4663$;
$\cot 25° = 2.145$; $\sec 25° = 1.103$; $\csc 25° = 2.366$

**3. (a)** $\tan 22.5° = \dfrac{\sqrt{2 - \sqrt{2}}}{\sqrt{2 + \sqrt{2}}}$, $\cot 22.5° = \dfrac{\sqrt{2 + \sqrt{2}}}{\sqrt{2 - \sqrt{2}}}$,

$\sec 22.5° = \dfrac{2}{\sqrt{2 + \sqrt{2}}}$, $\csc 22.5° = \dfrac{2}{\sqrt{2 - \sqrt{2}}}$;

**(b)** $\sin 67.5° = \dfrac{\sqrt{2 + \sqrt{2}}}{2}$, $\cos 67.5° = \dfrac{\sqrt{2 - \sqrt{2}}}{2}$,

$\tan 67.5° = \dfrac{\sqrt{2 + \sqrt{2}}}{\sqrt{2 - \sqrt{2}}}$, $\cot 67.5° = \dfrac{\sqrt{2 - \sqrt{2}}}{\sqrt{2 + \sqrt{2}}}$,

$\sec 67.5° = \dfrac{2}{\sqrt{2 - \sqrt{2}}}$, $\csc 67.5° = \dfrac{2}{\sqrt{2 + \sqrt{2}}}$

**5.** $\tan\left(x - \dfrac{\pi}{2}\right) = \dfrac{\sin\left(x - \dfrac{\pi}{2}\right)}{\cos\left(x - \dfrac{\pi}{2}\right)} = \dfrac{-\cos x}{\sin x} = -\cot x$

**7.** $\csc\left(x + \dfrac{\pi}{2}\right) = \dfrac{1}{\sin\left(x + \dfrac{\pi}{2}\right)} = \dfrac{1}{\cos x} = \sec x$

**9.** $\sin(x - \pi) = \sin x \cos\pi - \cos x \sin\pi$
$\qquad\qquad = (\sin x)(-1) - (\cos x)(0) = -\sin x$

**11.** $\tan(\theta + 270°) = -\cot\theta$

**13.** $\sin(\theta + 450°) = \cos\theta$    **15.** $\sec(\pi - \theta) = -\sec\theta$

**17.** $\tan(x - 4\pi) = \tan x$    **19.** $\cos\left(\dfrac{9\pi}{2} + x\right) = -\sin x$

**21.** $\csc(540° - \theta) = \csc\theta$    **23.** $\cot\left(x - \dfrac{3\pi}{2}\right) = -\tan x$

**25. (a)** $\cos\theta = -\dfrac{2\sqrt{2}}{3}$; $\tan\theta = -\dfrac{1}{2\sqrt{2}}$, or $-\dfrac{\sqrt{2}}{4}$;

$\cot\theta = -2\sqrt{2}$; $\sec\theta = -\dfrac{3}{2\sqrt{2}}$, or $-\dfrac{3\sqrt{2}}{4}$; $\csc\theta = 3$;

**(b)** $\sin\left(\dfrac{\pi}{2} - \theta\right) = -\dfrac{2\sqrt{2}}{3}$, $\cos\left(\dfrac{\pi}{2} - \theta\right) = \dfrac{1}{3}$,

$\tan\left(\dfrac{\pi}{2} - \theta\right) = -2\sqrt{2}$, $\cot\left(\dfrac{\pi}{2} - \theta\right) = -\dfrac{\sqrt{2}}{4}$,

$\sec\left(\dfrac{\pi}{2} - \theta\right) = 3$, $\csc\left(\dfrac{\pi}{2} - \theta\right) = -\dfrac{3\sqrt{2}}{4}$;

**(c)** $\sin(\pi + \theta) = -\dfrac{1}{3}$, $\cos(\pi + \theta) = \dfrac{2\sqrt{2}}{3}$,

$\tan(\pi + \theta) = -\dfrac{\sqrt{2}}{4}$, $\cot(\pi + \theta) = -2\sqrt{2}$,

$\sec(\pi + \theta) = \dfrac{3\sqrt{2}}{4}$, $\csc(\pi + \theta) = -3$;

**(d)** $\sin(\pi - \theta) = \dfrac{1}{3}$, $\cos(\pi - \theta) = \dfrac{2\sqrt{2}}{3}$,

$\tan(\pi - \theta) = \dfrac{\sqrt{2}}{4}$, $\cot(\pi - \theta) = 2\sqrt{2}$,

$\sec(\pi - \theta) = \dfrac{3\sqrt{2}}{4}$, $\csc(\pi - \theta) = 3$;

**(e)** $\sin(2\pi - \theta) = -\dfrac{1}{3}$, $\cos(2\pi - \theta) = -\dfrac{2\sqrt{2}}{3}$,

$\tan(2\pi - \theta) = \dfrac{\sqrt{2}}{4}$, $\cot(2\pi - \theta) = 2\sqrt{2}$,

$\sec(2\pi - \theta) = -\dfrac{3\sqrt{2}}{4}$, $\csc(2\pi - \theta) = -3$

**27.** **(a)** $\cos 27° = 0.89101$, $\tan 27° = 0.50952$,
$\cot 27° = 1.9626$, $\sec 27° = 1.1223$, $\csc 27° = 2.2027$;
**(b)** $\sin 63° = 0.89101$, $\cos 63° = 0.45399$, $\tan 63° = 1.9626$,
$\cot 63° = 0.50952$, $\sec 63° = 2.2027$, $\csc 63° = 1.1223$
**29.** $\sin 128° = 0.78801$, $\cos 128° = -0.61566$,
$\tan 128° = -1.2799$, $\cot 128° = -0.78129$,
$\sec 128° = -1.6243$, $\csc 128° = 1.2690$
**31.** $\sin^2 x$    **33.** $\sin x (\cos y + \tan y)$

**35.** $-\cos x (1 + \cot x)$    **37.** $\cos x + 1$    **39.** $\dfrac{\cos x - 1}{1 + \tan x}$

**41.** $\cos x - 1$    **43.** $-\tan^2 y$

**Margin Exercises, Section 7.4**
**1.** $24/25$
**2.** II, $\cos 2\theta = -119/169$, $\tan 2\theta = -120/119$,
$\sin 2\theta = 120/169$    **3.** $\cos \theta - 4 \sin^2 \theta \cos \theta$

**4.** $\sin^3 x = \sin x \dfrac{1 - \cos 2x}{2}$    **5.** $\dfrac{\sqrt{2 + \sqrt{3}}}{2}$    **6.** $\dfrac{1}{2 + \sqrt{3}}$,
or $2 - \sqrt{3}$

**7.** $\tan 2x$    **8.** $\dfrac{\frac{1}{2}(\cos^2 x - \sin^2 x)}{\sin x \cos x} = \cot 2x$

**Technology Connection, Section 7.4**
**TC 1.** B and D    **TC 2.** A and D

**Exercise Set 7.4, pp. 476–477**

**1.** $\sin 2\theta = \dfrac{24}{25}$, $\cos 2\theta = -\dfrac{7}{25}$, $\tan 2\theta = -\dfrac{24}{7}$, II

**3.** $\sin 2\theta = \dfrac{24}{25}$, $\cos 2\theta = \dfrac{7}{25}$, $\tan 2\theta = \dfrac{24}{7}$, I

**5.** $\sin 2\theta = \dfrac{24}{25}$, $\cos 2\theta = -\dfrac{7}{25}$, $\tan 2\theta = -\dfrac{24}{7}$, II

**7.** $8 \sin \theta \cos^3 \theta - 4 \sin \theta \cos \theta$, or $4 \sin \theta \cos^3 \theta - 4 \sin^3 \theta \cos \theta$, or $4 \sin \theta \cos \theta - 8 \sin^3 \theta \cos \theta$

**9.** $\dfrac{3 - 4 \cos 2\theta + \cos 4\theta}{8}$    **11.** $\dfrac{\sqrt{2 + \sqrt{3}}}{2}$    **13.** $2 + \sqrt{3}$

**15.** $\dfrac{\sqrt{2 + \sqrt{2}}}{2}$    **17.** $\dfrac{\sqrt{2 + \sqrt{2}}}{2}$    **19.** $\dfrac{\sqrt{2 + \sqrt{2}}}{2}$

**21.** $\dfrac{\sqrt{2 - \sqrt{2}}}{2}$    **23.** $0.6421$    **25.** $0.9844$    **27.** $0.1735$

**29.** $\cos x$    **31.** $\cos x$    **33.** $\sin x$    **35.** $\cos x$    **37.** $1$
**39.** $1$    **41.** $8$    **43.** $\sin 2x$
**45.** $(\sin x + \cos x)^2 = 1 + \sin 2x$

**47.** $\dfrac{2 \cot x}{\cot^2 x - 1} = \tan 2x$    **49.** $2 \sin^2 2x + \cos 4x = 1$

**51.** $\sin \theta = \sqrt{\dfrac{1}{2} + \dfrac{\sqrt{6}}{5}}$; $\cos \theta = \sqrt{\dfrac{1}{2} - \dfrac{\sqrt{6}}{5}}$;
$\tan \theta = \sqrt{\dfrac{5 + 2\sqrt{6}}{5 - 2\sqrt{6}}}$

**53.** $\sin \theta = -\dfrac{8}{17}$; $\cos \theta = \dfrac{15}{17}$; $\tan \theta = -\dfrac{8}{15}$

**55.** **(a)** $6062.76$ ft; **(b)** $6097$ ft;
**(c)** $N(\phi) = 6097 - 62 \cos^2 \phi$
**57.** Since $\cos^2 x - \sin^2 x = \cos 2x$, the graph of $f(x) = \cos^2 x - \sin^2 x$ is the same as the graph of $f(x) = \cos 2x$.

**Margin Exercises, Section 7.5**
**1.** *Method 1:*

| Left side | Right side |
|---|---|
| $\dfrac{\csc t - 1}{t \csc t}$ | $\dfrac{1 - \sin t}{t}$ |
| $\dfrac{\dfrac{1}{\sin t} - 1}{t \cdot \dfrac{1}{\sin t}}$ | |
| $\dfrac{\dfrac{1}{\sin t} - 1}{t \cdot \dfrac{1}{\sin t}} \cdot \dfrac{\sin t}{\sin t}$ | |
| $\dfrac{1 - \sin t}{t}$ | |

**2.** *Method 1:*

| Left side | Right side |
|---|---|
| $\sec \theta \csc \theta$ | $\tan \theta + \cot \theta$ |
| $\dfrac{1}{\cos \theta} \cdot \dfrac{1}{\sin \theta}$ | |
| $\dfrac{1}{\cos \theta \sin \theta} \cdot \dfrac{\sin^2 \theta + \cos^2 \theta}{1}$ | |
| $\dfrac{\sin \theta}{\cos \theta} + \dfrac{\cos \theta}{\sin \theta}$ | |
| $\tan \theta + \cot \theta$ | |

**3.** *Method 1:*

| Left side | Right side |
|---|---|
| $(\sin \theta + \cos \theta)^2$ | $1 - \sin 2\theta$ |
| | $1 - 2 \sin \theta \cos \theta$ |
| | $(\sin^2 \theta + \cos^2 \theta) - 2 \sin \theta \cos \theta$ |
| | $\sin^2 \theta - 2 \sin \theta \cos \theta + \cos^2 \theta$ |
| | $(\sin \theta - \cos \theta)^2$ |

**4.** *Method 1:*

| Left side | Right side |
|---|---|
| $(\sec u - \tan u)(1 + \sin u)$ | $\cos u$ |
| $\left(\dfrac{1}{\cos u} - \dfrac{\sin u}{\cos u}\right)(1 + \sin u)$ | |
| $\dfrac{1}{\cos u} + \dfrac{\sin u}{\cos u} - \dfrac{\sin u}{\cos u} - \dfrac{\sin^2 u}{\cos u}$ | |
| $\dfrac{1 - \sin^2 u}{\cos u}$ | |
| $\dfrac{\cos^2 u}{\cos u}$ | |
| $\cos u$ | |

**5.** *Method 1:*

| Left side | Right side |
|---|---|
| $\dfrac{\sin t}{1 + \cos t}$ | $\dfrac{1 - \cos t}{\sin t}$ |
| $\dfrac{\sin t}{1 + \cos t} \cdot \dfrac{1 - \cos t}{1 - \cos t}$ | |
| $\dfrac{(\sin t)(1 - \cos t)}{1 - \cos^2 t}$ | |
| $\dfrac{(\sin t)(1 - \cos t)}{\sin^2 t}$ | |
| $\dfrac{1 - \cos t}{\sin t}$ | |

**6.** *Method 2:*

| Left side | Right side |
|---|---|
| $\cot^2 x - \cos^2 x$ | $\cos^2 x \cot^2 x$ |
| $\dfrac{\cos^2 x}{\sin^2 x} - \cos^2 x$ | $\cos^2 x \cdot \dfrac{\cos^2 x}{\sin^2 x}$ |
| $\dfrac{\cos^2 x - \sin^2 x \cos^2 x}{\sin^2 x}$ | $\dfrac{\cos^4 x}{\sin^2 x}$ |
| $\dfrac{\cos^2 x\,(1 - \sin^2 x)}{\sin^2 x}$ | |
| $\dfrac{\cos^4 x}{\sin^2 x}$ | |

**7.** *Method 2:*

| Left side | Right side |
|---|---|
| $\dfrac{\sin 2\theta + \sin \theta}{\cos 2\theta + \cos \theta + 1}$ | $\tan \theta$ |
| $\dfrac{2 \sin \theta \cos \theta + \sin \theta}{(2 \cos^2 \theta - 1) + \cos \theta + 1}$ | $\dfrac{\sin \theta}{\cos \theta}$ |
| $\dfrac{\sin \theta\,(2 \cos \theta + 1)}{2 \cos^2 \theta + \cos \theta}$ | |
| $\dfrac{\sin \theta\,(2 \cos \theta + 1)}{\cos \theta\,(2 \cos \theta + 1)}$ | |
| $\dfrac{\sin \theta}{\cos \theta}$ | |

**8.**

| $\sin u \cdot \cos v$ | $\frac{1}{2}[\sin(u + v) + \sin(u - v)]$ |
|---|---|
| | $\frac{1}{2}[\sin u \cos v + \cos u \sin v$ |
| | $\quad + \sin u \cos v - \cos u \sin v]$ |
| | $\frac{1}{2}(2 \sin u \cos v)$ |
| | $\sin u \cdot \cos v$ |

**9.** $2 \sin 7\theta \cos 5\theta = \sin 12\theta + \sin 2\theta$

**Technology Connection, Section 7.5**
**TC 1.** C   **TC 2.** D   **TC 3.** A   **TC 4.** B

**Exercise Set 7.5, pp. 484–486**

**1.**

| $\csc x - \cos x \cot x$ | $\sin x$ |
|---|---|
| $\dfrac{1}{\sin x} - \cos x\,\dfrac{\cos x}{\sin x}$ | |
| $\dfrac{1 - \cos^2 x}{\sin x}$ | |
| $\dfrac{\sin^2 x}{\sin x}$ | |
| $\sin x$ | |

**3.**

| $\dfrac{1 + \cos \theta}{\sin \theta} + \dfrac{\sin \theta}{\cos \theta}$ | $\dfrac{\cos \theta + 1}{\sin \theta \cos \theta}$ |
|---|---|
| $\dfrac{1 + \cos \theta}{\sin \theta} \cdot \dfrac{\cos \theta}{\cos \theta} + \dfrac{\sin \theta \cdot \sin \theta}{\cos \theta \sin \theta}$ | $\dfrac{1 + \cos \theta}{\sin \theta \cos \theta}$ |
| $\dfrac{\cos \theta + \cos^2 \theta + \sin^2 \theta}{\sin \theta \cos \theta}$ | |
| $\dfrac{1 + \cos \theta}{\sin \theta \cos \theta}$ | |

**5.**

| $\dfrac{1 - \sin x}{\cos x}$ | $\dfrac{\cos x}{1 + \sin x}$ |
|---|---|
| $\dfrac{1 - \sin x}{\cos x} \cdot \dfrac{\cos x}{\cos x}$ | $\dfrac{\cos x}{1 + \sin x} \cdot \dfrac{1 - \sin x}{1 - \sin x}$ |
| $\dfrac{\cos x - \sin x \cos x}{\cos^2 x}$ | $\dfrac{\cos x - \sin x \cos x}{1 - \sin^2 x}$ |
| | $\dfrac{\cos x - \sin x \cos x}{\cos^2 x}$ |

**7.**

| $\dfrac{1 + \tan \theta}{1 + \cot \theta}$ | $\dfrac{\sec \theta}{\csc \theta}$ |
|---|---|
| $\dfrac{1 + \dfrac{\sin \theta}{\cos \theta}}{1 + \dfrac{\cos \theta}{\sin \theta}}$ | $\dfrac{\sin \theta}{\cos \theta}$ |
| | $\tan \theta$ |
| $\dfrac{\dfrac{\cos \theta + \sin \theta}{\cos \theta}}{\dfrac{\sin \theta + \cos \theta}{\sin \theta}}$ | |
| $\dfrac{\cos \theta + \sin \theta}{\cos \theta} \cdot \dfrac{\sin \theta}{\sin \theta + \cos \theta}$ | |
| $\dfrac{\sin \theta}{\cos \theta}$ | |
| $\tan \theta$ | |

**9.**

$$\frac{\sin x + \cos x}{\sec x + \csc x} \quad\bigg|\quad \frac{\sin x}{\sec x}$$

$$\frac{\sin x + \cos x}{\dfrac{1}{\cos x} + \dfrac{1}{\sin x}}$$

$$\frac{\sin x + \cos x}{\dfrac{\sin x + \cos x}{\sin x \cos x}}$$

$$(\sin x + \cos x)\cdot \frac{\sin x \cos x}{\sin x + \cos x}$$

$$\sin x \cos x$$

$$\frac{\sin x}{\sec x}$$

**11.**

$$\frac{1+\tan\theta}{1-\tan\theta}+\frac{1+\cot\theta}{1-\cot\theta} \quad\bigg|\quad 0$$

$$\frac{1+\dfrac{\sin\theta}{\cos\theta}}{1-\dfrac{\sin\theta}{\cos\theta}}+\frac{1+\dfrac{\cos\theta}{\sin\theta}}{1-\dfrac{\cos\theta}{\sin\theta}}$$

$$\frac{\dfrac{\cos\theta+\sin\theta}{\cos\theta}}{\dfrac{\cos\theta-\sin\theta}{\cos\theta}}+\frac{\dfrac{\sin\theta+\cos\theta}{\sin\theta}}{\dfrac{\sin\theta-\cos\theta}{\sin\theta}}$$

$$\frac{\cos\theta+\sin\theta}{\cos\theta}\cdot\frac{\cos\theta}{\cos\theta-\sin\theta}+$$

$$\frac{\sin\theta+\cos\theta}{\sin\theta}\cdot\frac{\sin\theta}{\sin\theta-\cos\theta}$$

$$\frac{\cos\theta+\sin\theta}{\cos\theta-\sin\theta}+\frac{\sin\theta+\cos\theta}{\sin\theta-\cos\theta}$$

$$\frac{\cos\theta+\sin\theta}{\cos\theta-\sin\theta}-\frac{\cos\theta+\sin\theta}{\cos\theta-\sin\theta}$$

$$0$$

**13.**

$$\frac{1+\cos 2\theta}{\sin 2\theta} \quad\bigg|\quad \cot\theta$$

$$\frac{1+2\cos^2\theta-1}{2\sin\theta\cos\theta} \quad\bigg|\quad \frac{\cos\theta}{\sin\theta}$$

$$\frac{\cos\theta}{\sin\theta}$$

**15.**

$$\sec 2\theta \quad\bigg|\quad \frac{\sec^2\theta}{2-\sec^2\theta}$$

$$\frac{1}{\cos 2\theta} \quad\bigg|\quad \frac{1+\tan^2\theta}{2-1-\tan^2\theta}$$

$$\frac{1}{\cos^2\theta-\sin^2\theta} \quad\bigg|\quad \frac{1+\tan^2\theta}{1-\tan^2\theta}$$

$$\frac{1+\dfrac{\sin^2\theta}{\cos^2\theta}}{1-\dfrac{\sin^2\theta}{\cos^2\theta}}$$

$$\frac{\cos^2\theta+\sin^2\theta}{\cos^2\theta-\sin^2\theta}$$

$$\frac{1}{\cos^2\theta-\sin^2\theta}$$

**17.**

$$\frac{\sin(\alpha+\beta)}{\cos\alpha\cos\beta} \quad\bigg|\quad \tan\alpha+\tan\beta$$

$$\frac{\sin\alpha\cos\beta+\cos\alpha\sin\beta}{\cos\alpha\cos\beta} \quad\bigg|\quad \frac{\sin\alpha}{\cos\alpha}+\frac{\sin\beta}{\cos\beta}$$

$$\frac{\sin\alpha\cos\beta+\cos\alpha\sin\beta}{\cos\alpha\cos\beta}$$

**19.**

$$1-\cos 5\theta\cos 3\theta-\sin 5\theta\cos 3\theta \quad\bigg|\quad 2\sin^2\theta$$

$$1-[\cos 5\theta\cos 3\theta+\sin 5\theta\cos 3\theta] \quad\bigg|\quad 1-\cos 2\theta$$

$$1-\cos(5\theta-3\theta)$$

$$1-\cos 2\theta$$

**21.**

$$\frac{\tan\theta+\sin\theta}{2\tan\theta} \quad\bigg|\quad \cos^2\frac{\theta}{2}$$

$$\frac{1}{2}\left[\frac{\dfrac{\sin\theta+\sin\theta\cos\theta}{\cos\theta}}{\dfrac{\sin\theta}{\cos\theta}}\right] \quad\bigg|\quad \frac{1+\cos\theta}{2}$$

$$\frac{1}{2}\left[\frac{\sin\theta(1+\cos\theta)}{\cos\theta}\cdot\frac{\cos\theta}{\sin\theta}\right]$$

$$\frac{1+\cos\theta}{2}$$

**23.**

$$\cos^4 x-\sin^4 x \quad\bigg|\quad \cos 2x$$

$$(\cos^2 x-\sin^2 x)(\cos^2 x+\sin^2 x) \quad\bigg|\quad \cos^2 x-\sin^2 x$$

$$\cos^2 x-\sin^2 x$$

**25.**

$$\frac{\tan 3\theta-\tan\theta}{1+\tan 3\theta\tan\theta} \quad\bigg|\quad \frac{2\tan\theta}{1-\tan^2\theta}$$

$$\tan(3\theta-\theta) \quad\bigg|\quad \tan 2\theta$$

$$\tan 2\theta$$

**27.**

| | |
|---|---|
| $\dfrac{\cos^3 x - \sin^3 x}{\cos x - \sin x}$ | $\dfrac{2 + \sin 2x}{2}$ |
| $\dfrac{(\cos x - \sin x)(\cos^2 x + \cos x \sin x + \sin^2 x)}{\cos x - \sin x}$ | $\dfrac{2 + 2\sin x \cos x}{2}$ |
| $1 + \cos x \sin x$ | $1 + \sin x \cos x$ |

**29.**

| | |
|---|---|
| $\sin(\alpha + \beta)\sin(\alpha - \beta)$ | $\sin^2 \alpha - \sin^2 \beta$ |
| $\begin{pmatrix}\sin\alpha\cos\beta + \\ \cos\alpha\sin\beta\end{pmatrix}\begin{pmatrix}\sin\alpha\cos\beta - \\ \cos\alpha\sin\beta\end{pmatrix}$ | $\begin{array}{l}1 - \cos^2\alpha - \\ (1 - \cos^2\beta)\end{array}$ |
| $\sin^2\alpha\cos^2\beta - \cos^2\alpha\sin^2\beta$ | $\cos^2\beta - \cos^2\alpha$ |
| $\begin{array}{l}\cos^2\beta(1-\cos^2\alpha) - \\ \cos^2\alpha(1-\cos^2\beta)\end{array}$ | |
| $\begin{array}{l}\cos^2\beta - \cos^2\alpha\cos^2\beta - \\ \cos^2\alpha + \cos^2\alpha\cos^2\beta\end{array}$ | |
| $\cos^2\beta - \cos^2\alpha$ | |

**31.**

| | |
|---|---|
| $\cos(\alpha + \beta) + \cos(\alpha - \beta)$ | $2\cos\alpha\cos\beta$ |
| $\begin{array}{l}\cos\alpha\cos\beta - \sin\alpha\sin\beta + \cos\alpha\cos\beta + \\ \sin\alpha\sin\beta\end{array}$ | |
| $2\cos\alpha\cos\beta$ | |

**33.**

| | |
|---|---|
| $\sin^2 x - \cos^2 x$ | $1 - 2\cos^2 x$ |
| $-(\cos^2 x - \sin^2 x)$ | $-(2\cos^2 x - 1)$ |
| $-\cos 2x$ | $-\cos 2x$ |

**35.**

| | |
|---|---|
| $\sin^2\theta$ | $\cos^2\theta(\sec^2\theta - 1)$ |
| | $\cos^2\theta \cdot \tan^2\theta$ |
| | $\cos^2\theta \cdot \dfrac{\sin^2\theta}{\cos^2\theta}$ |
| | $\sin^2\theta$ |

**37.**

| | |
|---|---|
| $\dfrac{\tan x}{\sec x - \cos x}$ | $\dfrac{\sec x}{\tan x}$ |
| $\dfrac{\dfrac{\sin x}{\cos x}}{\dfrac{1}{\cos x} - \cos x}$ | $\dfrac{\dfrac{1}{\cos x}}{\dfrac{\sin x}{\cos x}}$ |
| $\dfrac{\dfrac{\sin x}{\cos x}}{\dfrac{1 - \cos^2 x}{\cos x}}$ | $\dfrac{1}{\cos x} \cdot \dfrac{\cos x}{\sin x}$ |
| $\dfrac{\sin x}{1 - \cos^2 x}$ | $\dfrac{1}{\sin x}$ |
| $\dfrac{\sin x}{\sin^2 x}$ | |
| $\dfrac{1}{\sin x}$ | |

**39.**

| | |
|---|---|
| $\dfrac{\cos\theta + \sin\theta}{\cos\theta}$ | $1 + \tan\theta$ |
| $\dfrac{\cos\theta}{\cos\theta} + \dfrac{\sin\theta}{\cos\theta}$ | |
| $1 + \tan\theta$ | |

**41.**

| | |
|---|---|
| $\dfrac{\tan^2 x}{1 + \tan^2 x}$ | $\sin^2 x$ |
| $\dfrac{\tan^2 x}{\sec^2 x}$ | |
| $\dfrac{\sin^2 x}{\cos^2 x}$ | |
| $\dfrac{\dfrac{1}{\cos^2 x}}{\dfrac{1}{\cos^2 x}}$ | |
| $\sin^2 x$ | |

**43.**

| | |
|---|---|
| $\dfrac{\cot x + \tan x}{\csc x}$ | $\sec x$ |
| $\dfrac{\dfrac{\cos x}{\sin x} + \dfrac{\sin x}{\cos x}}{\dfrac{1}{\sin x}}$ | $\dfrac{1}{\cos x}$ |
| $\dfrac{\cos^2 x + \sin^2 x}{\sin x \cos x} \cdot \dfrac{\sin x}{1}$ | |
| $\dfrac{1}{\cos x}$ | |

**45.**

| | |
|---|---|
| $\dfrac{\tan\theta + \cot\theta}{\csc\theta}$ | $\sec\theta$ |
| $\dfrac{\dfrac{\sin\theta}{\cos\theta} + \dfrac{\cos\theta}{\sin\theta}}{\dfrac{1}{\sin\theta}}$ | $\dfrac{1}{\cos\theta}$ |
| $\dfrac{\sin^2\theta + \cos^2\theta}{\cos\theta\sin\theta} \cdot \dfrac{\sin\theta}{1}$ | |
| $\dfrac{1}{\cos\theta}$ | |

**47.**

| | |
|---|---|
| $\dfrac{\sin x}{1 + \cos x} + \dfrac{1 + \cos x}{\sin x}$ | $2\csc x$ |
| $\dfrac{\sin^2 x + (1 + \cos x)^2}{(1 + \cos x)\sin x}$ | $\dfrac{2}{\sin x}$ |
| $\dfrac{\sin^2 x + 1 + 2\cos x + \cos^2 x}{(1 + \cos x)\sin x}$ | |
| $\dfrac{2 + 2\cos x}{(1 + \cos x)\sin x}$ | |
| $\dfrac{2(1 + \cos x)}{(1 + \cos x)\sin x}$ | |
| $\dfrac{2}{\sin x}$ | |

**49.**

| | $\cos\theta$ |
|---|---|
| $\cos\theta(1+\csc\theta)-\cot\theta$ | |
| $\cos\theta\left(1+\dfrac{1}{\sin\theta}\right)-\dfrac{\cos\theta}{\sin\theta}$ | |
| $\cos\theta+\dfrac{\cos\theta}{\sin\theta}-\dfrac{\cos\theta}{\sin\theta}$ | |
| $\cos\theta$ | |

**51.**

| $\dfrac{\tan x+\cot x}{\sec x+\csc x}$ | $\dfrac{1}{\cos x+\sin x}$ |
|---|---|
| $\dfrac{\dfrac{\sin x}{\cos x}+\dfrac{\cos x}{\sin x}}{\dfrac{1}{\cos x}+\dfrac{1}{\sin x}}$ | |
| $\dfrac{\dfrac{\sin^2 x+\cos^2 x}{\cos x\sin x}}{\dfrac{\sin x+\cos x}{\cos x\sin x}}$ | |
| $\dfrac{\sin^2 x+\cos^2 x}{\sin x+\cos x}$ | |
| $\dfrac{1}{\cos x+\sin x}$ | |

**53.**

| $(\sec\theta-\tan\theta)(1+\csc\theta)$ | $\cot\theta$ |
|---|---|
| $\left(\dfrac{1}{\cos\theta}-\dfrac{\sin\theta}{\cos\theta}\right)\left(1+\dfrac{1}{\sin\theta}\right)$ | $\dfrac{\cos\theta}{\sin\theta}$ |
| $\left(\dfrac{1-\sin\theta}{\cos\theta}\right)\left(\dfrac{\sin\theta+1}{\sin\theta}\right)$ | |
| $\dfrac{1-\sin^2\theta}{\cos\theta\sin\theta}$ | |
| $\dfrac{\cos^2\theta}{\cos\theta\sin\theta}$ | |
| $\dfrac{\cos\theta}{\sin\theta}$ | |

**55.**

| $(\sec x+\tan x)(1-\sin x)$ | $\cos x$ |
|---|---|
| $\left(\dfrac{1}{\cos x}+\dfrac{\sin x}{\cos x}\right)(1-\sin x)$ | |
| $\left(\dfrac{1+\sin x}{\cos x}\right)(1-\sin x)$ | |
| $\dfrac{1-\sin^2 x}{\cos x}$ | |
| $\dfrac{\cos^2 x}{\cos x}$ | |
| $\cos x$ | |

**57.**

| $\cos^2\theta-\sin^2\theta$ | $\cos^4\theta-\sin^4\theta$ |
|---|---|
| | $(\cos^2\theta+\sin^2\theta)(\cos^2\theta-\sin^2\theta)$ |
| | $\cos^2\theta-\sin^2\theta$ |

**59.**

| $\dfrac{\cos\theta}{1-\cos\theta}$ | $\dfrac{1+\sec\theta}{\tan^2\theta}$ |
|---|---|
| | $\dfrac{1+\sec\theta}{\sec^2\theta-1}$ |
| | $\dfrac{1}{\sec\theta-1}$ |
| | $\dfrac{1}{\dfrac{1}{\cos\theta}-1}$ |
| | $\dfrac{1}{\dfrac{1-\cos\theta}{\cos\theta}}$ |
| | $\dfrac{\cos\theta}{1-\cos\theta}$ |

**61.**

| $\dfrac{\sin x-\cos x}{\cos^2 x}$ | $\dfrac{\tan^2 x-1}{\sin x+\cos x}$ |
|---|---|
| | $\dfrac{\dfrac{\sin^2 x}{\cos^2 x}-1}{\sin x+\cos x}$ |
| | $\dfrac{\dfrac{\sin^2 x-\cos^2 x}{\cos^2 x}}{\sin x+\cos x}\cdot\dfrac{1}{\sin x+\cos x}$ |
| | $\dfrac{\sin x-\cos x}{\cos^2 x}$ |

**63.**

| $1+\sec^4\theta$ | $\tan^4\theta+2\sec^2\theta$ |
|---|---|
| $1+(1+\tan^2\theta)^2$ | $\tan^4\theta+2(1+\tan^2\theta)$ |
| $1+1+2\tan^2\theta+\tan^4\theta$ | $\tan^4\theta+2\tan^2\theta+2$ |
| $\tan^4\theta+2\tan^2\theta+2$ | |

**65.**

| $\dfrac{1+\tan x}{1-\tan x}$ | $\dfrac{\cot x+1}{\cot x-1}$ |
|---|---|
| $\dfrac{1+\dfrac{1}{\cot x}}{1-\dfrac{1}{\cot x}}$ | |
| $\dfrac{\dfrac{\cot x+1}{\cot x}}{\dfrac{\cot x-1}{\cot x}}$ | |
| $\dfrac{\cot x+1}{\cot x-1}$ | |

**67.**

$$\dfrac{\sin\theta\,\tan\theta}{\sin\theta+\tan\theta} \qquad\qquad \dfrac{\tan\theta-\sin\theta}{\tan\theta\,\sin\theta}$$

$$\dfrac{\sin\theta\cdot\dfrac{\sin\theta}{\cos\theta}}{\sin\theta+\dfrac{\sin\theta}{\cos\theta}} \qquad\qquad \dfrac{\dfrac{\sin\theta}{\cos\theta}-\sin\theta}{\dfrac{\sin\theta}{\cos\theta}\cdot\sin\theta}$$

$$\dfrac{\dfrac{\sin^2\theta}{\cos\theta}}{\dfrac{\sin\theta\cos\theta+\sin\theta}{\cos\theta}} \qquad\qquad \dfrac{\dfrac{\sin\theta-\sin\theta\cos\theta}{\cos\theta}}{\dfrac{\sin^2\theta}{\cos\theta}}$$

$$\dfrac{\sin^2\theta}{\sin\theta\cos\theta+\sin\theta} \qquad\qquad \dfrac{\sin\theta-\sin\theta\cos\theta}{\sin^2\theta}$$

$$\dfrac{\sin\theta}{\cos\theta+1} \qquad\qquad \dfrac{1-\cos\theta}{\sin\theta}$$

$$\dfrac{\sin\theta}{\cos\theta+1}\cdot\dfrac{1-\cos\theta}{1-\cos\theta}$$

$$\dfrac{\sin\theta\,(1-\cos\theta)}{1-\cos^2\theta}$$

$$\dfrac{\sin\theta\,(1-\cos\theta)}{\sin^2\theta}$$

$$\dfrac{1-\cos\theta}{\sin\theta}$$

**69.**

$$\dfrac{\cos x+1}{\sin x}+\dfrac{\sin x}{\cos x+1} \qquad\qquad \dfrac{2}{\sin x}$$

$$\dfrac{(\cos x+1)^2+\sin^2 x}{\sin x\,(\cos x+1)}$$

$$\dfrac{\cos^2 x+2\cos x+1+\sin^2 x}{\sin x\,(\cos x+1)}$$

$$\dfrac{2\cos x+2}{\sin x\,(\cos x+1)}$$

$$\dfrac{2(\cos x+1)}{\sin x\,(\cos x+1)}$$

$$\dfrac{2}{\sin x}$$

**71.**

$$\sec\theta\,\csc\theta+\dfrac{\cot\theta}{\tan\theta-1} \qquad\qquad \dfrac{\tan\theta}{1-\cot\theta}-1$$

$$\dfrac{1}{\cos\theta}\cdot\dfrac{1}{\sin\theta}+\dfrac{\dfrac{\cos\theta}{\sin\theta}}{\dfrac{\sin\theta}{\cos\theta}-1} \qquad\qquad \dfrac{\dfrac{\sin\theta}{\cos\theta}}{1-\dfrac{\cos\theta}{\sin\theta}}-1$$

$$\dfrac{1}{\cos\theta\sin\theta}+\dfrac{\cos\theta}{\sin\theta}\cdot\dfrac{\cos\theta}{\sin\theta-\cos\theta} \qquad\qquad \dfrac{\sin\theta}{\cos\theta}\cdot\dfrac{\sin\theta}{\sin\theta-\cos\theta}-1$$

$$\dfrac{1}{\cos\theta\sin\theta}+\dfrac{\cos^2\theta}{\sin\theta\,(\sin\theta-\cos\theta)} \qquad\qquad \dfrac{\sin^2\theta-\cos\theta\,(\sin\theta-\cos\theta)}{\cos\theta\,(\sin\theta-\cos\theta)}$$

$$\dfrac{\sin\theta-\cos\theta+\cos^3\theta}{\cos\theta\sin\theta\,(\sin\theta-\cos\theta)} \qquad\qquad \dfrac{\sin^2\theta-\cos\theta\sin\theta+\cos^2\theta}{\cos\theta\,(\sin\theta-\cos\theta)}$$

$$\dfrac{\sin\theta-\cos\theta\,(1-\cos^2\theta)}{\cos\theta\sin\theta\,(\sin\theta-\cos\theta)} \qquad\qquad \dfrac{1-\cos\theta\sin\theta}{\cos\theta\,(\sin\theta-\cos\theta)}$$

$$\dfrac{\sin\theta-\cos\theta\cdot\sin^2\theta}{\cos\theta\sin\theta\,(\sin\theta-\cos\theta)}$$

$$\dfrac{1-\cos\theta\sin\theta}{\cos\theta\,(\sin\theta-\cos\theta)}$$

**73.**

$$\cot x+\csc x \qquad\qquad \dfrac{\sin x}{1-\cos x}$$

$$\dfrac{\cos x}{\sin x}+\dfrac{1}{\sin x} \qquad\qquad \dfrac{\sin x}{1-\cos x}\cdot\dfrac{1+\cos x}{1+\cos x}$$

$$\dfrac{\cos x+1}{\sin x} \qquad\qquad \dfrac{\sin x\,(1+\cos x)}{1-\cos^2 x}$$

$$\dfrac{\sin x\,(1+\cos x)}{\sin^2 x}$$

$$\dfrac{1+\cos x}{\sin x}$$

**75.**

$$\tan\theta+\cot\theta \qquad\qquad \dfrac{1}{\cot\theta\sin^2\theta}$$

$$\dfrac{\sin\theta}{\cos\theta}+\dfrac{\cos\theta}{\sin\theta} \qquad\qquad \dfrac{1}{\dfrac{\cos\theta}{\sin\theta}\cdot\sin^2\theta}$$

$$\dfrac{\sin^2\theta+\cos^2\theta}{\cos\theta\sin\theta} \qquad\qquad \dfrac{1}{\cos\theta\sin\theta}$$

$$\dfrac{1}{\cos\theta\sin\theta}$$

**77.**

$$2\cos^2 x-1 \qquad\qquad \cos^4 x-\sin^4 x$$

$$2\cos^2 x-(\sin^2 x+\cos^2 x) \qquad\qquad (\cos^2 x+\sin^2 x)(\cos^2 x-\sin^2 x)$$

$$\cos^2 x-\sin^2 x \qquad\qquad \cos^2 x-\sin^2 x$$

**79.**

| $(\cos x + \sin x)(1 - \sin x \cos x)$ | $\cos^3 x + \sin^3 x$ |
|---|---|
| | $(\cos x + \sin x)(\cos^2 x - \cos x \sin x + \sin^2 x)$ |
| | $(\cos x + \sin x)(1 - \sin x \cos x)$ |

**81.**

| $\cos u \cdot \sin v$ | $\frac{1}{2}[\sin(u+v) - \sin(u-v)]$ |
|---|---|
| | $\frac{1}{2}[\sin u \cos v + \cos u \sin v - \sin u \cos v + \cos u \sin v]$ |
| | $\frac{1}{2}(2 \cos u \sin v)$ |
| | $\cos u \cdot \sin v$ |

The other formulas follow similarly.

**83.** $\sin 3\theta - \sin 5\theta = 2\cos\dfrac{8\theta}{2}\sin\dfrac{-2\theta}{2}$

$\qquad\qquad\qquad = -2\cos 4\theta \sin\theta$

**85.** $\sin 8\theta + \sin 5\theta = 2\sin\dfrac{13\theta}{2}\cos\dfrac{3\theta}{2}$

**87.** $\sin 7u \sin 5u = \frac{1}{2}(\cos 2u - \cos 12u)$

**89.** $7\cos\theta \sin 7\theta = \frac{7}{2}[\sin 8\theta - \sin(-6\theta)]$

$\qquad\qquad\qquad = \frac{7}{2}(\sin 8\theta + \sin 6\theta)$

**91.** $\cos 55° \sin 25° = \frac{1}{2}(\sin 80° - \sin 30°)$

$\qquad\qquad\qquad = \frac{1}{2}\sin 80° - \frac{1}{4}$

**93.**

| $\sin 4\theta + \sin 6\theta$ | $\cot\theta(\cos 4\theta - \cos 6\theta)$ |
|---|---|
| $2\sin\dfrac{10\theta}{2}\cos\dfrac{-2\theta}{2}$ | $\dfrac{\cos\theta}{\sin\theta}\left(2\sin\dfrac{10\theta}{2}\sin\dfrac{2\theta}{2}\right)$ |
| $2\sin 5\theta\cos(-\theta)$ | $\dfrac{\cos\theta}{\sin\theta}(2\sin 5\theta\sin\theta)$ |
| $2\sin 5\theta\cos\theta$ | $2\sin 5\theta\cos\theta$ |

**95.**

| $\cot 4x(\sin x + \sin 4x + \sin 7x)$ | $\cos x + \cos 4x + \cos 7x$ |
|---|---|
| $\dfrac{\cos 4x}{\sin 4x}\left(\sin 4x + 2\sin\dfrac{8x}{2}\cos\dfrac{-6x}{2}\right)$ | $\cos 4x + 2\cos\dfrac{8x}{2}\cdot\cos\dfrac{6x}{2}$ |
| $\dfrac{\cos 4x}{\sin 4x}(\sin 4x + 2\sin 4x\cos 3x)$ | $\cos 4x + 2\cos 4x\cdot\cos 3x$ |
| $\cos 4x(1 + 2\cos 3x)$ | $\cos 4x(1 + 2\cos 3x)$ |

**97.**

| $\cot\dfrac{x+y}{2}$ | $\dfrac{\sin y - \sin x}{\cos x - \cos y}$ |
|---|---|
| $\dfrac{\cos\dfrac{x+y}{2}}{\sin\dfrac{x+y}{2}}$ | $\dfrac{2\cos\dfrac{x+y}{2}\sin\dfrac{y-x}{2}}{2\sin\dfrac{x+y}{2}\sin\dfrac{y-x}{2}}$ |
| | $\dfrac{\cos\dfrac{x+y}{2}}{\sin\dfrac{x+y}{2}}$ |

**99.**

| $\tan\dfrac{\theta+\phi}{2}(\sin\theta - \sin\phi)$ | $\tan\dfrac{\theta-\phi}{2}(\sin\theta + \sin\phi)$ |
|---|---|
| $\dfrac{\sin\dfrac{\theta+\phi}{2}}{\cos\dfrac{\theta+\phi}{2}}\left(2\cos\dfrac{\theta+\phi}{2}\sin\dfrac{\theta-\phi}{2}\right)$ | $\dfrac{\sin\dfrac{\theta-\phi}{2}}{\cos\dfrac{\theta-\phi}{2}}\left(2\sin\dfrac{\theta+\phi}{2}\cos\dfrac{\theta-\phi}{2}\right)$ |
| $2\sin\dfrac{\theta+\phi}{2}\cdot\sin\dfrac{\theta-\phi}{2}$ | $2\sin\dfrac{\theta+\phi}{2}\cdot\sin\dfrac{\theta-\phi}{2}$ |

**101.** $\ln|\sec x| = \ln\left|\dfrac{1}{\cos x}\right| = \ln\dfrac{1}{|\cos x|}$

$\qquad\qquad = \ln(|\cos x|)^{-1} = -\ln|\cos x|$

**103.** $\ln|\tan x| = \ln\left|\dfrac{\sin x}{\cos x}\right| = \ln\dfrac{|\sin x|}{|\cos x|}$

$\qquad\qquad = \ln|\sin x| - \ln|\cos x|$

**105.** $\ln e^{\sin t} = (\sin t)(\ln e) = \sin t\cdot 1 = \sin t$

**107.**

| $\ln|\csc\theta - \cot\theta|$ | $-\ln|\csc\theta + \cot\theta|$ |
|---|---|
| $\ln\left|\dfrac{1}{\sin\theta} - \dfrac{\cos\theta}{\sin\theta}\right|$ | $-\ln\left|\dfrac{1}{\sin\theta} + \dfrac{\cos\theta}{\sin\theta}\right|$ |
| $\ln\left|\dfrac{1 - \cos\theta}{\sin\theta}\right|$ | $-\ln\left|\dfrac{1 + \cos\theta}{\sin\theta}\right|$ |
| $\ln\left|\dfrac{1 - \cos\theta}{\sin\theta}\cdot\dfrac{1 + \cos\theta}{1 + \cos\theta}\right|$ | $\ln\left|\dfrac{1 + \cos\theta}{\sin\theta}\right|^{-1}$ |
| $\ln\left|\dfrac{1 - \cos^2\theta}{\sin\theta(1 + \cos\theta)}\right|$ | $\ln\left|\dfrac{\sin\theta}{1 + \cos\theta}\right|$ |
| $\ln\left|\dfrac{\sin^2\theta}{\sin\theta(1 + \cos\theta)}\right|$ | |
| $\ln\left|\dfrac{\sin\theta}{1 + \cos\theta}\right|$ | |

**109.** $\log(\cos x - \sin x) + \log(\cos x + \sin x)$
$\quad = \log[(\cos x - \sin x)(\cos x + \sin x)]$
$\quad = \log(\cos^2 x - \sin^2 x) = \log\cos 2x$

**111.** $\dfrac{1}{\omega C(\tan\theta + \tan\phi)}$

$$= \frac{1}{\omega C\left(\dfrac{\sin\theta}{\cos\theta} + \dfrac{\sin\phi}{\cos\phi}\right)}$$

$$= \frac{1}{\omega C\left(\dfrac{\sin\theta\cos\phi + \sin\phi\cos\theta}{\cos\theta\cos\phi}\right)}$$

$$= \frac{\cos\theta\cos\phi}{\omega C\sin(\theta + \phi)}$$

**Margin Exercises, Section 7.6**

**1.** 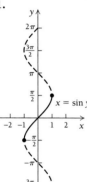    **2.**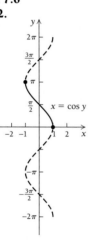

**3.** 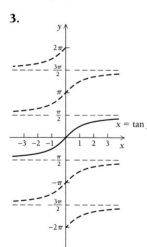    **4.**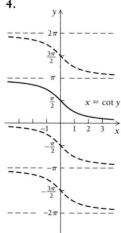

**5.** $\dfrac{\pi}{3}$, or $60°$   **6.** $\dfrac{3\pi}{4}$, or $135°$   **7.** $\dfrac{3\pi}{4}$, or $135°$

**8.** $-\dfrac{\pi}{4}$, or $-45°$   **9.** $50.76°$   **10.** $-5.86°$   **11.** $53.38°$

**12.** $-87.45°$

**Exercise Set 7.6, pp. 491–492**

**1.** $\dfrac{\pi}{4}$, or $45°$   **3.** $\dfrac{\pi}{4}$, or $45°$   **5.** $-\dfrac{\pi}{4}$, or $-45°$

**7.** $\dfrac{3\pi}{4}$, or $135°$   **9.** $\dfrac{\pi}{3}$, or $60°$   **11.** $\dfrac{\pi}{4}$, or $45°$

**13.** $-\dfrac{\pi}{6}$, or $-30°$   **15.** $\dfrac{3\pi}{4}$, or $135°$   **17.** $0$, or $0°$

**19.** $\dfrac{\pi}{2}$, or $90°$   **21.** $-\dfrac{\pi}{4}$, or $-45°$   **23.** $\dfrac{\pi}{3}$, or $60°$

**25.** $-\dfrac{\pi}{3}$, or $-60°$   **27.** $\dfrac{2\pi}{3}$, or $120°$   **29.** $\dfrac{\pi}{4}$, or $45°$

**31.** $\dfrac{2\pi}{3}$, or $120°$   **33.** $23°$   **35.** $-38.31°$   **37.** $36.97°$

**39.** $168.82°$   **41.** $20.17°$   **43.** $38.33°$   **45.** $31.09°$
**47.** $-170.83°$   **49.** $13.50°$   **51.** $-39.50°$
**53.** $-62.70°$   **55.** $-22.23°$   **57.** $170.44°$   **59.** $177.51$
**61.**

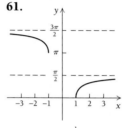

$y = \sec^{-1} x$

**63.** Let $x = \dfrac{\sqrt{2}}{2}$. Then $\sin^{-1}\dfrac{\sqrt{2}}{2} = \dfrac{\pi}{4} \approx 0.7854$, and

$\left(\sin\dfrac{\sqrt{2}}{2}\right)^{-1} \approx (0.6496)^{-1} \approx 1.5393$.

**65.** Let $x = 1$. Then $\tan^{-1} 1 = \dfrac{\pi}{4} \approx 0.7854$, and

$(\tan 1)^{-1} \approx (1.5574)^{-1} \approx 0.6421$.

**67.** Let $x = \dfrac{\sqrt{2}}{2}$. Then $\tan^{-1}\dfrac{\sqrt{2}}{2} \approx 0.6155$ and

$\dfrac{\sin^{-1}\dfrac{\sqrt{2}}{2}}{\cos^{-1}\dfrac{\sqrt{2}}{2}} = \dfrac{\pi/4}{\pi/4} = 1$.   **69.** $\tan\theta = \dfrac{50}{b}$, or $\theta = \tan^{-1}\dfrac{50}{b}$

**71.** (a) $3.1415927$; (b) $\pi$

**Margin Exercises, Section 7.7**

**1.** $\dfrac{1}{2}$   **2.** $1$   **3.** $\dfrac{\sqrt{2}}{2}$

**4.** The expression cannot be evaluated.   **5.** $\dfrac{2\pi}{3}$   **6.** $-\dfrac{\pi}{4}$

**7.** $\dfrac{\pi}{6}$   **8.** $\dfrac{2\pi}{3}$   **9.** $\dfrac{\sqrt{2}}{2}$   **10.** $\dfrac{\sqrt{3}}{2}$   **11.** $\dfrac{\pi}{3}$   **12.** $\dfrac{\pi}{2}$

**13.** $1$   **14.** $\dfrac{3}{\sqrt{b^2 + 9}}$   **15.** $\dfrac{2}{\sqrt{5}}$   **16.** $\dfrac{t}{\sqrt{1 - t^2}}$   **17.** $1$

**18.** $\dfrac{1}{3}$

**Exercise Set 7.7, pp. 497–498**

**1.** 0.3　**3.** −4.2　**5.** $\dfrac{\pi}{3}$　**7.** $-\dfrac{\pi}{4}$　**9.** $\dfrac{\pi}{5}$　**11.** $-\dfrac{\pi}{3}$

**13.** $\dfrac{\sqrt{3}}{2}$　**15.** $\dfrac{1}{2}$　**17.** 1　**19.** $\dfrac{\pi}{6}$　**21.** $\dfrac{\pi}{3}$　**23.** $\dfrac{\pi}{2}$

**25.** $\dfrac{x}{\sqrt{x^2+4}}$　**27.** $\dfrac{\sqrt{x^2-9}}{3}$　**29.** $\dfrac{\sqrt{b^2-a^2}}{a}$　**31.** $\dfrac{3}{\sqrt{11}}$

**33.** $\dfrac{1}{3\sqrt{11}}$　**35.** $-\dfrac{1}{2\sqrt{6}}$　**37.** $\dfrac{1}{\sqrt{1+y^2}}$　**39.** $\dfrac{1}{\sqrt{1+t^2}}$

**41.** $\dfrac{\sqrt{1-y^2}}{y}$　**43.** $\sqrt{1-x^2}$　**45.** $\dfrac{p}{3}$　**47.** $\dfrac{1}{2}$

**49.** $\dfrac{\sqrt{2+\sqrt{3}}}{2}$　**51.** $\dfrac{24}{25}$　**53.** $\dfrac{119}{169}$　**55.** $\dfrac{3+4\sqrt{3}}{10}$

**57.** $-\dfrac{\sqrt{2}}{10}$　**59.** $xy+\sqrt{(1-x^2)(1-y^2)}$

**61.** $y\sqrt{1-x^2}-x\sqrt{1-y^2}$　**63.** 0.9861

**65.** $\sin\theta=x;\ \cos\theta=\sqrt{1-x^2};\ \tan\theta=\dfrac{x}{\sqrt{1-x^2}};$

$\cot\theta=\dfrac{\sqrt{1-x^2}}{x};\ \sec\theta=\dfrac{1}{\sqrt{1-x^2}};\ \csc\theta=\dfrac{1}{x}$

**67.** $\sin\theta=\dfrac{x}{\sqrt{1+x^2}};\ \cos\theta=\dfrac{1}{\sqrt{1+x^2}};\ \tan\theta=x;$

$\cot\theta=\dfrac{1}{x};\ \sec\theta=\sqrt{1+x^2};\ \csc\theta=\dfrac{\sqrt{1+x^2}}{x}$

**69.**

| | |
|---|---|
| $\sin^{-1}x+\cos^{-1}x$ | $\dfrac{\pi}{2}$ |
| $\sin(\sin^{-1}x+\cos^{-1}x)$ | $\sin\dfrac{\pi}{2}$ |
| $[\sin(\sin^{-1}x)][\cos(\cos^{-1}x)]+$ $[\cos(\sin^{-1}x)][\sin(\cos^{-1}x)]$ | 1 |
| $x\cdot x+\sqrt{1-x^2}\cdot\sqrt{1-x^2}$ | |
| $x^2+1-x^2$ | |
| 1 | |

**71.**

| | |
|---|---|
| $\sin^{-1}x$ | $\tan^{-1}\dfrac{x}{\sqrt{1-x^2}}$ |
| $\sin(\sin^{-1}x)$ | $\sin\left(\tan^{-1}\dfrac{x}{\sqrt{1-x^2}}\right)$ |
| $x$ | $x$ |

**73.**

| | |
|---|---|
| $\arcsin x$ | $\arccos\sqrt{1-x^2}$ |
| $\sin(\arcsin x)$ | $\sin(\arccos\sqrt{1-x^2})$ |
| $x$ | $x$ |

**75.** $\theta=\arctan\dfrac{y+h}{x}-\arctan\dfrac{y}{x}$

**Margin Exercises, Section 7.8**

**1.** $60°+360°k,\ 300°+360°k;\ \dfrac{\pi}{3}+2k\pi,\ \dfrac{5\pi}{3}+2k\pi$

**2.** $\dfrac{\pi}{6},\dfrac{5\pi}{6},\dfrac{7\pi}{6},\dfrac{11\pi}{6};\ 30°,150°,210°,330°$

**3.** $\dfrac{\pi}{6},\dfrac{5\pi}{6},\dfrac{7\pi}{6},\dfrac{11\pi}{6}$　**4.** $40°,320°$　**5.** $140°,220°$

**6.** $120°,240°,75.52°,284.48°$　**7.** $120°,240°,90°,270°$
**8.** $241.66°,118.34°$

**Technology Connection, Section 7.8**
**TC 1.** $-40.1°,102.0°$　**TC 2.** $0.79,0.98$　**TC 3.** $\phi$
**TC 4.** $2.13,3.85$

**Exercise Set 7.8, p. 503**

**1.** $\dfrac{\pi}{3}+2k\pi,\dfrac{2\pi}{3}+2k\pi$　**3.** $\dfrac{\pi}{4}+2k\pi,-\dfrac{\pi}{4}+2k\pi$

**5.** $20.17°+k\cdot360°,\ 159.83°+k\cdot360°$

**7.** $236.67°,123.33°$　**9.** $\dfrac{4\pi}{3},\dfrac{5\pi}{3}$　**11.** $123.69°,303.69°$

**13.** $\dfrac{\pi}{6},\dfrac{5\pi}{6},\dfrac{7\pi}{6},\dfrac{11\pi}{6}$　**15.** $\dfrac{\pi}{6},\dfrac{5\pi}{6},\dfrac{7\pi}{6},\dfrac{11\pi}{6}$

**17.** $\dfrac{\pi}{6},\dfrac{5\pi}{6},\dfrac{3\pi}{2}$　**19.** 0　**21.** $0,\dfrac{\pi}{6},\dfrac{5\pi}{6},\pi,\dfrac{7\pi}{6},\dfrac{11\pi}{6}$

**23.** $\dfrac{\pi}{6},\dfrac{5\pi}{6}$　**25.** $109.47°,250.53°,120°,240°$

**27.** $\dfrac{\pi}{6},\dfrac{5\pi}{6},\pi$　**29.** $0,\pi,\dfrac{\pi}{2},\dfrac{3\pi}{2}$　**31.** $198.27°,341.73°$

**33.** $70.12°,128.31°,250.12°,308.31°$
**35.** $37.22°,169.35°,217.22°,349.35°$
**37.** $60°,120°,240°,300°$　**39.** $60°,240°$
**41.** $30°,60°,120°,150°,210°,240°,300°,330°$

**43.** $0,\pi$　**45.** $\dfrac{\pi}{2}$

**47. (a)**

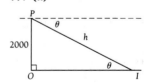

The angle of elevation is equal to the angle of depression.

Thus, $\sin\theta=\dfrac{2000}{h}$. **(b)** $41.81°$

**Margin Exercises, Section 7.9**

**1.** $\dfrac{\pi}{2},\pi$　**2.** $\dfrac{\pi}{2},\pi$　**3.** $\dfrac{\pi}{4},\dfrac{3\pi}{4},\dfrac{5\pi}{4},\dfrac{7\pi}{4}$　**4.** $\dfrac{\pi}{2},\dfrac{3\pi}{2},\dfrac{7\pi}{6},\dfrac{11\pi}{6}$

**5.** $\dfrac{\pi}{12},\dfrac{5\pi}{12},\dfrac{13\pi}{12},\dfrac{17\pi}{12}$　**6.** $\dfrac{\pi}{3},\pi,\dfrac{5\pi}{3}$

**Exercise Set 7.9, pp. 507–508**

**1.** $0, \pi$   **3.** $\dfrac{3\pi}{4}, \dfrac{7\pi}{4}$   **5.** $\dfrac{\pi}{2}, \dfrac{3\pi}{2}, \dfrac{\pi}{6}, \dfrac{5\pi}{6}$

**7.** $\dfrac{\pi}{2}, \dfrac{3\pi}{2}, \dfrac{\pi}{4}, \dfrac{3\pi}{4}, \dfrac{5\pi}{4}, \dfrac{7\pi}{4}$   **9.** $0, \dfrac{\pi}{2}, \pi, \dfrac{3\pi}{2}$   **11.** $0$

**13.** $0, \dfrac{\pi}{2}, \pi, \dfrac{3\pi}{2}$   **15.** $\dfrac{\pi}{6}, \dfrac{5\pi}{6}, \pi$   **17.** $\dfrac{\pi}{6}, \dfrac{5\pi}{6}, \dfrac{7\pi}{6}, \dfrac{11\pi}{6}$

**19.** $63.43°, 243.43°, 101.31°, 281.31°$

**21.** $\dfrac{\pi}{6}, \dfrac{\pi}{2}, \dfrac{5\pi}{6}, \dfrac{7\pi}{6}, \dfrac{3\pi}{2}, \dfrac{11\pi}{6}$   **23.** $\dfrac{2\pi}{3}, \dfrac{4\pi}{3}$   **25.** $\dfrac{\pi}{4}, \dfrac{7\pi}{4}$

**27.** $\dfrac{\pi}{12}, \dfrac{5\pi}{12}$   **29.** $\dfrac{\pi}{6}, \dfrac{3\pi}{2}$   **31.** $0.9669, 1.853, 4.108, 4.995$

**33.** $0.7297, 2.412, 3.665, 5.760$   **35.** $t \approx 1.24, 6.76$
**37.** $16.5°$   **39.** $1$   **41.** $1.13, 5.66, -0.622, -0.516,$ etc.
**43.** $0.1923$

**Review Exercises: Chapter 7, pp. 508–509**
**1.** [7.3] $\cot(x - \pi) = \cot x$   **2.** [7.5] $1$   **3.** [7.5] $\csc^2 x$
**4.** [7.3] $-\sin x$   **5.** [7.3] $\sin x$   **6.** [7.3] $-\cos x$
**7.** [7.1] $\tan x = \pm \sqrt{\sec^2 x - 1}$   **8.** [7.1] $\csc x$   **9.** [7.1] $1$
**10.** [7.1] $\dfrac{\sqrt{\tan x \sec x}}{\sec x}$   **11.** [7.1] $\dfrac{\tan x}{\sqrt{\sec x \tan x}}$
**12.** [7.3] $\sin(90° - \theta) = 0.7314$; $\cos(90° - \theta) = 0.6820$;
$\tan(90° - \theta) = 1.0724$; $\cot(90° - \theta) = 0.9325$;
$\sec(90° - \theta) = 1.4663$; $\csc(90° - \theta) = 1.3673$

**13.** [7.2] $\cos x \cos \dfrac{3\pi}{2} - \sin x \sin \dfrac{3\pi}{2}$

**14.** [7.2] $\dfrac{\tan 45° - \tan 30°}{1 + \tan 45° \tan 30°}$

**15.** [7.2] $\cos(27° - 16°)$ or $\cos 11°$

**16.** [7.2] $\dfrac{-\sqrt{6} - \sqrt{2}}{4}$   **17.** [7.2] $2 - \sqrt{3}$

**18.** [7.4] $\sin 2\theta = \dfrac{24}{25}$, $\cos 2\theta = -\dfrac{7}{25}$, $\tan 2\theta = -\dfrac{24}{7}$, $2\theta$ is

in quadrant II   **19.** [7.4] $\dfrac{1}{2}\sqrt{2 - \sqrt{2}}$   **20.** [7.4] $2\cot \theta$

**21.**

| | |
|---|---|
| $\tan 2\theta$ | $\dfrac{2\tan \theta}{1 - \tan^2 \theta}$ |
| $\dfrac{\sin 2\theta}{\cos 2\theta}$ | $\dfrac{2\dfrac{\sin \theta}{\cos \theta}}{\dfrac{\cos^2 \theta}{\cos^2 \theta} - \dfrac{\sin^2 \theta}{\cos^2 \theta}}$ |
| $\dfrac{2\sin \theta \cos \theta}{\cos^2 \theta - \sin^2 \theta}$ | $\dfrac{2\dfrac{\sin \theta}{\cos \theta} \cdot \dfrac{\cos^2 \theta}{\cos^2 \theta - \sin^2 \theta}}{}$ |
| | $\dfrac{2\sin \theta \cos \theta}{\cos^2 \theta - \sin^2 \theta}$ |

**22.** [7.5]

| | |
|---|---|
| $\dfrac{\sec x - \cos x}{\tan x}$ | $\sin x$ |
| $\dfrac{\dfrac{1}{\cos x} - \cos x}{\dfrac{\sin x}{\cos x}}$ | |
| $\dfrac{1 - \cos^2 x}{\cos x} \cdot \dfrac{\cos x}{\sin x}$ | |
| $\dfrac{1 - \cos^2 x}{\sin x}$ | |
| $\dfrac{\sin^2 x}{\sin x}$ | |
| $\sin x$ | |

**23.** [7.6] $\dfrac{\pi}{6}$   **24.** [7.6] $81°$   **25.** [7.6] $-45°$, or $-\dfrac{\pi}{4}$

**26.** [7.6] $150°$, or $\dfrac{5\pi}{6}$   **27.** [7.7] $\dfrac{7}{8}$   **28.** [7.7] $\dfrac{\pi}{3}$

**29.** [7.7] $\dfrac{\pi}{6}$   **30.** [7.7] $\dfrac{b}{\sqrt{b^2 + 25}}$   **31.** [7.8] $0, \pi$

**32.** [7.8] $\dfrac{\pi}{3}, \dfrac{5\pi}{3}$   **33.** [7.9] $\dfrac{\pi}{4}, \dfrac{3\pi}{4}, \dfrac{5\pi}{4}, \dfrac{7\pi}{4}$

**34.** [7.9] $0.5117, 1.845, 3.6533, 4.9861$

**35.** [7.8] $\dfrac{\pi}{6}, \dfrac{5\pi}{6}, \dfrac{7\pi}{6}, \dfrac{11\pi}{6}$   **36.** [7.8] No solution in $[0, 2\pi)$.

**37.** [7.4]

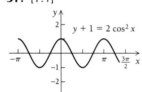

**38.** [7.6]

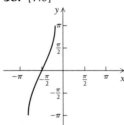

$f(x) = 2\sin^{-1}\left(x + \dfrac{\pi}{2}\right)$

**Test: Chapter 7, pp. 509–510**
**1.** [7.2] $\cos \pi \cos x + \sin \pi \sin x$

**2.** [7.2] $\dfrac{\tan 83° + \tan 15°}{1 - \tan 83° \tan 15°}$   **3.** [7.2] $\sin 35°$

**4.** [7.2] $\dfrac{\sqrt{2} - \sqrt{6}}{4}$   **5.** [7.2] $2 - \sqrt{3}$   **6.** [7.3] $\csc x$

**7.** [7.5] 1   **8.** [7.5] $\csc^2 x$   **9.** [7.3] $\cos x$   **10.** [7.3] $\sin x$
**11.** [7.3] $\sin x$   **12.** [7.1] $\csc x = \pm\sqrt{1 + \cot^2 x}$
**13.** [7.1] $\sec x$   **14.** [7.1] 1   **15.** [7.1] $\dfrac{\sqrt{\sec x \csc x}}{\csc x}$

**16.** [7.4] $\sin 2\theta = -\dfrac{120}{169}$; $\cos 2\theta = -\dfrac{119}{169}$; $\tan 2\theta = \dfrac{120}{119}$;

quadrant III   **17.** [7.4] $2 + \sqrt{3}$   **18.** [7.4] $2 \sin 4x$

**19.** [7.5]

| | |
|---|---|
| $\dfrac{1 - \cos 2\theta}{\sin 2\theta}$ | $\tan \theta$ |
| $\dfrac{1 - (1 - 2\sin^2 \theta)}{2\sin \theta \cos \theta}$ | $\dfrac{\sin \theta}{\cos \theta}$ |
| $\dfrac{2\sin^2 \theta}{2\sin \theta \cos \theta}$ | |
| $\dfrac{\sin \theta}{\cos \theta}$ | |

**20.** [7.6] $-\dfrac{\pi}{4}$   **21.** [7.6] $42.96°$   **22.** [7.6] $\dfrac{\pi}{6}$

**23.** [7.6] $\dfrac{3\pi}{4}$   **24.** [7.7] $\dfrac{\sqrt{3}}{2}$   **25.** [7.7] $\dfrac{\pi}{4}$

**26.** [7.8] $\dfrac{7\pi}{6}, \dfrac{11\pi}{6}$   **27.** [7.9] $\dfrac{\pi}{2}, \dfrac{3\pi}{2}, \dfrac{\pi}{4}, \dfrac{3\pi}{4}, \dfrac{5\pi}{4}, \dfrac{7\pi}{4}$

**28.** [7.8] 0.8730, 1.805, 4.015, 4.947
**29.** [7.3] $\sin(90° - \theta) = 0.8910$, $\cos(90° - \theta) = 0.4540$,
$\tan(90° - \theta) = 1.963$, $\cot(90° - \theta) = 0.5095$,

$\sec(90° - \theta) = 2.203$, $\csc(90° - \theta) = 1.122$   **30.** [7.8] $\dfrac{\pi}{2}$

## CHAPTER 8

### Margin Exercises, Section 8.1
**1.** $C = 29°$, $c = 5.9$, $b = 11.1$   **2.** $\sin A = 2.8$; no solution
**3.** $A = 43°$, $C = 104°$, $c = 36$; or $A = 137°$, $C = 10°$, $c = 6$
**4.** $C = 18°$, $A = 124°$, $a = 27$   **5.** $22.5 \text{ in}^2$

### Exercise Set 8.1, pp. 517–519
**1.** $C = 17°$, $a = 26$, $c = 11$   **3.** $A = 121°$, $a = 33$, $c = 14$
**5.** $B = 68.8°$, $a = 32.2 \text{ cm}$, $b = 32.3 \text{ cm}$
**7.** $A = 110.36°$, $a = 5 \text{ mi}$, $b = 3 \text{ mi}$
**9.** $C = 103°41'$, $a = 1804 \text{ km}$, $b = 5331 \text{ km}$
**11.** $B = 32°$, $a = 1752 \text{ in.}$, $c = 720 \text{ in.}$
**13.** $B = 57.4°$, $C = 86.1°$, $c = 40$, or $B = 122.6°$,
$C = 20.9°$, $c = 14$   **15.** $B = 19.0°$, $C = 44.7°$, $b = 6.2$
**17.** $A = 74°26'$, $B = 44°24'$, $a = 33.3$   **19.** No solution
**21.** $B = 14.7°$, $C = 135.0°$, $c = 28.04 \text{ cm}$   **23.** No solution
**25.** $B = 71°26'$, $A = 56°46'$, $a = 3.668 \text{ km}$; or $B = 108°34'$,
$A = 19°38'$, $a = 1.473 \text{ km}$   **27.** No solution
**29.** $C = 54°36'$, $B = 83°34'$, $b = 134.1 \text{ mi}$; or $C = 125°24'$,
$B = 12°46'$, $b = 29.8 \text{ mi}$   **31.** $8.2 \text{ ft}^2$   **33.** $12 \text{ yd}^2$
**35.** $596.98 \text{ ft}^2$   **37.** $76 \text{ m}$   **39.** $51 \text{ ft}$   **41.** $1467 \text{ km}$
**43.** From $A$: 35 mi; from $B$: 66 mi   **45.** 10.6 km or 2.4 km
**47.** See Student's Solution Manual or Instructor's Solution
Manual.

**49.**

$A = bh$, $h = a\sin\theta$, so $A = ab\sin\theta$

**51.** Area $\triangle ADB$ + Area $\triangle BDC$ = Area $\triangle ADC$. Using
Theorem 2, we have

$$\frac{1}{2}xy\sin\alpha + \frac{1}{2}yz\sin\beta = \frac{1}{2}xz\sin(\alpha + \beta).$$

Multiplying both sides by $\dfrac{2}{x \cdot y \cdot z}$ gives

$$\frac{\sin\alpha}{z} + \frac{\sin\beta}{x} = \frac{\sin(\alpha + \beta)}{y}.$$

### Margin Exercises, Section 8.2
**1.** $a = 41$, $B = 21°$, $C = 37°$
**2.** $A = 108.21°$, $B = 22.33°$, $C = 49.46°$
**3.** Law of cosines; $A = 108.21°$, $B = 22.33°$, $C = 49.46°$
**4.** Law of sines; $C = 18°$, $A = 124°$, $a = 54$
**5.** Law of cosines; $a = 122$, $B = 21.9°$, $C = 35.7°$
**6.** Law of sines; $A = 43°$, $C = 104°$, $c = 107$
**7.** Law of sines; $B = 66°49'$, $C = 69°36'$, $c = 6.1$; or
$B = 113°11'$, $C = 23°14'$, $c = 2.6$
**7.** Law of sines; no solution

### Exercise Set 8.2, pp. 523–525
**1.** $a = 15$, $B = 24°$, $C = 126°$
**3.** $b = 25$, $A = 20°$, $C = 27°$
**5.** $b = 75 \text{ m}$, $A = 94°51'$, $C = 12°29'$
**7.** $c = 51.2 \text{ cm}$, $A = 10.82°$, $B = 146.90°$
**9.** $a = 25.5 \text{ yd}$, $B = 38°2'$, $C = 45°45'$
**11.** $c = 45.17 \text{ mi}$, $A = 89.3°$, $B = 42.0°$
**13.** $A = 36.18°$, $B = 43.53°$, $C = 100.29°$
**15.** $A = 24.15°$, $B = 30.75°$, $C = 125.10°$   **17.** No solution
**19.** $A = 128.71°$, $B = 22.10°$, $C = 29.19°$
**21.** Law of sines; $C = 98°$, $a = 96.7$, $c = 101.9$
**23.** Law of cosines; $A = 73.71°$, $B = 51.75°$, $C = 54.54°$
**25.** Law of cosines; $c = 44$, $B = 42°$, $A = 91°$
**27.** Law of sines; $A = 61°20'$, $a = 5.633$, $b = 6.014$
**29.** 28 nautical mi, S $54°38'$ E   **31.** 37 nautical mi
**33.** 59 ft   **35.** $67.98°$, $67.98°$, $44.04°$   **37.** 46 ft
**39.** (a) 16 ft; (b) $121 \text{ ft}^2$   **41.** $3424 \text{ yd}^2$   **43.** 12 m, 24 m
**45.** $60.08°$, $40.78°$, $79.14°$   **47.** 50 ft   **49.** 4.7 cm
**51.** See Student's Solution Manual or Instructor's Solution
Manual.   **53.** 64 in.   **55.** 162 ft
**57.** $A = \frac{1}{2}a^2 \sin\theta$; when $\theta = 90°$
**59.** Let $S^2 = a^2 + b^2 + c^2$. Then
$$a^2 = b^2 + c^2 - 2bc\cos A$$
$$b^2 = c^2 + a^2 - 2ca\cos B$$
$$(+)\ \frac{c^2 = a^2 + b^2 - 2ab\cos C}{S^2 = S^2 + S^2 - 2(bc\cos A + ac\cos B + ab\cos C).}$$
$\therefore\ S^2 = a^2 + b^2 + c^2 = 2(bc\cos A + ac\cos B + ab\cos C)$
**61.** 10,106 ft

### Margin Exercises, Section 8.3
**1.** 15.5 N, $20°$   **2.** $75°$, 171 km/h   **3.** E: 71, S: 71
**4.** 19 km/h from S $32°$ E   **5.** (a) 4 up, 21 left;
(b) 21.4, $11°$ with horizontal   **6.** $\langle 12, -8 \rangle$   **7.** $\langle 12, 4 \rangle$

**8.** $(8.5, 159°)$   **9.** $\langle 14.2, -4.9 \rangle$   **10.** (a) $\langle 1, 7 \rangle$;
(b) $\langle 7, -3 \rangle$; (c) $\langle -23, 21 \rangle$; (d) $17.1$

**Exercise Set 8.3, pp. 533–534**
**1.** $57.0, 38°$   **3.** $18.4, 37°$   **5.** $20.9, 59°$   **7.** $68.3, 18°$
**9.** $13.6$ N, $65°$   **11.** $655$ N, $21°$   **13.** $22$ ft/sec, $34°$
**15.** $726$ lb, $47°$   **17.** $174$ nautical mi, S $15°$ E
**19.** An angle of $12°$ upstream
**21.** Vertical $118$, horizontal $92$
**23.** S. $192$ km/h, W. $161$ km/h   **25.** $43$ N, S $36°$ W
**27.** (a) N. $28$, W. $7$; (b) $28.9$, N $14°$ W   **29.** $\langle 5, 16 \rangle$
**31.** $\langle 4.8, 13.7 \rangle$   **33.** $\langle -662, -426 \rangle$   **35.** $(5, 53°)$
**37.** $(18, 304°)$   **39.** $(5, 233°)$   **41.** $(18, 124°)$
**43.** $\langle 3.5, 2 \rangle$   **45.** $\langle -5.7, -8.2 \rangle$   **47.** $\langle 17.3, -10 \rangle$
**49.** $\langle 70.7, -70.7 \rangle$   **51.** $\langle 19, 36 \rangle$   **53.** $\langle 14, 8 \rangle$   **55.** $18$
**57.** $17.9$   **59.** $35.8$   **61.** $\langle 0, 0 \rangle$
**63.** To show: If $k$ is a scalar, then $k(\mathbf{u} + \mathbf{v}) = k\mathbf{u} + k\mathbf{v}$.
Proof: Let $\mathbf{u} = \langle a, b \rangle$ and $\mathbf{v} = \langle c, d \rangle$.
Then $k(\mathbf{u} + \mathbf{v}) = k(\langle a, b \rangle + \langle c, d \rangle)$
$$= k\langle a + c, b + d \rangle$$
$$= \langle k(a + c), k(b + d) \rangle$$
$$= \langle ka + kc, kb + kd \rangle$$
$$= \langle ka, kb \rangle + \langle kc, kd \rangle$$
$$= k\langle a, b \rangle + k\langle c, d \rangle$$
$$= k\mathbf{u} + k\mathbf{v}.$$

**65.**
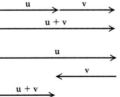
If the vectors are *not* collinear (or parallel), they form a triangle as shown. From plane geometry (Euclid), we know that (any) one side of a triangle is *less than* the sum of the other two sides.

If the vectors *are* collinear, then either
$$|\mathbf{u} + \mathbf{v}| = |\mathbf{u}| + |\mathbf{v}|$$
or
$$|\mathbf{u} + \mathbf{v}| < |\mathbf{u}| + |\mathbf{v}|.$$

**Margin Exercises, Section 8.4**
**1.**
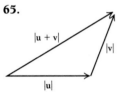

**2.** Many answers possible. *A*: $(4, 30°)$, $(4, 390°)$, $(-4, 210°)$, etc.; *B*: $(5, -60°)$, $(5, 300°)$; *C*: $(2, 150°)$, $(2, -210°)$; *D*: $(3, 225°)$, $(3, -135°)$; *E*: $(5, 60°)$, $(-5, -120°)$   **3.** (a) $(3\sqrt{2}, 45°)$; (b) $(4, 270°)$;
(c) $(6, 120°)$; (d) $(4, 330°)$   **4.** (a) $\left( \frac{5}{2}\sqrt{3}, \frac{5}{2} \right)$;

(b) $(5, 5\sqrt{3})$; (c) $\left( -\frac{5}{\sqrt{2}}, -\frac{5}{\sqrt{2}} \right)$; (d) $(4\sqrt{3}, 4)$

**5.** $2r\cos\theta + 5r\sin\theta = 9$   **6.** $r^2 + 8r\cos\theta = 0$
**7.** $x^2 + y^2 = 49$   **8.** $y = 5$   **9.** $x^2 + y^2 - 3x = 5y$
**10.**   **11.**
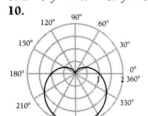
$r = 1 - \sin\theta$

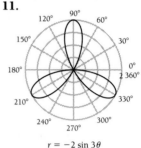
$r = -2\sin 3\theta$

**Technology Connection, Section 8.4**
**TC 1.**   **TC 2.**
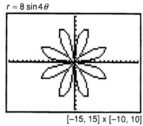
$r = 8\sin 4\theta$
$[-15, 15] \times [-10, 10]$

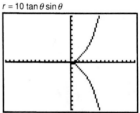
$r = 10\tan\theta\sin\theta$
$[-15, 15] \times [-10, 10]$

**TC 3.**   **TC 4.**
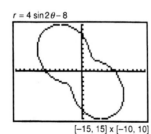
$r = 4\sin 2\theta - 8$
$[-15, 15] \times [-10, 10]$

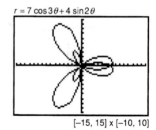
$r = 7\cos 3\theta + 4\sin 2\theta$
$[-15, 15] \times [-10, 10]$

**Exercise Set 8.4, pp. 539–540**
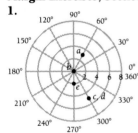

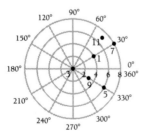

**25.** $(4\sqrt{2}, 45°)$   **27.** $(5, 90°)$   **29.** $(4, 0°)$
**31.** $(6, 60°)$   **33.** $(2, 30°)$   **35.** $(6, 30°)$
**37.** $\left( \frac{4}{\sqrt{2}}, \frac{4}{\sqrt{2}} \right)$ or $(2.8, 2.8)$   **39.** $(0, 0)$

**41.** $\left(-\dfrac{3}{\sqrt{2}}, -\dfrac{3}{\sqrt{2}}\right)$   **43.** $(3, -3\sqrt{3})$   **45.** $(5\sqrt{3}, 5)$

**47.** $\left(\dfrac{5\sqrt{3}}{2}, -\dfrac{5}{2}\right)$   **49.** $r(3\cos\theta + 4\sin\theta) = 5$

**51.** $r\cos\theta = 5$   **53.** $r^2 = 36$

**55.** $r^2(\cos^2\theta - 4\sin^2\theta) = 4$   **57.** $x^2 + y^2 = 25$

**59.** $y = x$   **61.** $y = 2$   **63.** $x^2 + y^2 = 4x$

**65.** $x^2 - 4y = 4$   **67.** $x^2 - 2x + y^2 - 3y = 0$

**69.**
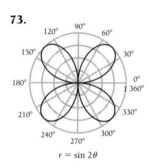
$r = 4\cos\theta$

**71.**
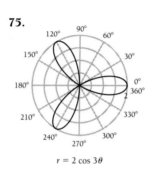
$r = 1 - \cos\theta$

**73.**
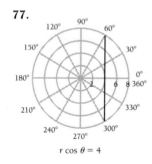
$r = \sin 2\theta$

**75.**
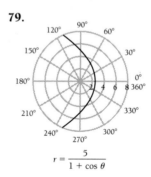
$r = 2\cos 3\theta$

**77.**
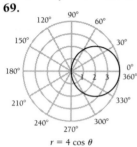
$r\cos\theta = 4$

**79.**
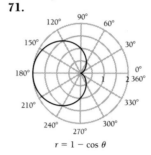
$r = \dfrac{5}{1 + \cos\theta}$

**81.**
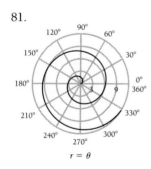
$r = \theta$

**83.**
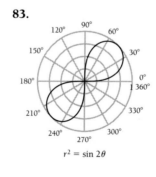
$r^2 = \sin 2\theta$

**85.**
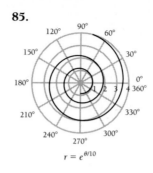
$r = e^{\theta/10}$

**87.** $y^2 = -4x + 4$

**89.**
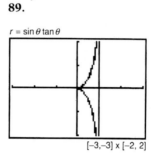
$r = \sin\theta\tan\theta$
$[-3, -3] \times [-2, 2]$

**91.**
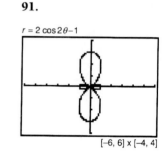
$r = 2\cos 2\theta - 1$
$[-6, 6] \times [-4, 4]$

**93.**
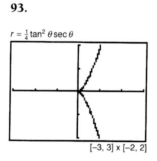
$r = \frac{1}{4}\tan^2\theta\sec\theta$
$[-3, 3] \times [-2, 2]$

**Margin Exercises, Section 8.5**

**1.**
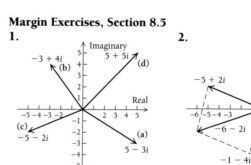

**2.**

**3.** (a) 5; (b) 13   **4.** $1 - i$   **5.** $\sqrt{3} - i$   **6.** $\sqrt{2}\operatorname{cis} 315°$

**7.** $6\operatorname{cis} 225°$   **8.** $20\operatorname{cis} 55°$   **9.** $4\operatorname{cis}\dfrac{5\pi}{4}$   **10.** $4\operatorname{cis} 90°$

**11.** $2\operatorname{cis}\dfrac{\pi}{4}$   **12.** $\sqrt{2}\operatorname{cis} 285°$

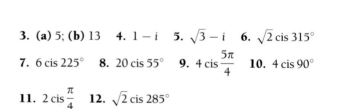

**Exercise Set 8.5, pp. 546–547**

**1.**

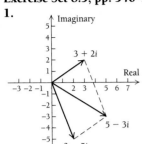

**3.**

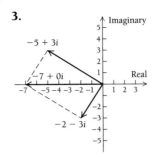

**5.**

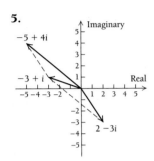

**7.**

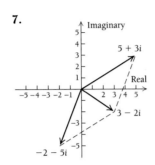

**9.** $\dfrac{3\sqrt{3}}{2} + \dfrac{3}{2}i$ **11.** $-10i$ **13.** $2 + 2i$ **15.** $-2 - 2i$

**17.** $\sqrt{2}\text{ cis }315°$ **19.** $20\text{ cis }330°$ **21.** $5\text{ cis }180°$

**23.** $4\text{ cis }0°$, or $4$ **25.** $8\text{ cis }120°$ **27.** $\text{cis }270°$, or $-i$

**29.** $2\text{ cis }270°$, or $-2i$ **31.** $z = a + bi$,

$|z| = \sqrt{a^2 + b^2}; -z = -a - bi, |-z| =$

$\sqrt{(-a)^2 + (-b)^2} = \sqrt{a^2 + b^2}, \therefore |z| = |-z|$

**33.** $|(a + bi)(a - bi)| = |a^2 + b^2| = a^2 + b^2;$

$|(a + bi)^2| = |a^2 + 2abi - b^2| = |a^2 - b^2 - 2abi| =$

$\sqrt{(a^2 - b^2)^2 + (2ab)^2} = \sqrt{a^4 + 2a^2b^2 + b^4} = a^2 + b^2$

**35.** $z \cdot w = (r_1 \text{ cis }\theta_1)(r_2 \text{ cis }\theta_2) = r_1 r_2 \text{ cis }(\theta_1 + \theta_2),$

$|z \cdot w| = \sqrt{[r_1 r_2 \cos(\theta_1 + \theta_2)]^2 + [r_1 r_2 \sin(\theta_1 + \theta_2)]^2} =$

$\sqrt{(r_1 r_2)^2} = |r_1 r_2|, |z| = \sqrt{(r_1 \cos\theta_1)^2 + (r_1 \sin\theta_1)^2} =$

$\sqrt{r_1^2} = |r_1|, |w| = \sqrt{(r_2 \cos\theta_2)^2 + (r_2 \sin\theta_2)^2} = \sqrt{r_2^2} = |r_2|.$

Then $|z| \cdot |w| = |r_1| \cdot |r_2| = |r_1 r_2| = |z \cdot w|$

**37.**

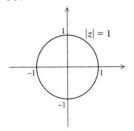

**39.** $\cos\theta - i\sin\theta$

**Margin Exercises, Section 8.6**

**1.** $32\text{ cis }270°$, or $-32i$ **2.** $16\text{ cis }120°$, or $-8 + 8i\sqrt{3}$

**3.** $1 + i, -1 - i$ **4.** $\sqrt{5} + i\sqrt{5}, -\sqrt{5} - i\sqrt{5}$

**5.** $1\text{ cis }60°, 1\text{ cis }180°, 1\text{ cis }300°$; or $\dfrac{1}{2} + \dfrac{\sqrt{3}}{2}i, -1, \dfrac{1}{2} - \dfrac{\sqrt{3}}{2}i$

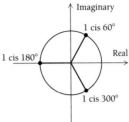

**Technology Connection, Section 8.6**

**TC 1.** $\pm 1.495349, \pm 1.495349i$

**TC 2.** $-2.154435, 1.0772173 \pm 1.8657952i$

**TC 3.** $\pm 1, \pm 0.80901699 \pm 0.58778525i,$

$\pm 0.30901699 \pm 0.95105652i$

**TC 4.** $1.1486984, 1.0493882 \pm 0.46721771i,$

$0.76862923 \pm 0.85364924i, 0.35496731 \pm 1.0924771i,$

$-0.1200717 \pm 1.1424057i, -0.5743492 \pm 0.99480196i,$

$-0.9293165 \pm 0.67518795i, -1.123597 \pm 0.23882782i$

**Exercise Set 8.6, p. 551**

**1.** $8\text{ cis }\pi$ **3.** $64\text{ cis }\pi$ **5.** $8\text{ cis }270°$

**7.** $-8 - 8\sqrt{3}i$ **9.** $-8 - 8\sqrt{3}i$ **11.** $i$ **13.** $1$

**15.** $\text{cis }45°, \text{cis }225°$; or $\dfrac{\sqrt{2}}{2} + i\dfrac{\sqrt{2}}{2}, -\dfrac{\sqrt{2}}{2} - i\dfrac{\sqrt{2}}{2}$

**17.** $2\text{ cis }157.5°, 2\text{ cis }337.5°$ **19.** $\sqrt{2}\text{ cis }60°,$

$\sqrt{2}\text{ cis }240°$; or $\dfrac{\sqrt{2}}{2} + \dfrac{\sqrt{6}}{2}i, -\dfrac{\sqrt{2}}{2} - \dfrac{\sqrt{6}}{2}i$

**21.** $\text{cis }30°, \text{cis }150°, \text{cis }270°$; or $\dfrac{\sqrt{3}}{2} + \dfrac{1}{2}i, -\dfrac{\sqrt{3}}{2} + \dfrac{1}{2}i, -i$

**23.** $2\text{ cis }0°, 2\text{ cis }90°, 2\text{ cis }180°, 2\text{ cis }270°$; or $2, 2i,$
$-2, -2i$

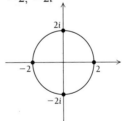

**25.** $\text{cis }36°, \text{cis }108°, \text{cis }180°, \text{cis }252°, \text{cis }324°$

**27.** $2\text{ cis }30°, 2\text{ cis }120°, 2\text{ cis }210°, 2\text{ cis }300°$;
or $\sqrt{3} + i, -1 + \sqrt{3}i, -\sqrt{3} - i, 1 - \sqrt{3}i$

**29.** $\sqrt[3]{4}\text{ cis }110°, \sqrt[3]{4}\text{ cis }230°, \sqrt[3]{4}\text{ cis }350°$ **31.** $\text{cis }0°,$

$\text{cis }120°, \text{cis }240°$; or $1, -\dfrac{1}{2} + \dfrac{\sqrt{3}}{2}i, -\dfrac{1}{2} - \dfrac{\sqrt{3}}{2}i$

**33.** $\text{cis }36°, \text{cis }108°, \text{cis }180°, \text{cis }252°, \text{cis }324°$

**35.** $\sqrt[5]{2}\text{ cis }42°, \sqrt[5]{2}\text{ cis }114°, \sqrt[5]{2}\text{ cis }186°, \sqrt[5]{2}\text{ cis }258°,$

$\sqrt[5]{2}\text{ cis }330°$

**37.** $3\text{ cis }45°, 3\text{ cis }135°, 3\text{ cis }225°, 3\text{ cis }315°$;

or $\dfrac{3\sqrt{2}}{2} + \dfrac{3\sqrt{2}}{2}i, -\dfrac{3\sqrt{2}}{2} + \dfrac{3\sqrt{2}}{2}i, -\dfrac{3\sqrt{2}}{2} - \dfrac{3\sqrt{2}}{2}i,$

$\dfrac{3\sqrt{2}}{2} - \dfrac{3\sqrt{2}}{2}i$

**39.** $-1.366 + 1.366i$, $0.366 - 0.366i$
**41.** $\pm 1.8612097$, $\pm 1.8612097i$
**43.** $\pm 1.2311444$, $\pm 0.99601675 \pm 0.72364853i$, $\pm 0.38044455 \pm 1.1708879i$

### Review Exercises: Chapter 8, p. 552

**1.** [8.1] $A = 46°$, $C = 99°$, $c = 34$; or $A = 134°$, $C = 11°$, $c = 7$   **2.** [8.1] $A = 34.2°$, $a = 0.622$, $c = 0.511$
**3.** [8.2] 8.0   **4.** [8.1] 419 cm
**5.** [8.2] $A = 92.1°$, $B = 33.0°$, $C = 54.8°$
**6.** [8.1] 13.72 cm²   **7.** [8.3] 76 mph; N 20° E
**8.** [8.1] 33.0 m²   **9.** [8.3] 23 lb down, 17 lb left; 28.6 lb, 54° down from horizontal (to left)
**10.** [8.3] $\langle -18.8, 6.8 \rangle$   **11.** [8.3] $(3.6, 124°)$
**12.** [8.3] **(a)** $\langle 25, 0 \rangle$; **(b)** 7.1   **13.** [8.4] $(4, 270°)$
**14.** [8.4] $(2, 330°)$   **15.** [8.4] $(-2, -2\sqrt{3})$
**16.** [8.4] $(-2, 2\sqrt{3})$   **17.** [8.4] $r + 2\cos\theta - 3\sin\theta = 0$
**18.** [8.4] $x^2 + y^2 = -4y$
**19.** [8.4]

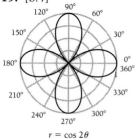

$r = \cos 2\theta$

**20.** [8.4]

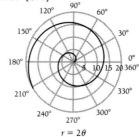

$r = 2\theta$

**21.** [8.4]

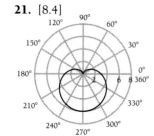

$r = 3(1 - \sin\theta)$

**22.** [8.5]

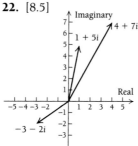

**23.** [8.5] $-\sqrt{2} + \sqrt{2}i$   **24.** [8.5] $\sqrt{2}$ cis 45°

**25.** [8.5] 70 cis 50°   **26.** [8.5] $\dfrac{\sqrt{2}}{2}$ cis 15°

**27.** [8.6] $4\sqrt{2}$ cis 135°   **28.** [8.6] $-\dfrac{81}{2} + \dfrac{81\sqrt{3}}{2}i$

**29.** [8.6] $\sqrt[6]{2}$ cis 15°; $\sqrt[6]{2}$ cis 135°; $\sqrt[6]{2}$ cis 255°
**30.** [8.6] $3, \frac{3}{2}(-1 \pm \sqrt{3}i)$   **31.** [8.6] 2 cis 45°, 2 cis 225°; or $\sqrt{2} + \sqrt{2}i$, $-\sqrt{2} - \sqrt{2}i$
**32.** [8.1] $50.52°$, $129.48°$   **33.** [8.3] $\langle \frac{36}{13}, \frac{15}{13} \rangle$
**34.** [8.5] $y^2 = 4x + 4$

### Test: Chapter 8, p. 553

**1.** [8.1] $A = 115°$, $a = 88.4$, $c = 31.8$   **2.** [8.2] 12
**3.** [8.1] 6 ft, 5 ft   **4.** [8.1] 3.3 cm²   **5.** [8.3] 22.8 N, 61°
**6.** [8.3] $(21.7, 12.5)$   **7.** [8.3] $(11.7, 149°)$

**8.** [8.3] **(a)** $\langle -1, -14 \rangle$; **(b)** 8.1   **9.** [8.3]
**(a)** N. 7, W. 5; **(b)** 8.6, N 36° W   **10.** [8.4] $\left( -\dfrac{3\sqrt{3}}{2}, -\dfrac{3}{2} \right)$
**11.** [8.4] $x^2 + y^2 = 625$   **12.** [8.6] $2\sqrt{2}$ cis 45°, $2\sqrt{2}$ cis 225°; or $2 + 2i$, $-2 - 2i$
**13.** [8.5]

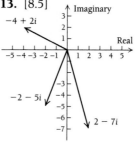

**14.** [8.5] $\dfrac{3\sqrt{3}}{2} - \dfrac{3}{2}i$   **15.** [8.5] 2 cis 120°

**16.** [8.5] 10 cis 90°   **17.** [8.5] $\dfrac{1}{2}$ cis 80°

**18.** [8.6] 8 cis 270°   **19.** [8.6] 2 cis 60°, 2 cis 180°, 2 cis 300°; or $1 + \sqrt{3}i$, $-2$, $1 - \sqrt{3}i$
**20.** [8.6]

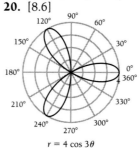

$r = 4\cos 3\theta$

**21.** [8.4]

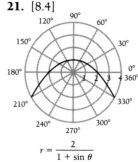

$r = \dfrac{2}{1 + \sin\theta}$

**22.** [8.4] $\left( -\dfrac{4}{\sqrt{53}}, \dfrac{14}{\sqrt{53}} \right)$